Ernest Irving Freese's
Geometric Transformations
The Man, the Manuscript, the Magnificent Dissections!

Ernest Irving Freese's
Geometric Transformations
The Man, the Manuscript, the Magnificent Dissections!

Greg N. Frederickson
Purdue University, USA

World Scientific

NEW JERSEY · LONDON · SINGAPORE · BEIJING · SHANGHAI · HONG KONG · TAIPEI · CHENNAI · TOKYO

Published by

World Scientific Publishing Co. Pte. Ltd.

5 Toh Tuck Link, Singapore 596224

USA office: 27 Warren Street, Suite 401-402, Hackensack, NJ 07601

UK office: 57 Shelton Street, Covent Garden, London WC2H 9HE

Library of Congress Cataloging-in-Publication Data

Names: Frederickson, Greg N. (Greg Norman), 1947–   author. | Freese, Ernest Irving, –1957. Geometric transformations.
Title: Ernest Irving Freese's Geometric transformations : the man, the manuscript, the magnificent dissections! /
    by Greg N. Frederickson (Purdue University, USA).
Other titles: Geometric transformations
Description: New Jersey : World Scientific, 2017. | Includes bibliographical references and index.
Identifiers: LCCN 2017042754| ISBN 9789813220461 (hardcover : alk. paper) | ISBN 9789813220478 (pbk : alk. paper)
Subjects: LCSH: Freese, Ernest Irving, –1957 | Geometric dissections. | Mathematical recreations. | Shapes.
Classification: LCC QA95 .F6825 2017 | DDC 793.74--dc23
LC record available at https://lccn.loc.gov/2017042754

British Library Cataloguing-in-Publication Data
A catalogue record for this book is available from the British Library.

For any available supplementary material, please visit
http://www.worldscientific.com/worldscibooks/10.1142/10460#t=suppl

Printed in Singapore

To Harry Lindgren (1912-1992),
Robert Reid (1926-2016),
and Gavin Theobald (1961-)

—amazing dissectors and great friends!

Preface

Researching and writing this book has been a blast! How often is one privileged to recover a revelatory document that had seemed lost forever? How could I be prepared for a manuscript so inventive and artistic, produced with vision and passion as the life of its creator slipped away?

It is not enough to rescue such a gem, to clean and ready it for publication. One needs to understand how it came to be. What manner of man, with just eight years of formal education, became obsessed with its completion? A man who had published over a hundred articles in a variety of periodicals, and also a well-received book. A man who had ridden the rails exploring the continental United States, and then worked his passage around the world on steamers. From what tradition and circumstances had he sprung?

How had this adventurer distinguished himself from those who had earlier explored geometrical puzzles? How did his solutions stack up against those who came later? In the interim, others had found about three dozen improvements. By how much had Ernest Freese anticipated the problems posed and the methods used by those who had not seen his manuscript? Check out how he dissected sequences of squares, how he dissected many-sided polygons, and how he completed the tessellation.

As I pondered this man's delightful creations, I endeavored to put them to the test: Could I find improved versions of his solutions, ones that used fewer pieces? Would you believe four dozen? And I divined some of his particular preferences: The elegant use of hinges beguiled him, and yet he hadn't realized that a surprising number (more than thirty) of his constructions were indeed hingeable. Moreover, he had only partially-hinged other constructions of his, a number of which I have since modified to be completely hingeable.

It was as though I had been drawn into an imaginary conversation with someone who could now never respond. Oh, the tricks that I wish I could show *him*! Perhaps a future reader of this book will someday be in a similar situation. Will he, or she, want to share an unbridled enthusiasm back through time? What wonderful ideas will *I* then be missing out on?

How to Read This Book

This book consists of several different parts: The first chapter will immerse you in a brief history of geometric dissections, displaying some stunning dissections that should motivate you to proceed further. It will also introduce pioneers who have made startlingly beautiful contributions to geometric dissections. That chapter also introduces our main character, Ernest Irving Freese, who is probably not yet so well-known. As you read about his remarkable life in the second chapter, you will learn what kinds of adventures were possible a hundred years ago. I have built the remaining chapters around sections of Freese's manuscript. Each of those chapters contains a sequence of illustrated plates, preceded by a chapter introduction and a commentary on those plates.

At first glance, you may wonder how to approach this book. Of course you could start at the beginning and read straight through to the end. The first chapter, highlighting a brief history of geometric dissections, is certainly a good place to start. But before reading about Freese's life, you might benefit from examining some of the plates in his manuscript. This could give you motivation for learning about his extraordinary life. When you sample the plates in his manuscript, you might find it enlightening to quickly read some of the corresponding commentary in that chapter. You could thus see how well Freese had done on certain types of dissections, compared with those dissectionists who explored a topic before him and those who came after him. This may work well if you are only mildly curious, but not so well if you would like to pit your skills against the masters.

So here is an approach if you can't resist a challenge: First read the brief history of geometric dissections in Chapter 1, then browse through Freese's index in Chapter 4 to identify the types of dissection problems that he investigated. Or hop into his manuscript at random to <u>see</u> what kinds of problems Freese found irresistible. Then you may be ready to try your hand on problems that sound intriguing, to see if you can improve on any of Freese's solutions. Finally, check the commentary in the relevant chapter to learn the current state of knowledge.

Remember that there is no guarantee that the best solution will appear in either Freese's manuscript or a chapter commentary. No one seems to know how to prove that a dissection has the fewest possible number of pieces. If you are clever enough, you may find something better. On the other hand, you need not find any new dissections in order to have a ball with this book. Simply revel in the creativity and imagination of people who have indulged in this glorious activity over the preceding centuries, including Ernest Irving Freese!

Acknowledgments.

I am indebted to Vanessa Kibbe, a niece of Ernest Freese. She plowed through the disorganized mass of her uncle's records that accumulated over his lifetime and lay untouched for almost half a century after his death. It was amazing to view, at last, the manuscript that I had given up almost all hope of finding. I would also like to acknowledge William A. (Bill) Freese and Dixie B. Barry, who shared their memories of their father with me before they passed away. I thank Judy Lindgren for surveying a mountain of her father's correspondence for me. One of those letters led me to the Los Angeles address of Freese's house, without which I would never have made contact with Bill Freese and his cousin Vanessa. I thank Jon Grinspan for sharing his insight on how young people at the turn of the twentieth century grappled with their challenges. Thanks to Christoph Irmscher, who shared his knowledge of the history of acrostics. I thank Prof. John Fritsch at Purdue University, who helped me identify biographical information on Ernest's wife Winifred.

It is my pleasure to acknowledge Dana Roth, reference librarian at the Millikan Library of Caltech, for determining the acquisition history of relevant volumes of the *Messenger of Mathematics*. I thank Dan Trinkle at Purdue for showing me how to convert bitmaps into a suitable format, how to incorporate postscript files, how to process photographic images, and how to solve various other software-related mysteries. I thank Kristyn Childres at Purdue for preparing a photographic image. I thank Dic Sonneveld for supplying unasked, but much appreciated, many details of Ernest Freese's life. I thank Shana, from Random Acts of Genealogical Kindness, in Colorado, for sending me an entry from the US Census of 1900. I thank Dan McLaughlin, reference librarian at the Pasadena Public Library, who located an early article about Ernest Freese and his bicycling exploits. I thank Gavin Theobald and Robert Reid for their wonderful dissections that I have included in this volume. Thanks also to Gavin, whose suggestions helped to improve my manuscript, and whose many last-minute improvements have been so exciting!

I also thank friends and family who listened patiently as I waxed enthusiastic about this project over the course of many years.

Contents

Preface   vii

1. The Rich History of Geometric Dissections   1

2. The "Wild Adventures" of Ernest Irving Freese   11

3. Techniques, Special Properties, Hardness   31

4. Freese's Title Page and Descriptive Index   35

5. Isosceles Triangles   (Freese's Plates 1–7)   37

6. Equilateral Triangles   (Plates 8–26)   41

7. Squares, Crosses, Rectangles   (Plates 27–46)   53

8. Pentagons and Pentagrams   (Plates 47–63)   67

9. Hexagons and Hexagrams   (Plates 64–88)   85

10. Octagons and Octagrams (Plates 89–98)   105

11. Enneagons (Nonagons)   (Plates 99–103)   113

12. Decagons and Decagrams   (Plates 104–111)   117

13. Dodecagons and Dodecagrams   (Plates 112–133)   125

14. Many-sided Polygons   (Plates 134–141)   137

15. Miscellaneous Figures   (Plates 142–153)   147

16. More Crosses   (Plates 154–160)   155

17. More Miscellaneous Figures   (Plates 161–178)   163

18. Mixed Polygons to One   (Plates 179–188)   175

19. Special Triangles   (Plates 189–200)   181

20. From the Past, Into the Future   189

Bibliography   191

Index   195

Chapter 1

The Rich History of Geometric Dissections

A geometric dissection is a cutting of a geometric figure into pieces that we can rearrange to form another figure. As visual demonstrations of relationships such as the Pythagorean theorem, dissections have had a surprisingly rich history, reaching back to Islamic mathematicians a millennium ago and Greek mathematicians more than two millennia ago. Over the years these curious puzzles have charmed geometrically-inclined people, including the star of this book, Ernest Irving Freese, an architect who lived and worked in Los Angeles. Shortly before his death in 1957, Freese completed a 200-page manuscript that presented a wealth of geometric dissections, many of his own invention. His tract constituted the first book-length document devoted in large measure to geometric dissections. Yet aside from some sample pages that circulated amongst a lucky few, it seemed for decades as though the bulk of the manuscript might have been lost.

However, with determination, with good fortune, and with the assistance of key people, I was able to track down this prize! Laid out with a draftsman's keen eye and labeled with an architect's exquisite hand lettering, the plates will take your breath away. Although he had no formal training in mathematics, Freese found within himself the ingenuity and resourcefulness to create novel geometric dissections. This current book showcases a carefully restored manuscript, electronically cleaned to remove stains and smudges that had accumulated over the years. To understand what motivated a self-educated man to create such a dazzling document near the end of his life, we begin with a brief history of dissections, followed by a biography of Freese in the next chapter. Then come a series of chapters, each consisting of a contiguous sequence of plates from Freese's manuscript and introduced with appropriate commentary. The commentary highlights recent improvements in the dissections that Freese had presented.

Resurrecting a 60-year-old all-but-forgotten manuscript is admittedly an audacious project, mirroring both the audacity of the man and the audacity of his manuscript. We will focus on mathematical recreations that are primarily geometric dissections, but we will also touch on biography, history, culture, and art. It's the biography of an uncommon 'common man', the history of a 'recreational area' of a serious field, an everyday culture that stretches from self-education to transformer toys, and a sort of art that has both geometric and kinetic overtones. (Just glance again at this book's cover art, and imagine the pieces swinging on their hinges!)

Let's begin by picking up the history of geometric dissections several centuries ago. In 1778 Jean Montucla introduced geometric dissections in a revision of Jacques Ozanam's book on mathematical and physical recreations. In 1821 John Jackson included dissections in his book on "rational amusement." During the same time period, John Lowry (1814), William Wallace (1831), Farkas Bolyai (1832), and Prussian Lieutenant (Karl) Gerwien (1833) showed that for any two simple polygons of equal area, there exists a dissection from one to the other that uses a finite number of pieces. The basic idea is to decompose each polygon into triangles and then dissect the set of triangles into pieces that fill out a square. Projecting the (dissected) square for one polygon onto the (dissected) square for the other polygon produces the overlapping cuts that define the dissection for the two polygons. Unfortunately, this method usually results in an ugly dissection with a mind-numbing surfeit of pieces.

Luckily, many people strove to identify attractive dissections with fewer pieces. Édouard Lucas incorporated dissection material in his 1883 book. In England, Henry Perigal (1873,) Henry M. Taylor (1905,) and William Macaulay (1914, 1919) wrote articles for periodicals such as the *Messenger of Mathematics* and the *Mathematical Gazette*.

1.1: Henry Ernest Dudeney - 1910

As mathematical puzzles, geometric dissections enjoyed great popularity a century ago, in newspaper and magazine columns written by the American Sam Loyd and the Englishman Henry Ernest Dudeney (Figure 1.1). Loyd and Dudeney chose as their goal the minimization of the number of pieces in such dissections. Their puzzles captivated readers, converting many amateur mathematicians into enthusiasts, while professional mathematics tended towards abstraction and beyond the reach of all but specialists. Singularly notable were Loyd's book, *Cyclopedia of Puzzles* (1914), and Dudeney's books, *The Canterbury Puzzles* (1907) and *Amusements in Mathematics* (1917).

As geometric dissections attracted more attention, the topic appeared with greater frequency in books devoted to mathematical recreations. Those included Emile Fourrey's 1907 book, Coxeter's revision of W. W. Rouse Ball's book in 1939, Geoffrey Mott-Smith's 1946 book, and Cundy and Rollett's 1952 book.

Let's cherry-pick a few stunning examples to understand why these puzzles became so popular. First, marvel at the remarkable 5-piece dissection of a regular octagon to a square in Figure 1.2. In both the octagon and the square, a small square anchors the center, with the four remaining, identical pieces arranged around it. Both dissected polygons display lovely 4-fold rotational symmetry. This grand dissection originated at least 700 years ago in an anonymous Persian manuscript, *Interlocks of Similar or Complementary Figures*. In (1926), in his column "Perplexities" in the *Strand*

Magazine, vol. 72, page 316, Henry Dudeney reported the dissection's (re)discovery by Geoffrey Thomas Bennett, a mathematician at the University of Cambridge.

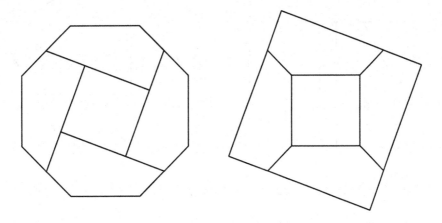

1.2: Dissection of a regular octagon to a square

In 1951 Harry Lindgren provided a wondrous explanation for how to derive this dissection. We can tile the plane with pairs consisting of a regular octagon and a square of the same side length. We can also tile the plane with pairs of two squares, one of area equal to that of the regular octagon and the other of side length equal to that of the octagon. Figure 1.3 denotes the first tiling with solid lines and the second tiling with dashed lines. We overlay the two tilings so that the centers of the small squares coincide with the centers of the octagons and the large squares. It's amazing how the dissection just pops out at you! Harry Lindgren called this method *completing the tessellation*, where tessellation is another term for tiling, and we can create a tiling for the octagon by "completing" it with small squares.

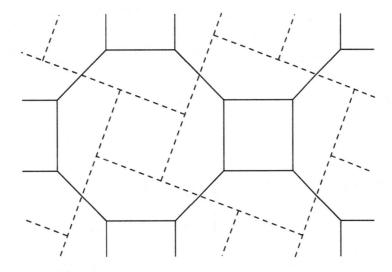

1.3: Superposition of tessellations for the dissection of an octagon to a square

One of the most symmetrical dissections is Henry Perigal's dissection of two not necessarily congruent squares to one, which we see in Figure 1.4. We leave the smaller square as is and cut

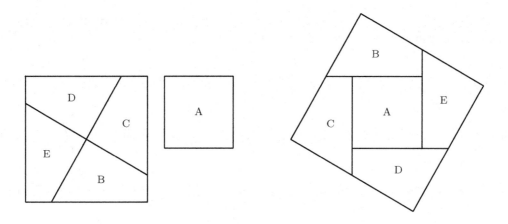

1.4: Dissection of two not necessarily congruent squares to one square

the larger square into four congruent pieces. When we tilt the largest square appropriately, we can easily shift the pieces from one figure to another by translation with no rotation. We call a dissection with this property *translational*. Thus square piece A shifts from being by itself to being in the middle of the largest square, and pieces B, C, D, and E each shift without rotation from the leftmost square to the rightmost square. Philip Kelland had noted such a property for a specific dissection in 1864, and Hugo Hadwiger and Paul Glur characterized all such dissections in 1951.

Perigal published his nifty dissection in 1873, although one could argue that Abū'l-Wafā, a tenth-century mathematician and astronomer, had used the same technique in his dissection of three congruent squares to one. Figure 1.5 shows how we can derive the dissection by taking a tessellation based on pairs of smaller and larger squares, represented with solid line segments, and superposing a tessellation consisting of the resulting squares, represented with dashed line segments. The dots represent some of the points of 2-fold rotational symmetry in both of the tessellations.

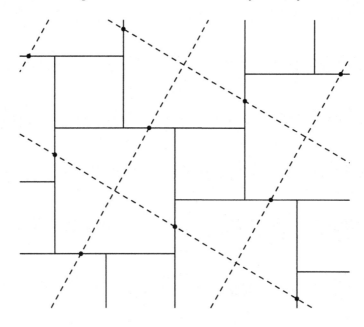

1.5: Superposition of tessellations for dissection of two noncongruent squares

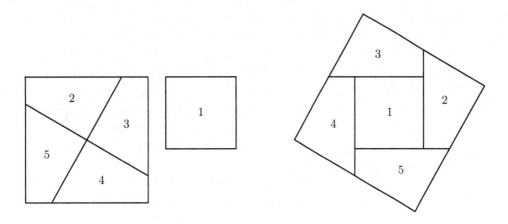

1.6: Relabeled dissection of two noncongruent squares to one square

In the latter part of the twentieth century David Singmaster saw how to hinge together the dissection pieces, so that we can rotate the pieces around on their hinges to form the desired figures. We redraw the dissection and relabel the pieces as in Figure 1.6 and then show how to hinge the pieces in Figure 1.7. If we start with piece 1 astride piece 5 in Figure 1.7, we can then swing the three remaining pieces clockwise until they enclose piece 1, producing the rightmost square in Figure 1.6. If instead we swing pieces 3, 4, and 5 clockwise around piece 2, we get the leftmost square in Figure 1.6. How magical is Perigal's dissection – not only translational but also hingeable!

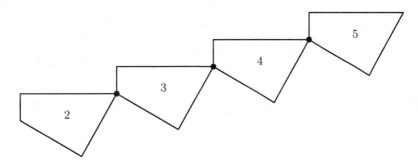

1.7: Hinged pieces for dissection of two noncongruent squares

Another outstanding dissection is that of an equilateral triangle to a square in Figure 1.8. Henry Dudeney posed the problem on April 6, 1902 in the *Weekly Dispatch*, and supplied the startling 4-piece solution (Figure 1.8) four weeks later in that same periodical. While many people credit Dudeney with the discovery, Dudeney apportioned some degree of credit to one of his most prolific correspondents, Charles William McElroy. You can read an analysis of those curious circumstances in my 2002 book.

In 1964 Harry Lindgren described a simple way to derive this dissection. With reference to Figure 1.9, take an infinite strip of equilateral triangles (shown with solid line segments) and crosspose with it an infinite strip of squares (shown with dashed line segments). Do this so that when a solid line segment crosses a dashed line segment, either both line segments coincide with the boundaries of the strips or the crossing point is a point of 2-dimensional rotational symmetry

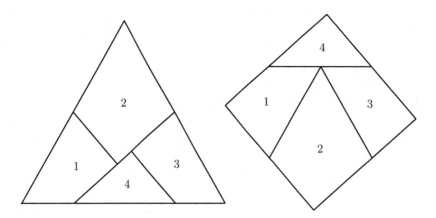

1.8: Dissection of an equilateral triangle to a square

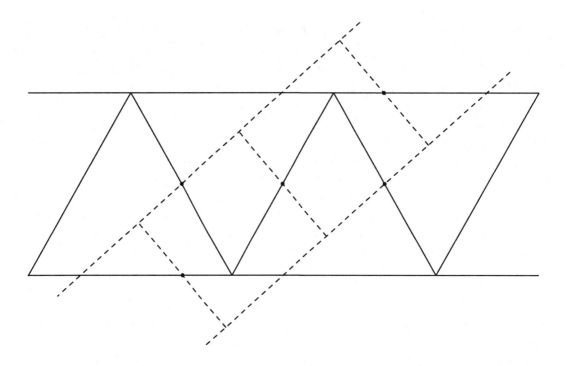

1.9: Crossposition for dissection of an equilateral triangle to a square

(identified by a small dot) in one strip. I explain this method, which I call a T-strip method, in my 1997 book.

With respect to special properties, how does the triangle-to-square dissection stack up against Perigal's dissection? Unfortunately the triangle-to-square dissection is not translational. Indeed, it is not possible for any dissection of an equilateral triangle to a square to be translational, as Hadwiger and Glur proved. However, a remarkable feature of the triangle-to-square dissection is that it is hingeable, as Dudeney noted in his 1907 book. We see one of four possible hingings of the pieces in Figure 1.10. To obtain the equilateral triangle, hold piece 1 in a fixed position and swing the rest of the chain clockwise around that piece. To obtain the square, hold piece 1 and then swing the rest of the chain counterclockwise around that piece. The dissection is hingeable

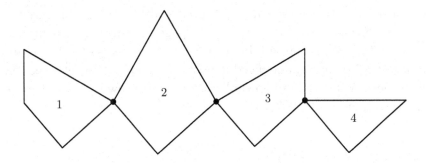

1.10: Hinged pieces for an equilateral triangle to a square

because you can place a hinge at each point for which a line segment in one strip crosses a line segment in the other strip. Now that is cool!

Freese was fascinated by dissections that are hingeable, either completely or partially, and noted positions for hinges in eight of the plates in his manuscript. This put him ahead of Harry Lindgren, whose 1964 book illustrated just three explicit examples of hinged dissections. In 2002, almost half a century after Freese completed his manuscript, I published a whole book of hingeable dissections. On page 3 of that book, I asked, "Given any pair of figures that are of equal area and bounded by straight line segments, is it always possible to find a swing-hingeable dissection of them?" Within a decade, the question was answered in the affirmative by Timothy Abbott, Zachary Abel, David Charlton, Erik Demaine, Martin Demaine, and Scott Kominers (2012).

While the mathematical proof affirming the above question is an impressive achievement, hinged dissections as constructed by the techniques of that proof involve in general a truly sizable number of pieces. It would seem that to fully appreciate the magic of hinged dissections, we would do well to find hinged dissections with relatively few pieces and preferably some degree of symmetry. At their best, such examples of hinged dissections can be fairly described as a form of kinetic art, which has prompted gasps and spontaneous applause from conference audiences. They are perhaps the ultimate transformer toys for geometry geeks and motion planners! Thus with that early prod by Freese, and encouraged by the great progress made on hinged dissections in recent decades, I will identify many hingeable dissections when I comment on various individual dissections in Freese's manuscript.

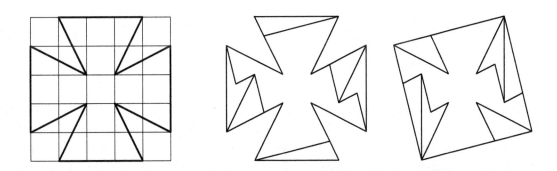

1.11: Definition of a Maltese Cross and its dissection to a square

An astounding dissection burst onto the scene in the mid-1920's. It is a 7-piece dissection of a Maltese Cross to a square. Henry Dudeney had proposed this dissection problem in his puzzle column in 1920 in *The Strand Magazine*, volume 60, pages 368 and 452, defining the cross on a 5 × 5 grid, as we see on the left in Figure 1.11. Dudeney supplied his own 13-piece dissection soon afterward, but in his (1926) book he announced a most unexpected 7-piece dissection (Figure 1.11), conceived by a mysterious A. E. Hill. It's so sublime, we might view it as art! I proposed a possible derivation for this dissection in my 1997 book.

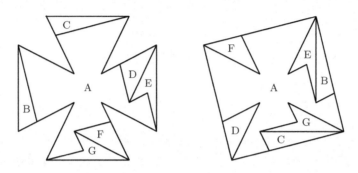

1.12: Translational dissection of a Maltese Cross to a square

As impressive as Hill's dissection is, it is neither translational nor hingeable. Dudeney's original dissection is translational, but those 13 pieces are just too many! However, I managed to find a 7-piece dissection that is translational, though not as symmetrical as Hill's beauty. Figure 1.12 spotlights the 7-piece dissection from my 1997 book, now labeled to show translationality. Check it out!

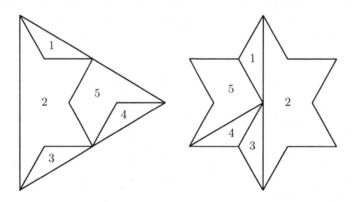

1.13: Dissection of a 6-pointed star to an equilateral triangle

In 1946 Geoffrey Mott-Smith published five different 5-piece dissections of a hexagram to an equilateral triangle. One of his five dissections used the set of pieces in Figure 1.13, although Mott-Smith arranged the pieces differently. The key observation for his five dissections is that the side length of the equilateral triangle equals the diameter of the hexagram, thus making it an easy dissection problem. As luck would have it, none of Mott-Smith's five arrangements of the pieces are hingeable, whereas the arrangement in Figure 1.13 is hingeable, as we see in Figure 1.14. However, the dissection is not translational, and in fact there can exist no translational dissection for this pair of figures.

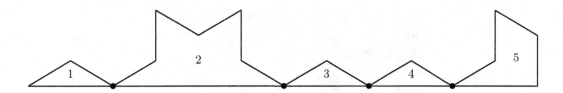

1.14: Hinged pieces for the dissection of a 6-pointed star to an equilateral triangle

Soon recreational geometers began searching for elegant methods to dissect congruent copies of regular polygons to a larger copy of the same polygon, for various polygons of more than four sides. Figure 1.15 reveals C. Dudley Langford's seductive 12-piece dissection of five congruent regular pentagons to one, published in 1956. The key insight is that the ratio of the length of the pentagon's diagonal to the length of the pentagon's side is the golden ratio $\phi = (1+\sqrt{5})/2 \approx 1.618$. Since the sum of ϕ and its inverse is $\sqrt{5}$, we get an effective way to construct the side of the large pentagon, using pieces from the five small pentagons. I wager that you'll be smitten by this "pentacular" dissection, even though it is neither hingeable nor translational.

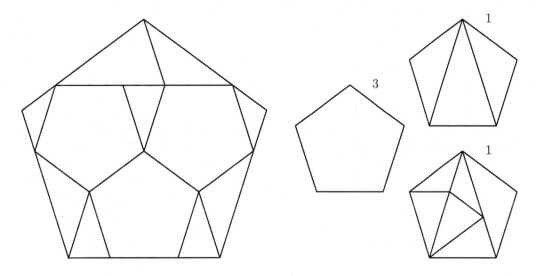

1.15: Dissection of five regular pentagons to one regular pentagon

In the early 1960's Harry Lindgren (Figure 1.16) gathered together the dissections of others plus more than a decade of his own work, to produce the first published book devoted exclusively to geometric dissections. He had started writing articles on geometric dissections in 1951, sending them to venues such as *Australian Mathematics Teacher*, *American Mathematical Monthly*, *Mathematical Gazette*, and *Recreational Mathematics Magazine*. His projected book, *Geometric Dissections*, got a big boost from Martin Gardner's math games column in the November 1961 issue of the *Scientific American*, and garnered much praise when it appeared in 1964.

Among his many treats, Lindgren introduced a new class of dissections that involved star figures. He noted pairs of these stars and/or polygons that, due of their special geometry, realized trigonometric relationships that made for elegant dissections. One amazing example is the dissection of a pentagram to a regular decagon in Figure 1.17. The special relationship for these two figures, when of equal area, is that the ratio of the side length of the pentagram to the side length of the regular

1.16: Harry Lindgren - early 1960's

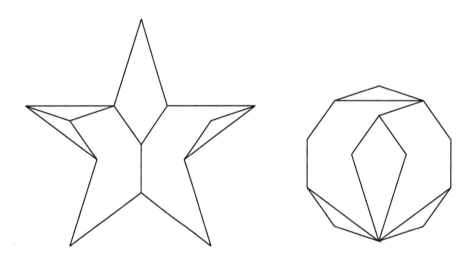

1.17: Dissection of a pentagram to a regular decagon

decagon is equal to $2\cos(\pi/10)$.

A welcome feature of Lindgren's book was his Appendix F, in which he identified sources and credits for the dissections of others. A puzzling aspect of this appendix was his reference to a then-mysterious person whom he called Irving L. Freese. Lindgren credited Freese with nine unpublished dissections, and claimed that Freese had lived in Los Angeles and died in 1957, but provided no further information about him. Freese seemed almost as elusive as that A. E. Hill whom Henry Dudeney had once mentioned.

Chapter 2

The "Wild Adventures" of Ernest Irving Freese

2.1: Ernest Irving Freese - later in life

When I started writing my 1997 book on geometric dissections, I learned more about Freese. To assist me, Harry Lindgren's daughter Judy had searched through her father's correspondence to identify letters that mentioned dissections. The letters indicated that Freese's name was actually Ernest Irving Freese, that he had written a manuscript consisting of 200 plates, and that Freese had been an architect who lived and worked in Los Angeles. One letter to Lindgren noted Freese's address before he passed away in 1957. Several of Freese's correspondents had encouraged his widow Winifred to share her husband's opus, but she had never responded.

As I researched my first book, I became discouraged that I might not be able to locate Freese's manuscript. In my frustration, I lampooned the situation at the beginning of my Chapter 17, imagining a private eye from southern California who was trying to track down the lost treatise.

Little did I imagine that I would morph into that detective and eventually lay my hands on that phantom document. Hoping against hope that Winifred was still alive, I mailed a letter to the address where she and her husband Ernest had resided almost four decades earlier.

2.2: Ernest, Bill, and Winifred Freese, early 1940s

Although Mrs. Freese had by that time also passed away, her son Bill was living in that very same house and answered my letter! Unfortunately, the son did not know if any of his father's work on dissections had survived, but he said that he would look. After not hearing from him for a year, I wrote another letter, addressed this time to him, but got no further response from him. However, in 2002, when Bill entered a hospice for cancer patients, his cousin Vanessa Kibbe, who was his designated heir, started to clean out the house. She found the unopened letter from me to Bill, which alerted her to the existence and significance of the manuscript. Within a few months she had located that legendary tome! In 2003 I flew out to Los Angeles to make a copy of the manuscript as well as many other documents that Vanessa had located and set aside for me to review. What a treasure hunt!

We learn about the completion of Ernest's manuscript from his announcement in a letter to his friend Dorman Luke on July 20, 1957:

> For the past 4 months I've let everything go hang – work, friends, correspondence,

2.3: Freese's house at 6247 Pine Crest Drive, Los Angeles, 1940s

or what have you – BUT, it's finished – the first book on Geometric Transformations (Dissective Geometry) ever produced. 200 plates comprising about 400 original examples. Cost $28.$\underline{^{00}}$ just to get a blueprint copy of the dwgs. Well, it's off my chest – but who will publish it? Probably <u>nobody</u>!

A Quick Biography

Why was Freese inspired so late in his life to assemble such a compendium? Let's study a letter from Paul Peter Kiessig, a friend who had known Freese for almost half a century. Kiessig was born in 1887 in California and passed away in 1967 in Vista, San Diego, California. On March 12, 1958, several months after Freese's death, Kiessig sat down in Vista to write a long letter to Freese's widow Winifred, recalling Freese's exploits and accomplishments:

Paul Peter Kiessig's letter

Dear Winifred: At long last I'll try to put down what I learned of Ernest's adventurous meanderings thru his allotted participation in life.

I don't know whether he ever knew his mother. I never heard him mention her. Meeting his father in time, I became fond of the kindly old man. But apparently he either was in no position when Ernest was a young boy to care for him or Ernest just struck out into world confident that he could meet it barehanded.

At about 15 I believe he got a job as a waterboy on the Flagler railroad then being built to connect the Florida Keys. The work of the engineers in the drafting room fascinated him. He watched them whenever the opportunity presented. They took a fancy to him – answered his questions – and when they observed a ready grasp of the math basis, they tutored him. That is the only technical education he ever received. But they had given him a star to ride and he had found his orbit. An inherent curiosity and flair kept him on the trail of mathematics and engineering. He was nothing less than a prodigy. It's a matter of unending wonderment to me that some few are born with such keen perception and that affinity for their special field that they just seem to absorb by instinct akin to extrasensory powers that which is obscure indeed even with expert guidance to the rest of us.

About 1909 I was working for Sellon and Hemmings architects in Sacramento. Sellon was State Architect. I worked at the Capitol and later transferred to the private practice of the two. The other boys told me of Freese. I saw some of his drawings – Lovely pen and ink perspectives – confident drawings with personality in the craftsmanship. You could identify his hand whenever you saw the drawings.

They all spoke of him with great respect and admiration and as yet I only knew him thru his works. He showed up one day they said fresh off of a freight train and asked Hemmings for a job. Hemmings was skeptical but needed help badly so with fingers crossed took him on probation. In no time Freese had the whole program and office in the palm of his hand. The indispensible [sic] man. Then one day out of the blue he came in in the morning and went right into Hemmings' office. "Well I'm leaving I'd like to have my check" He might as well have [hit] him with a club. "What's wrong"? "Nothing I just got an urge to move on." "If it's a matter of money we can fix that." "No the pay is O.K. the office is O.K. I just have an urge." And so he departed. Things were chaos for a bit and the firm swore that they would rather fail than ever give Ernest another job.

About 2 yrs later Hemmings' door opens and there stands Freese grinning. They welcomed him joyously with open arms – and Ernest put his gifted hand to their service once again. Thats when I met him – prob 1909. I was only a tracer and office boy and yet I was almost the same age. The Sacramento Hotel was the one job of that period that Ernest did that I remember.

I didn't see him after Sacto. days for several years. Then I ran into him at Hunt and Burns, Los Angeles. Of course he was chief draftsman. I was trying to make a living importing and selling drawing instruments. But I had so little capital that it was hard going and I believe about twice I borrowed ten dollars from him just to stay alive. So you can imagine how pleased I was to lend him 100.00 in after years when the need was his.

Finally I had to abandon my importing venture and he gave me a job in the Hunt + Burns office. He was a good chief – rarely got rough or impatient with those under him but he seemed to resent authority above him. Maybe it was just the irk that a dreamer or a master craftsman in any line feels at restraint or being fenced in by stipulations. But his reaction to direction or interference from above could be swift and savage. He got along pretty well with Hunt who confined himself largely to the business end. But he regarded poor old Burns as a bumbling incompetent and did not conceal his feeling. B. stayed pretty well out of his way but had occasion to come into the drafting room one day. I don't know the details but before the whole crew Ernest told poor old Burns "Get out of this drafting room and don't ever come in again."

"Many a shaft at random sent finds mark the archer little meant." I often wondered just what effect this humiliation might have had on this bumbling old pomposo's self respect and peace of mind. Perhaps it hurt me more than it did him.

I can only recall about three of the boys of that crew; Cannon, Harry Moore (Marry Hoore) and Wayne Lyons (who was [gay]). Freese was generous and helpful with them. Cannon he coached and tutored until he became an adequate engineer –

2.4: Sumner P. Hunt, Los Angeles architect - 1913

upgrading his whole life – which wasn't easy as Cannon's marriage was an unfortunate one. Moore had reached his capacity and solicited no aid. I don't know what happened to Lyons. There was also a capable educated, fine old spinster secretary (cant recall her name.) She had tact and was a good and interesting human being.

It wasn't too long after Ernest quit them that the firm disintegrated completely – as we all knew it would. I believe it was early in this period that Ernest married his first wife. She was pretty in a way with a good figure but you were aware at once that she was shallow and frivolous. I'm not sure that Ernest at that time was good husband material – maybe a little splenetic and impatient of stupidity and vacuity. Anyway she was soon two timing him and this split up was probably the best thing that ever happened.

As you probably realize most men especially those hoboing or without families or roots "take their fun where they find it." Society makes her rules but mother nature seems not always in accord. But I was never aware of any dalliance on Freese's part tho this is no certification.

Before and after Sellon and Hemmings days Ernest indulged his restless travels and search for adventure. On one hobo interlude he met Jack London and got to know him quite well. Ernest too did a little circumnavigating. I have only a few skip facts to relate of this phase of his adventuring. It wasn't easy at that time to find arch. employment in the Orient. This help was usually contracted for from Europe and had to be blue ribbon gentlemen. But Freese did work in offices in the Philippines (Manila) and Hong Kong I believe. You can bet they regretted losing him. Who didn't?

The last episode I have to relate is I'm sorry very sketchy but it was a star performance and red blooded adventure. Of course he worked his passage on ships but whether he just worked from port to port, as was possible at that time, or whether he jumped ship at Madras I do not know. But after seeing what he wished in Madras he registered with the police as required for anyone taking off for the interior – and disappeared. After 30 days the police sent his father a notice saying "Your son left Madras and went into the wilderness thirty days ago. We have heard nothing of him since and fear for his safety." About a month or so later he arrived in Bombay. He had walked all the way across India – which would have been a major achievement even in a civilized area. I don't recall whether it was from Madras or Bombay that he sent bales of postcards photos etc to his father for safekeeping but these were all lost in transit. Egypt was also on his itinerary.

We felt a common interest in both having hoboed it which was less rare in that day before the young discovered that the world was made for them and owed them a fast buck.

Well Winifred I realize that most of this account is not news to you but that it's possible that some of it might be.

Unfortunately not all genius clicks but click or not there were lots of people who found out it was there. Incidentally there are quite a few professionals who were transformed by him from run of the mill draftsmen to confident engineers.

I would like to observe also that you were the real haven of his spirit. I was apprehensive of the difference in your ages but it proved you who gave him love and peace of mind and encouragement – the realizations of the goal [of] his restless search. You are the best thing that ever happened to Ernest.

A More Comprehensive Account of Freese's Life

Ernest Irving Freese was born on February 5, 1886 in Minneapolis, Minnesota. Freese's father William Henry Freese was a cooper who had migrated west from Maine in 1882, married Freese's mother Bertha Reeves in Minneapolis in 1885, and moved on to Denver, Colorado, and later to California. Ernest quit school after the eighth grade and left home to travel around the country. As Paul Kiessig wrote, about 1901 Ernest came to work on the Flagler railway and was befriended by the engineers. Around 1904 or 1905 Freese worked at the architectural firm of Sellon and Hemmings in Sacramento until he up and quit one day. That coincided with a period in which his father had started to work as a grocery clerk in Pasadena.

Twenty years after his arrival in Pasadena, Ernest recalled how he came to live in the Los Angeles area, in an article "Judges Complete Gigantic Task of Selecting Winners of Twin-Matching Contest" which the *Los Angeles Sunday Times* published in December 13, 1925 on page B6:

> The Times, in a way, was responsible for keeping me in Los Angeles for the last twenty years. I came here about twenty years ago, just a boy out of art school. I landed in Los Angeles broke. And I drew a cartoon—the first cartoon that I had ever drawn in my life—and sold

it to the Times for $25. It kept me in food until I got a job as a draftsman—and I stayed here.

Yet staying there permanently did not preclude Freese from striking out on many adventures. On Nov. 17, 1905, the *Pasadena Daily News* published an article, "Starts Out as a Tramp," with the subheadlines "A Pasadena Boy," "Young Ernest Freeze [sic] Leaves Pasadena," and "'Jumped Out' Last Night on His Third Wild Adventure." The article sketched his wild adventures:

> When the Santa Fe's Overland pulled out for the east last night it carried one passenger who was not paying any fare. That was Ernest Freeze [sic], a Pasadena boy who is well known to many. Freeze [sic] was comfortably ensconsed [sic] on top of one of the baggage cars.
>
> In a spirit of adventure and because he wants to see more of the country, Freeze [sic] is starting out on a prospective tour of about half a year. He is heading for New Orleans, and then contemplates going to Florida. He doesn't propose to pay any railroad fare, and he has made similar trips before without ever paying a cent to the railway corporations it is reasonable to presume that he may be lucky enough to get through all right again.
>
> It was only a few months ago that young Freeze [sic] returned to Pasadena after a six months' tour of the country. He is skilled in illustrative work and architectural drawing, and he at once went to work here. He has been employed at different times in two of the well known architect offices of Pasadena.
>
> Though Freeze [sic] travels as a tramp, or rather as tramps do, he says he has never been arrested. He does not mix up with the hoboes any more than he can help, as they are a rough and often times desperate lot. Except some of the states in the old south, the young fellow has been in all parts of the country and now intends to see the places where he has never been before. Sometimes he works to help pay expenses. Freeze [sic] is big and strong, though not yet twenty years old, and he isn't afraid to turn his hand to anything. His strong physique stands him in good stead in the rough life. Many have seen him around the Pasadena Y. M. C. A. gym, and frequently he has taken part in the exhibitions. If he doesn't lose a leg or two around the cars, or doesn't get all mangled up and find burial in some pauper's grave yard as an unknown tramp, he may yet live to look back on his travels and adventures with a reminiscent enjoyment. One of his trips out of Pasadena was made with Arthur Clark.

Apparently Freese made it to Florida, because in the material uncovered years after his death was a 4 1/4 inch by 5 1/2 inch handwritten card that stated

> Camp 7   Dec 23 05
> Mr. F A Barrett
>   Paymaster
> Bearer is John Reynolds
>   HJGault
>   Res Engr

The remainder of the card contains handwritten comments by Freese stating in the top margin, "Working on Key Largo of Key West Extension of Florida East Coast R.R.," in the left margin, "My 'identification' paper!," and a circle around the name John Reynolds, linked down to a signature

"Ernest Irving Freese." This was Freese's ID card made out by Homer J. Gault to Freese under the pseudonym John Reynolds!

On February 27, 1906, the *Pasadena Daily News* featured Ernest Freese in an article entitled "Unique Hobo Strikes Town." An excerpt:

> The passenger train rolled into Pasadena yesterday, and two bright-looking, energetic-appearing young men rolled out from beneath the body of one of the coaches, where they had been riding on the rods for several hours. Both these young fellows were under twenty years of age, and although they were and are tramps or hoboes in every sense of the word, as the definition of these words are understood by the ordinary run of people, they were clean looking, hustling young men apparently full of life and activity, but unwilling to become what they term "wage slaves."
>
> "The News printed something about me last November and I have a few lines here I would like to see printed," said one of the strangers half an hour later in the News office. Following is the communication that was handed in:
>
> Dear Editor: The "Tramp" spoken of in the November 17 issue of the News last year has returned. He wishes to say, apropos the above mentioned article, that he has railroaded through every state in the Union and Florida, and that he has not yet "lost a leg or two around the cars." Neither did he "get all mangled up and find burial in some paupers' graveyard as an unknown tramp." Oh, no—he wishes to say that he has learned things—among others, that hoboes, instead of being a "rough and oftimes desperate lot," are more often incapacitated or discouraged workers whom an ill-ordered capitalistic system has cast adrift to live as parasites on their more "fortunate" brothers—even as the idle rich live as parasites. And furthermore, he, the tramp, is now contemplating another "wild adventure," namely he is going to take a zig-zag ramble around the globe, from California to California, eastward. He is going for the "fun of it"—not as a peddler of religion or glad tidings, but just for the pure, simple joy of it. And he will start on this world ramble on the first day of July, 1906. I guess that will hold you for a while. (?)
>
> Yours for "wild adventures,"
> ERN FREESE,
> (The Tramp.)

Freese had trimmed this article out of the newspaper and pasted it in a diary that he called "The Road." He then wrote the following in an expansive script, with tiny x's as periods and also as the dots over the i's.

> Off the Road + home again. I say "home" - but I mean the place from which I started - for "any old place I can hang my hat is home sweet home to me." My old boss - C. W. Buchanan - stopped me on the street today + gave me the glad hand before I'd been in town 24 hours. Said I could start in the morning. So I guess I'll dig in for a while and be "good." Here I am - on top once more - and less than a month ago I was away down. The wheel has turned. Well, it'll seem kind of good to juggle a pencil again + turn imagination loose on paper. There is certainly a difference between a hobo on the Road + an Architectural Draftsman. Between juggling a pencil + shovelling coal for a ride on a crimpy night. When I'm a hobo – I do as the hoboes do. When I'm a "respectable young man" I do as other "respectable" young men do – that is – if I care to. That is adaptability.

2.5: Ernest Freese and his bicycle - June 1906

On June 27, 1906, the *Pasadena Evening Star* carried an article, "Pasadena Men Will go on Bicycle World Tour":

> On Monday morning next at 9 o'clock two young Pasadena men will start on a bicycle tour of the world which will embrace a ride of 7000 miles in the United States and a tour of European countries. These men are Arthur S. Armstrong, who has been employed in the office of Architect F. S. Allen, and his chum, Ernest I. Freeze [sic]. Both came from Portland, Maine, and this will be the objective point of their excursion from this city.
>
> Suitably equipped with an army camping outfit the long ride will commence. However, the young men plan to live after civilized methods and will only resort to extreme roughing it when obliged to do so.
>
> They do not plan to reach Portland for a year and on the way will pause at various places where they will work for brief periods at their profession, architectural drafting. In fact, out of the next twelve months they plan to spend at least a fourth of their time at work.
>
> With them they will take cameras, drawing materials, etc., and a careful diary

of their experiences will be kept for they may write illustrated articles on their experiences.

Their route will be via the Pacific Coast to 'Frisco, thence north to Portland, Ore., east to Seattle and Bozeman; through the Yellowstone to Denver, Salt Lake City and south to Texas and New Orleans. From New Orleans the bicyclists will go to Florida and then follow the Atlantic Coast north to their destination.

At Portland, Me., they plan to stop for half a year and then they will resume their travels, crossing the ocean and touring Europe and foreign countries. The two men expect that their entire trip will last not less than four years.

Armstrong and Freeze [sic] reside at No. 322 Summit avenue and the purpose of their trip is to gain world-experience and regain their health.

Freese arrived in Portland, Maine, as confirmed in the 1907 city directory, which listed him as a "draughtsman." His entry in the 1908 city directory indicated that he had moved to Texas. When he revisited the city in December of 1909, he sent a postcard to his father, displaying a picture of the Billiard Room of the Elks Home in Portland, on which he claimed that he had worked for the architect Austin W. Pease.

It is unclear what Freese did between 1907 and 1909, because his next verifiable location was a listing in the 1909 city directory of Sacramento, California, as a "draughtsman" in the office of Sellon and Hemmings. According to Freese's annotations on a map of the world, he claimed to have made "tramp trips" in the "States." The routes were from California to Seattle, through the Dakotas, between Chicago and San Francisco via Salt Lake City and Denver, between Los Angeles and Montreal, between St. Louis and Washington, D.C., between Los Angeles and Florida via New Orleans, with a side trip from New Mexico to Mexico City, between New York City and Key West, Florida along the East Coast.

Early in 1909 Freese embarked on an around-the-world trip, departing from San Francisco and journeying west to the Orient, through the Strait of Malacca, to Madras (now Chennai) and across India to Bombay (now Mumbai), through the Suez Canal, to Algiers, across the Atlantic Ocean to Baltimore, then New York City and finally Portland, Maine. It is not clear exactly when he started his around-the-world trip, but he claimed to have started from San Francisco. And he claimed to have returned home soon after New Years Day, 1910. His voyage is documented by postcards sent to his father with the following postmarks:

Apr 26: Honolulu

Jul 28: Tokyo

Jul 29: Yokohama, Japan

Aug 2: Nagasaki

Aug 3: Shanghai, China

Aug 6-7: Hong Kong

Aug 10: (arrived in) Manila on S.S. Rubl'

Sep 1: (still in) the Philippines

Sep 7: Singapore

Sep 10: Kuala Lumpur, Malay peninsula

Sep 17: Penang, Malay peninsula

Sep 24-29: Madras (Chennai), India

Oct 3: Bombay (Mumbai), India

Oct 9: (departed) Bombay (Mumbai) on the British steamship Strathblane, across Indian Ocean and through Suez Canal, working for his passage

Oct 25: Port Said - stop for coaling

Nov 1: Algiers - stop for coaling

Nov 23: (arrived in) Baltimore, paid wages at the rate of 4 pounds/month

Nov 27: New York City

Dec 20: Portland, Maine

Freese worked as a crew member of a ship for at least a portion of his trip, from Bombay (Mumbai), India to Baltimore, Maryland, USA. His circumnavigation of the world evokes the undertakings of explorers and adventurers from the time of Magellan, as recounted by Joyce E. Chaplin in her 2012 book. Was Freese influenced by contemporaries who had worked their way around the globe? One such young adventurer, Harry A. Franck, worked and tramped eastward around the globe, as he recounted in his 1910 book. How common were such working trips at that time?

2.6: William Henry Freese (father)

Coincidentally, Freese's mother Bertha Freese must still have been alive at the start of his trip, because he mailed a postcard addressed to her from Penang in the Malay peninsula. Also, there would not seem to have been time for Freese to walk from Madras (Chennai) to Bombay (Mumbai), so perhaps he took more than one trip around the world. Indeed, there was a copy of a letter dated November 18, 1910 from the chief police office in Kuala Lumpur to Freese's father, inquiring if he had heard whether Freese had concluded his tour of the Malay Peninsula. Perhaps Paul Kiessig had misremembered a few details of Freese's adventures decades after the fact.

Again Freese was listed in the 1911 city directory of Portland, Maine as a draughtsman, this time having moved to California. Also in 1911, two years after his father started working as a meat cutter and clerk in Los Angeles, Freese surfaced in Los Angeles, working as a draftsman. And by the next year he was working for the architectural firm of Hunt & Burns, claiming in 1912 to be an architect there. In 1919 Freese was listed as an engineer in the San Diego city directory. He was back with Hunt & Burns in 1921, and was last listed with them in 1923. After that, he seems to have been in practice on his own.

Paul Peter Kiessig's appraisal of Ernest Freese's talent while at Sellon and Hemmings in Sacramento was echoed during Freese's time with Hunt and Burns in Los Angeles. In the November 1922 issue (Vol. 22, no. 5) of the *The Building Review*, the respected San Francisco architect Harris Allen published an article, "Southern California Architects: Hunt and Burns—Experts on Clubhouses," which appeared on pages 57-60, 63-65, and Plates 53-68. On page 64, Allen wrote: "It is interesting to compare the sketches of [the Virginia Country Club (Los Angeles)] and the Southwest Museum with the photographs of executed buildings. These clean and brilliant drawings are the work of Mr. Ernest Irving Freese, to whose able technique in handling Hunt and Burns' office work much credit is due."

Young Ernest Freese was audacious as he charted the course of his life. He took off on tramps across America, circumnavigated the globe, and did not marry until age 32. Yet he was perhaps more in tune with his times than we might at first suppose. Jon Grinspan, an historian who has studied the behavior of young adults in the late nineteenth century, pointed out in 2014 that when life at home seemed stalled, young adults headed out on "wander years," and the institution of marriage changed. The average age of marriage rose abruptly to 26 – shocking, at a time when worldwide life expectancy was below 50! It was not until 1990 that America returned to a marriage age of 26.

Freese had less enthusiasm for, and perhaps less access to, an evolving educational system that would eventually allow many more young people to attend college. Again, as pointed out by Grinspan in 2015, youths of 19th century America took more personal responsibility for their education and often pursued a do-it-yourself approach. Reaching the middle class on their own initiative, through self-study, seemed a viable and promising approach in uncertain times.

Perhaps there is no better indicator of how Freese deliberately set off on his own than the astounding number of magazine articles that he wrote throughout his career.

Freese's Publications

Considering Freese's early hobo tramps around the United States, his cheeky interaction with the Pasadena newspapers, and his around-the-world tour, we should not be surprised that he wrote articles about his travel adventures. He first found an audience in periodicals devoted to motorcycling, bicycling, and outing. On a trip to Hawaii Freese brought his motorcycle, with a sidecar for his first wife Philippa (Cuny). After they motored off the beaten track in those islands, he wrote about the early Hawaiian customs that they encountered. He published three articles in the periodical *Art and Archaeology*.

Once Freese started publishing articles, it seemed as though he couldn't stop. He published over 150 articles, on a wide range of topics, with one series of articles leading to a book! As a measure of his ambition in this activity, Freese listed his occupation as "architect, writer" when registering for the U.S. military draft in 1917. I shall identify the periodicals in which Freese published, the number of articles and the time span during which he published in each periodical. These lists exemplify the initiative and adventure evident during much of his life. Freese based the articles in the first six periodicals on his explorations of the United States, including the territory of Hawaii:

Motorcycling: 5 articles, Aug. 1914 - Oct. 1914.

Pacific Motorcyclist: 1 article, March 1915.

Motorcycle & Bicycle Illustrated: 5 articles, Jan. 1915 - Apr. 1918.

Motorcycling and Bicycling: 8 articles, Jan. 1915 - Oct. 1918.

Outing: 1 article, Sept. 1918.

Art and Archaeology: 3 articles, Aug. 1919 - Jan.-Feb. 1924.

In a letter to the A. N. Marquis Company, Freese claimed to have had up through a high school education. Yet both his younger daughter Dixie and his younger son William wrote that their father had stopped after the eighth grade. He thereafter entered the employ of an architect, starting in 1904. From 1904 through 1911 he was an architectural and structural draftsman, and also a construction superintendent. He mastered differential and integral calculus, and the theory of structural engineering, strictly by self-study. From 1911 to 1917 he worked for the architectural firm of Hunt and Burns in Los Angeles, being in charge of their architectural and structural design. Starting in 1913, he was a licensed architect of California.

Freese's most substantial effort in publishing was in the areas of architecture and architectural engineering. His articles ranged from planning the orientation, layout, and choice of materials for houses, to how-to manuals of construction, and to illustrations of a wide variety of home designs from different regions of the country and around the world. A charming article, "A bungle-ode," which appeared in the *Architect and Engineer of California* in 1918, was a tongue-in-cheek homage to his soon-to-be-wed sweetheart, who fancied living in a bungalow!

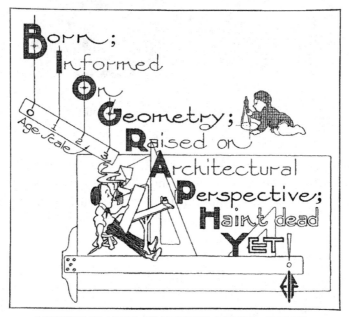

AUTOBIOGRAPHY OF ERNEST IRVING FREESE

2.7: Ernest Irving Freese's acrostic, 1930

Following the publication in *Pencil Points* of his 5-part series of articles on perspective projection, and preceding the publication of his book *Perspective Projection* by Pencil Points Press, *Pencil Points* published his "BIOGRAPHY," a pithy description of key events in Freese's life in the form of an acrostic. An acrostic is an ancient and popularly admired literary form, as discussed by Joshua Scodel (1991).

2.8: Ernest Irving Freese - mid-life

Freese targeted his most ambitious writings for the periodical *Pencil Points*, which published a series of his articles, "Perspective projection," which he then expanded into an influential book of the same title. *Pencil Points* followed that with Freese's extensive series of articles, "Geometry of architectural drafting," running through 1932, and Freese's periodic column, "Freese's Corner," which addressed issues in mathematics and geometry related to drafting and ran until mid-1932.

House Beautiful: 3 articles, Nov. 1914 - Sept. 1915.

Bungalow Magazine: 2 articles, Feb. 1916 - Apr. 1916.

Building: 1 article, Apr. 1916.

House and Garden: 6 articles, Jan. 1916 - June 1918.

Keith's Magazine: 1 article, May 1918.

Metal Worker, Plumber & Steam Fitter: 1 article, Sept., 1918.

Everyday Engineering: 3 articles, Nov. 1918 - Feb. 1919.

National Builder: 6 articles, Nov. 1918 - July 1919.

Building Age: 14 articles, July 1918 - Aug. 1920.

Architect and Engineer of California: 10 articles, June 1912 - Jan. 1934.

California Home Owner: 29 articles, Aug. 1923 - Sept. 1925.

Southwest Builder & Contractor: 1 article, Jan. 1926.

Western Construction News: 1 article, Oct. 1930.

Los Angeles Times: 17 articles, May 1920 - March 1936.

Pencil Points: series of 5 articles, "Perspective projection," 1929.

Perspective Projection: A Simple and Exact Method of Making Perspective Drawings, by Ernest Irving Freese, architect, New York: The Pencil Points Press, 1930 (Review in *American Architect*, 138:60, Dec. 1930).

Pencil Points: 22 articles, "Geometry of architectural drafting," 1929–32.

Pencil Points column, "Freese's Corner," 10 articles, 1931-1932.

Pencil Points: 3 articles, Oct. 1930 - March 1935.

American Architect: 11 articles, Sept. 1920 - May 1936.

American Architect and Architecture: 2 articles, Aug. 1936 - Oct. 1936.

American Builder and Building Age: 2 articles, Jan. 1935 - March 1936.

Southwest Builder & Contractor: 15 articles, "Decoding the Codes," 1939–40.

At 5:54pm on March 10, 1933, a magnitude 6.4 earthquake struck in Long Beach, California. The earthquake caused damage of approximately $50,000,000 in 1933 dollars throughout the city and adjacent communities. The destruction was greatest for poorly designed and unreinforced brick structures, and 120 people lost their lives. In just seconds, the earthquake hit 120 schools in the vicinity of Long Beach hard and destroyed 70 of them. Experts projected that had the earthquake struck during school hours, thousands of children and teachers would have perished.

Infuriated by the shoddy work that had put so many children at risk, Freese participated in a newspaper campaign to make sure that no cost would be spared to make the schools safe. He waged the campaign in the newspaper of the historic neighborhood of Highland Park, in Los Angeles. Freese was anything but diplomatic towards the politicians who tried to remedy the problem with as little money as possible, and his impolitic approach turned many of them against his proposals.

The issue of safe schools resonated with Freese not only because he was an architect, but because he was the father of three school-age children. Freese had married Philippa Cuny in 1918. She had then given birth in quick succession to Patricia, Irving, and Dixie, born in 1919, 1921, and 1922, respectively. Freese's marriage started to unravel around the time that he was battling the politicians, with Philippa and the children moving out in 1936, soon followed by a divorce. The marriage breakup was so contentious that son Irving changed his name to Phil. In 1937 Freese married Winifred Muriel Anderson, who gave birth to son William in 1938.

Highland Park News-Herald: 10 articles, Jan. 1934 - Sept. 1934.

Later in his life Freese focused on mental challenges such as brain teasers based on mathematics. While finishing his manuscript on geometric dissections, he communicated with Dorman Luke, H. Martyn Cundy, and Jekuthiel Ginsburg.

John Martin's Book: 1 article, May 1928.

Pathfinder: 1 article, March 1939.

Scripta Mathematica: 1 article, 1956.

Some Illuminating Details

Before we home in on the main attraction (Freese's manuscript), let's collect a few more details that will bring our man into sharper focus.

Practicing nudists

Freese and his second family were practicing nudists. Included with materials documenting various aspects of Freese's life is a set of six photographs of Freese and friends posing outdoors in the buff. Freese's handwriting on an envelope containing the photos stated "Pictures taken at Fraternity Elysia (Glassey's Ranch)" with dates 1936 and 1939.

Fraternity Elysia was founded by two pioneers in the movement for social nudism, Hobart and Lura Glassey, in La Tuna Canyon in 1935. Their camp was a successful business, with paying members granted access to the private colony. In 1938 it was the setting for a nudist movie, "The Unashamed," which was widely shown. One famous member of Fraternity Elysia was Caltech seismologist Charles Richter, for whom the Richter scale, the standard measure of earthquake intensity, was named.

Even though the camp was successful, it was dogged by tragedy. In 1938 Hobart Glassey died at Elysia in a fall that broke his neck. Then in 1939, actress Dawn Hope Noel, committed suicide there. To some people these events suggested a "corruption of mind" caused by nudism, and negative publicity about the camp led to concern over the morals of Los Angeles. By 1940, this attitude had prompted local authorities to initiate a series of raids on Elysia.

There seems to be no record that Freese ran afoul of the law. Rather, Freese's involvement may well have been a relatively harmless example of his adventurous and unconventional approach to life. Freese's nude photos are tame, projecting an almost art-museum sort of innocence.

After the Second World War

On October 1, 1948, Freese posted a letter to Wheeler Sammons of the A. N. Marquis Company in response to a request for biographical data. In addition to some standard facts and a lengthy list of his publications, he included a few pithy remarks:

> I am a firm disbeliever in college education ... except possibly as a pastime.
>
> Married twice, the last time happily!
>
> Belong to no society, association, club, church or political organization; in fact am most cordially hated by most of them! Strictly a "lone wolf" who never knows which side of the fence his bread is buttered on, nor cares.
>
> Special activities: Solving problems whose answers are not in the books. Telling school boards and city fathers how to build buildings to resist earthquakes ... and getting buried deep for my temerity.
>
> Incidentally I was the first California architect to formulate a theory of earthquake-resisting construction, and put it into practice long before building codes demanded such measures.
>
> I was finally forced to conclude that technical writing is a damnable way to try and get butter to go with your bread.

After 1945, Freese was on his own with respect to architectural projects. Luckily, he had purchased a significant parcel of land on Pine Crest Drive. To provide income to live on, he would periodically sell off a parcel of that land, and serve as architect for a bungalow on the site. He and Winifred also entered various contests, using the winnings to supplement their income.

Math Recreations Books that Influenced Freese

In 2003 a search through Freese's former home turned up several books that had fostered his interest in geometric dissections. They are:

- *Mathematical Puzzles for Beginners and Enthusiasts*, by Geoffrey Mott-Smith, Blakiston Company: Philadelphia, 1946.
- *Mathematical Recreations & Essays*, by W. W. Rouse Ball, revised by H. S. M. Coxeter, 11th edition, Macmillan and Co.: London, 1939.
- *Amusements in Mathematics*, by Henry Ernest Dudeney, Thomas Nelson and Sons, Ltd.: London, 1917.

Gift inscriptions in these books indicated something of a family affair. The first book was a birthday gift to Freese from his son Bill in 1947, the second a gift to him from his wife Winnie in 1948, and the third a Christmas gift from Winnie in 1952. In each book Freese neatly noted, in red pencil, some thoughts about geometric dissections.

In Mott-Smith's book, Freese underscored Mott-Smith's observation that a 2-piece "step-cut" dissection transforms a $(n-1)^2 \times n^2$ rectangle into a $n(n-1)$-square. Freese also claimed that he had found a 6-piece dissection of a regular pentagon to a square that was different from the dissection of Dudeney's that Mott-Smith had given.

Freese underlined sentences in the "Minimum Dissections," "Puzzle Dissections," and "Macaulay's Four-Part Dissections" sections of Coxeter's revision of the Rouse Ball book. These sentences emphasized the goal of minimizing the number of pieces. Freese inserted a careful analysis of the 5-piece dissection of a regular octagon to a square. He also made special note of the 4-piece dissection of a Greek Cross to a square, the 4-piece dissection of an isosceles right triangle to a Greek Cross, the 6-piece dissection of a regular pentagon to a square, and the 4-piece dissection of an equilateral triangle to a square.

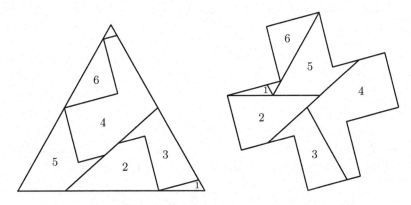

2.9: Dudeney's dissection of a Greek Cross to an equilateral triangle

In Dudeney's book, Freese noted how Dudeney's dissection of a Greek Cross to an equilateral triangle (in Figure 2.9) is similar to the 4-piece dissection of an equilateral triangle to a square. Freese also identified the position of 2 hinges, but seemed unaware that the dissection is fully hingeable. In my 2002 book I established the full hingeability, which is on display in Figure 2.10, where the circled inset for piece 6 shows the neck of that piece under 5-fold magnification. The irregular shape of the pieces might seem a flaw, but the use of an extra hinge to connect pieces 2 through 5 into a cycle that forces those four pieces to move in unison is a point of endearing artistry that more than compensates for any possible flaw. Freese also noted that he had given a 5-piece (unhingeable) solution to the problem in Plate 19 of his manuscript.

Freese's Contacts with 'Professional' Mathematicians

When Freese worked on his geometric dissections manuscript, he corresponded with several mathematicians, including Dorman Luke, H. Martyn Cundy, and Jekuthiel Ginsburg. Luke, whom we have already encountered on page 12, studied, drew, and constructed models of various polyhedra, and had been acknowledged by H. S. M. Coxeter in the preface of the first edition of his book *Regular Polytopes* in 1947. In 1952, H. Martyn Cundy and A. P. Rollett included Luke's

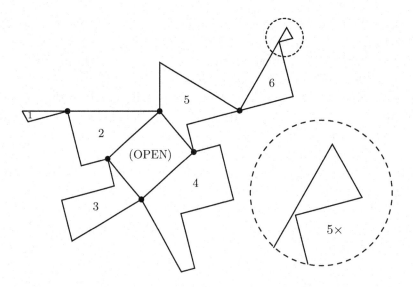

2.10: Hinged pieces for Dudeney's Greek Cross to an equilateral triangle [GNF]

stellated forms in their book *Mathematical Models.* Luke published "Stellations of the rhombic do-decahedron" in the *Mathematical Gazette*, vol. 41, pp. 189-194, in 1957, and Magnus J. Wenninger featured the so-called "Dorman Luke construction" prominently in his 1983 book *Dual Models.*

H. Martyn Cundy was the assistant editor of the *Mathematical Gazette.* Freese had sent Cundy a selection of plates from his manuscript, and Cundy had recommended that Freese get in touch with Harry Lindgren. When Dorman Luke informed Cundy of Freese's death, Cundy wrote a letter to Winifred encouraging her to send a copy of the manuscript to Lindgren.

2.11: Jekuthiel Ginsburg with Albert Einstein, ca. 1950

Freese had also been in touch with Jekuthiel Ginsburg, a gentle, absent-minded Polish-Jewish emigre professor of mathematics at Yeshiva College in New York City. At Yeshiva, Ginsburg had been chairman of the mathematics department and later the director of the mathematics institute.

He had founded the journal *Scripta Mathematica*, served as its longtime editor, and published one of Freese's brain teasers the year before Freese passed away. Unbeknownst to Winifred, Ginsburg had passed away three weeks before her husband. Thus Ginsburg never received the following disheartening letter that Winifred wrote on January 28, 1958. It underscores the importance to Freese of his interaction with mathematicians and his expectations with respect to his manuscript.

> Dear Mr. Ginsberg:
>
> I am sorry to have to tell you that my husband died last October 23rd. There is not really any excuse for my delay in writing this letter except that I wasn't very happy about writing it. Your correspondence meant a great deal to Ernest, not only because of the satisfaction he had from knowing some of his work was valued by you and considered worth publishing, but from the enjoyment he got out of communicating with someone of your caliber. He had very few such contacts and none, I think, that he valued as highly.
>
> This brings me to the subject of his work. His Dissective Geometry, of which he spoke to you, had occupied a good deal of his time for some years past, and should be looked over by someone capable of evaluating it. It is in a loose-leaf notebook, but is not by any means to be considered notes; all the pages are in finished form. Would you care to look it over, or is there someone in this area you think I should show it to? Mr. Cundy of the Mathematical Gazette in England expressed a wish to see it if I had made no other arrangements, but I see no reason to send it out of the country if it can serve any purpose here. Of course it would be very gratifying to have it published (though Ernest always realized that it was not a commercial product) but if it is not published, and has material in it that should be made available to others, I would be very glad for that to be arranged.

We have already seen that Freese's wife Winifred had played a crucial role in encouraging Freese's work on dissections, by virtue of giving him two math puzzle books that contained geometric dissections, and probably the third one as well, by arranging the gift from their 10-year-old son. That this encouragement came from someone who brought a formidable intellect into the family is emphasized by the fact that she had earned a BA from Scripps College in 1934 and an MA from Radcliffe College in 1935, and was the assistant editor of the Huntington Library Quarterly from 1962 to 1981. Yet the lack of any response to Winifred's letter by Jekuthiel Ginsburg helps to explain her ultimate failure to promote her husband's manuscript. Finally, a full sixty years later, we see here Freese's manuscript published!

And in the meantime, many people worked on geometric dissections, unaware, for the most part, of which dissection problems Freese had worked on and what magnificent dissections he had found. In the following chapters we shall learn how dissectionists reinvented some of Freese's dissection problems and even surpassed some of his precocious solutions. Two people who made great strides beyond what Freese had done are Robert Reid and Gavin Theobald. If you can't wait until you encounter their contributions in the following chapters, just look up the references to some of their marvelous dissections in the index for this book.

2.12: Robert Reid (on the right) visiting me in Indiana, 1997

2.13: Gavin Theobald at his home in Loughborough, England, 1999

Chapter 3

Techniques, Special Properties, Hardness

When Ernest Freese started to write his manuscript, there seemed to be no standard terminology for the various dissection techniques. Thus he created his own nomenclature for them. Below is a short summary of the terms, an explanation of what he might have meant, and a listing of the plates in which these terms appear.

General Grid Method - Take one set of figures that together tile the plane, and take another set of figures that also tile the plane, where both sets have the same total area. Overlay the two tessellations so that the overlay repeats in a periodic fashion, thus obtaining a dissection of one set to the other set. We see a great example in Plate 35, where Freese showed how to produce two different dissections by overlaying the two tessellations in different ways. Lindgren had called this a *tessellation method* in his 1964 book. Read more about this method in Chapter 10 of my 1997 book.

Normal Grid Method - Freese confusingly applied this name to Plate 76, which could be called the Raking Grid Method, and then later applied the name to Plate 89, in which he dissected one figure to another figure by adding a particular third figure to each so that tilings can be made. Lindgren had called this technique *completing a tessellation*. Two other examples of this technique result in Plates 113 and 116, although Freese just called the method superposed tessellations. Read more about this in Chapter 13 of my 1997 book. Unfortunately Freese was not consistent in the use of his terms.

Offset Grid Method - Cut a figure into pieces, and rearrange the pieces to form a pattern whose replication forms an infinite strip. Do the same for a second figure of equal area. Then crosspose one strip on top of the other so that the area of overlap is precisely the area of each figure. Plates 65, 104, and 105 show wonderful examples of this technique. Based on certain conditions the strip technique is either the *plain-strip* technique (*P-strip*) or the *twin-strip* technique (*T-strip*). Read more about these methods in Chapters 11 and 12 of my 1997 book.

Raking Grid Method - Take a regular polygon that tiles the plane, such as an equilateral triangle, a square, or a regular hexagon. Create a tiling using this figure, and take a tiling that uses a larger version of the same figure, such that we can overlay one tiling on top of the other to get a dissection of n figures to m figures, where n and m are integers. Plates 17, 18, and the lower parts of Plates 71, 73, and 80 are great examples of this technique. Lindgren also called this a *tessellation method*. Read more about this method in Chapter 6 of my 1997 book.

Rhombic Method - Dissect several congruent copies of a regular polygon to a larger copy of the same polygon, using decompositions of these polygons into rhombuses and isosceles triangles. Plates 93, 118, 137, 139, and 141 are attractive examples of this technique. Read the first few pages of Chapter 17 of my 1997 book to understand this method, with the remainder of the chapter containing a more general version of the *internal structure* of polygons.

Ring Expansion - With this technique we surround a rotationally symmetric figure (normally a regular n-sided polygon) in a symmetric fashion by (usually) pairs of pieces cut from one or more symmetric figures. Plates 50, 55, 56, 70, 83, 84, 94, 96, 100, 101, 103, 106, 107, 135, and 161 display a range of such examples.

Sliced Grid Method - Cut a parallelogram into three pieces that you rearrange to form another parallelogram with the same angles but a different base and different height. Plates 15, 16, 36, and 136 provide nice examples. Read more about this method on pages 32-33 of my 1997 book. Lindgren called this method a *parallelogram slide*, or *P-slide* in his 1964 book.

Superposed Tessellations - In Plate 116 Freese used two tessellations to convert a regular dodecagon to two equilateral triangles, by adding a small regular hexagon to each group. This is an example of the *completing the tessellation*, which we have mentioned already in the second example of Freese's normal grid method.

When Freese came to write his manuscript, the primary goal of dissections, as emphasized by Coxeter in his (1939) revision of the Rouse Ball classic book, was to minimize the number of pieces. In casting about for further goals that a dissection might realize, Freese also focused on achieving identical cuts in congruent polygons. (See for example Plates 98 and 126.) In a few dissections he chose to hinge pieces together. Although Freese never mentioned it, in some dissections he oriented figures so that we can move the pieces from one figure to another without rotation. (See for example the top right of Plate 20, as well as Plates 31 and 37.) It is possible that he took such dissections for granted. Finally, Freese also played with star-like figures, enjoying the contrast to regular polygons, which are clearly convex. Thus we should probably note dissections with the above special attributes:

Hinged Dissections - Let's examine the notion of a hinged dissection, which Freese never really defined. In all of his hinged examples, he dissected one geometric figure to another, with pieces connected by hinges. The canonical example is the legendary dissection of an equilateral triangle to a square, for which Freese showed a hinging in Plate 8. However, Freese did give dissections of $n > 1$ figures to another figure, such as that of an equilateral triangle and a square of equal area to a regular hexagon, in Plate 74. In that particular case, we might reasonably say that Freese's dissection is hingeable, because we hinge together all pieces from the equilateral triangle, and we also hinge together all pieces from the square. This gives us the two "assemblages" of pieces that we see in Figure I74b. Furthermore, Freese gave dissections of $n > 1$ figures to a set of $m > 1$ figures, such as three congruent regular hexagons to four congruent hexagons, in Plate 75. In this case, any hinging of pieces should necessarily result in $m + n - 1$ assemblages, where we should hinge all pieces in any given assemblage together. With $n = 3$ and $m = 4$, this gives us the $4 + 3 - 1 = 6$ assemblages of pieces that we see in Figure I75a. You can read a more complete discussion of these points in the first chapter of my 2002 book, but my examples in this current book are by and large self-explanatory, and also great fun!

Translational Dissections - A dissection possesses *translation with no rotation*, or is *translational*, if we can move the pieces from one figure to another without rotating any of them. We may also talk about a dissection of a set of figures to a different set of figures. This property of translationality stands in contrast to the property of hingeability, which involves rotation rather than translation. As early as 1864 Philip Kelland identified a dissection that was both hingeable and translational. We might surmise that for any pair of polygonal figures of the same area, we could find a translational dissection with a finite number of pieces, much as we can for finding a hingeable dissection. However, in 1951 Hugo Hadwiger and Paul Glur proved that this is not the case. Indeed, there is no translational dissection of a square to an equilateral triangle in any finite number of pieces. And yet there are many dissection problems for which there are translational dissections, including many that appear to have a minimum number of pieces over all dissections, translational or not, for that problem.

Freese did not seem to have been aware of this issue. Hadwiger and Glur used the German expression "translative Zerlegungsgleichheit" to indicate a dissection with translation but no rotation. It seems that my (1978-1979) article was the first publication in English to specifically discuss several translational dissections. Although Freese did not acknowledge this issue, he did show, perhaps serendipitously, certain dissections in a way that highlighted their translational nature. In Plate 64, he tilted the square so that its pieces are oriented in the same way as the pieces in the hexagon. In Plate 65, not only did he tilt the two squares so that their pieces matched the orientation of the pieces in the hexagon, but he also numbered all pieces so that the correspondence was clear. Similar examples appear in the lower parts of Plates 67 and 69, and in the middle dissection of Plate 72.

On the other hand, Freese presented many translational dissections in a way that indicated that he ignored or did not recognize the presence of that property. He often drew polygons so that they sat on a side, even though this would work against the display of their translational property. We can see this in Plates 66 and 79, for example. In the case of dissections of multiple copies of one figure to another, Freese might draw the multiple copies and yet not tilt them, leaving it unclear whether the dissection is translational. One such example is in the top part of Plate 75. Thus Freese's layout of dissections may have been subconscious rather than deliberate, missing opportunities to emphasize the favorable properties.

Turning Over Pieces - It is sometimes possible to use fewer pieces if we perform a dissection for which we must turn over some of the pieces. Should we view the corresponding special property as "turning over pieces" or as "not turning over pieces"? For such questions we have no absolute answers, only opinions! In his manuscript, Freese presented only dissections for which he did not turn over pieces. Furthermore, he went out of his way several times (in Plates 88, 134, and 169) to avoid turning over pieces, even if it meant using more pieces. In particular, to avoid turing over a piece in Plate 134, he cut a right triangle into two isosceles triangles, which he then could reassemble to form the mirror-image right triangle. In a letter to Dorman Luke dated November 4, 1956, Freese asserted his view:

> You are entirely correct in repeating my criticism of Cundy's "dissection" of the dodecagon (Cundy – pages 22 + 23) wherein he states that "six of the pieces must be turned over".

Thus if we want to commend ourselves for a sense of fair play, then maybe we should not be satisfied dissecting two congruent 15-sided regular polygons to one as in in Figure I135a, in which we turn over seven pieces. Instead, we should go on to produce Figure I135b, in which we do not turn over pieces.

Stars and Pseudo-stars - Readers may be excited to see stars in the form of pentagrams, hexagrams, octagrams, decagrams, and dodecagrams in the index for Freese's manuscript. As discussed in Harry Lindgren's 1964 book and my 1997 book, for integers $p > 4$ and q with $1 < q < p/2$, we have a *star* $\{p/q\}$, where each vertex is connected to its qth nearest vertex in a clockwise direction. And as discussed in my 2002 book, a *pseudo-star* $\{p/q\}$ need not have q be an integer but just a real number in the range $1/2 \leq q < p/2$. It is true that Freese considered only the $\{8/2\}$ as an octagram, the $\{10/2\}$ as a decagram, and the $\{12/2\}$ as a dodecagram, and thus ignored the $\{8/3\}$, $\{10/3\}$, $\{10/4\}$, $\{12/3\}$, $\{12/4\}$, and $\{12/5\}$ as stars worthy of dissection. Other than hexagrams and octagrams, no other stars seem to play a role in earlier dissections. However the $\{8/3\}$ makes an appearance in Plate 171 as the "co-octagram." Also, the pseudo-stars $\{5/1.5\}$ and $\{9/3.5\}$ make appearances in Plate 171 as the "co-pentagram" and "co-nonagram," respectively. The co-pentagram makes a repeat appearance six plates later, in Plate 177, there referred to as a "concave pentagon." And, as I discuss in the commentary for Plates 26 and 61, Freese also dissected pseudo-stars that he called the "stellated nonagon" and "stellated decagon." Thus he did break new ground in dissecting star-like figures, even though he did not take on some of the stars that Lindgren dissected in his 1964 book.

So far, we have encountered some mind-boggling dissections. We have also seen a few techniques that have helped people discover dissections that appear to be minimal. You may well wonder how difficult it is to identify such dissections. A century ago the puzzlists Sam Loyd and Henry Dudeney asked for dissections with the fewest possible number of pieces. Except for a very few special cases, we have no convincing proof that a known dissection is minimal. Not even for that 4-piece dissection of an equilateral triangle to a square, in Figure 1.9, or for that 5-piece dissection of a regular octagon to a square, in Figure 1.2! And it is hard, in a mathematical sense, to determine the minimum number of pieces for some specified dissection problem. In 2016 Jeffrey Bosboom, Erik D. Demaine, Martin L. Demaine, Jayson Lynch, Pasin Manurangsi, Mikhail Rudoy, and Anak Yodpinyanee showed that it is NP-hard to get within a constant factor of minimum. A problem that is NP-hard is at least as hard as the notorious traveling salesman problem and thousands of similar combinatorial optimization problems that require optimal solutions.

Okay, but suppose that we don't worry about having a minimum solution and just would like a hinged dissection. You were probably amazed by the hinged dissection of an equilateral triangle to a square in Figure 1.10, or the hinged dissection of 6-pointed star to an equilateral triangle, in Figure 1.14. Yes, but given some arbitrary dissection and the position of hinges in it, how can you determine if you can swing the pieces around to move from one specified configuration to another specified configuration? Note that the pieces must stay on a flat surface as you swing them and are not allowed to go through each other! That problem appears, mathematically speaking, to be even harder. In 2003 Robert Hearn, Erik Demaine, and I showed that this problem is PSPACE-hard. (For certain 2-person games, determining whether the first player has a winning strategy is PSPACE-hard. Computer scientists conjecture that PSPACE-hard problems are even harder than NP-hard problems. You can find formal definitions of NP-hard and PSPACE-hard in appropriate textbooks, such as the 2013 book by Michael Sipser.) Altogether, the problems associated with geometric dissections can be extremely challenging, which makes the dissections in this book even more awe-inspiring.

Now that we have identified organization, nomenclature, and various features to be found in Freese's manuscript, let's dive into both the manuscript and my commentary. Freese started gently, considering dissections of isosceles triangles. And so shall we, as we shadow his work on these symmetrical objects.

Chapter 4

Freese's Title Page and Descriptive Index

When an author sets out to write a treatise, book, or manuscript, he or she usually employs some sort of organizing principle. Ordering events by time works well for a history or biography. For a manuscript about geometric dissections, the author might organize each chapter around dissections that use a particular technique or around dissections that involve a certain figure or class of figures. For example, Harry Lindgren organized one chapter of his 1964 book around dissections created by using tessellations and another chapter around stars. In my 1997 book, I had much the same plan. On the other hand, Ernest Freese didn't write chapters, although on his title page he indicated that he arranged his plates primarily by increasing number of sides of polygons.

I have chosen a similar approach in this book, handling the plates in numerical order, and grouping plates in a similar fashion to Freese's ordering. This grouping does not produce what readers might view as true chapters, because Freese didn't really tie these sections together with explanatory text. Nonetheless, I have called these groupings "chapters" and have placed appropriate "textual glue" at the beginning and end of each such grouping. Following the introduction, commentary, and conclusion for each chapter, I then include Freese's plates for that grouping.

I have diverged a bit from Freese's plan, which he had outlined on his title page. Rather than making separate short chapters for rectangles and for simple crosses, I swept those up into a lengthy chapter dominated by squares. When it seemed reasonable to mention a few dissections of 7-sided figures, I smuggled them into chapters dealing with 5-sided and 8-sided figures. And when confronted with the prospect of short chapters, each devoted to one of several many-sided polygons, I took the plunge and amalgamated them into just one chapter with an appropriate chapter title!

And true (perhaps) to Freese's spirit, I have stuck my neck out by not acquiescing with certain traditional editorial concerns: I didn't want to clutter Freese's beautiful plates by adding the page numbers and running heads that you might find in many books. Thus Freese's plates have no page numbers. Only the pages that I wrote have page numbers, and they proceed sequentially. Although this is a bit nonstandard, readers are smart enough to deal with it.

Thus if you wish to find a particular plate, given its plate number, go to my table of contents and find the chapter that contains that particular plate number. Get the first page number for that chapter and the first page number for the next chapter. Your desired plate will be somewhere between those two pages. When you search for a plate number, you will open the book to either a plate or a page, and either way you will know whether to look forward or backward to narrow your search. I probably didn't need to explain this to you, but I just don't want to have an editor worry that readers can't handle it when Freese's plates don't sport page numbers. You can do it!

Following are Freese's title page and his twenty pages of index, embedded with all manner of unexpected nuggets. Happy prospecting!

GEOMETRIC TRANSFORMATIONS ★

BY
ERNEST IRVING FREESE

· · · · · · · · · ·

A GRAPHIC RECORD of explorations and discoveries
in the diversional domain of DISSECTIVE GEOMETRY

· · · · · · · · · ·

Under the following group-headings, all transformations
that have to do with the particular figure heading the
group will be found immediately in the DESCRIPTIVE INDEX
on the INDEX PAGES given opposite each such heading.

GROUP HEADING	INDEX PAGE
ISOSCELES TRIANGLE	1, 2.
EQUILATERAL TRIANGLE	2, 3, 4.
RECTANGLE	4.
SQUARE	4, 5, 6, 7, 8.
GREEK CROSS, LATIN CROSS, QUADRATE CROSS, STEPPED CROSS, SWASTIKA.	8.
PENTAGON & PENTAGRAM	8, 9.
HEXAGON & HEXAGRAM	9, 10, 11.
OCTAGON & OCTAGRAM	11, 12.
NONAGON	12.
DECAGON & DECAGRAM	12, 13.
DODECAGON & DODECAGRAM	13, 14.
PENTADECAGON	14.
HEXADECAGON	15.
ICOSAGON	15.
ICOSITETRAGON	15.
MISCELLANEOUS TRANSFORMATIONS	15 to 20, incl.
★ COMPRISING **200 PLATES** OF EXPOSITORY EXAMPLES.	

NOTES: All polygons are "regular" if not otherwise evident.

All dissections are made with the least possible number
of pieces except when made under other specific conditions.

DESCRIPTIVE INDEX of

GEOMETRIC TRANSFORMATIONS

BY
ERNEST IRVING FREESE

The indexing proceeds with the increase in the number of sides of the polygon at the head of each group of transformations.

Transformations of the
ISOSCELES TRIANGLE

PLATE 1:

The UNIQUE ISOSCELES TRIANGLE that may be dissected in four different ways to form differing PAIRS of equiareal ISOSCELES TRIANGLES.

PLATE 2:

The only two ISOSCELES TRIANGLES for each of which the pieces of a 6-part dissection will form either of 2 different PAIRS of equiareal ISOSCELES TRIANGLES.

PLATE 3:

An ISOSCELES RIGHT TRIANGLE = an EQUILATERAL TRIANGLE.
An ISOSCELES RIGHT TRIANGLE = a SECTOR of a DODECAGON.
An ISOSCELES RIGHT TRIANGLE = a SEGMENT of a DODECAGON.

PLATE 4:

An ISOSCELES TRIANGLE = a PENTAGON.
An ISOSCELES TRIANGLE = a HEXAGON.
An ISOSCELES TRIANGLE = an OCTAGON...two examples.

PLATE 5:

2 similar ISOSCELES TRIANGLES with respective sides in the ratio of 1:2 = a PENTAGON.
4 similar ISOSCELES TRIANGLES with respective sides in the ratio of 1:2:3:4 = a TRIACONTAGON (POLYGON of 30 sides).

PLATE 6:

Any ISOSCELES TRIANGLE = a PAIR of equiareal ISOSCELES TRIANGLES.

PLATE 7:

An ISOSCELES TRIANGLE WITH 30° BASE ANGLES = an EQUILATERAL ONE.
An ISOSCELES TRIANGLE WITH 30° BASE ANGLES = 3 EQUILATERAL ONES.
3 similar equiareal ISOSCELES TRIANGLES = 2 EQUILATERAL ONES.

PLATE 8:

General method of SQUARING an ISOSCELES TRIANGLE by cutting it into 4 pieces and turning it "inside out."

PLATE 9:

4 similar equiareal ISOSCELES TRIANGLES = 9 similar equiareal ISOSCELES TRIANGLES.

PLATE 27:

ISOSCELES TRIANGLES derived from special dissections of SQUARES.
2 similar equiareal ISOSCELES TRIANGLES derived from 1 SQUARE.

PLATE 28:

Two different ISOSCELES TRIANGLES and a SQUARE that are each formed by the SAME 5 PIECES.
An ISOSCELES TRIANGLE forming a DODECAGON SECTOR also forms a SQUARE from the SAME 3 PIECES. (For 4 PIECES see PLATE 8)
An ISOSCELES TRIANGLE forming a DODECAGON SEGMENT also forms a SQUARE from the SAME 4 PIECES.

PLATE 125:

An ISOSCELES RIGHT TRIANGLE = a DODECAGON.

Transformations of the
EQUILATERAL TRIANGLE

PLATE 3:

An EQUILATERAL TRIANGLE = an ISOSCELES RIGHT TRIANGLE.

PLATE 6:

2 EQUILATERAL TRIANGLES with respective sides in a given ratio = 1 EQUILATERAL TRIANGLE.

PLATE 7:

An EQUILATERAL TRIANGLE = an ISOSCELES TRIANGLE with base angles of 30°.
3 equiareal EQUILATERAL TRIANGLES = an ISOSCELES TRIANGLE with base angles of 30°.
2 equiareal EQUILATERAL TRIANGLES = 3 equiareal ISOSCELES TRIANGLES of the same base and altitude.

PLATE 8:

A simple method of SQUARING an EQUILATERAL TRIANGLE by cutting it into 4 pieces and turning it "inside out".
(See foot of PLATE 79 for a 5-piece SQUARING.)

PLATE 9:

An EQUILATERAL TRIANGLE = 2 equiareal SQUARES.
4 equiareal EQUILATERAL TRIANGLES = 9 equiareal ONES.

PLATE 10:

An EQUILATERAL TRIANGLE = 3 equiareal SQUARES.
9 equiareal EQUILATERAL TRIANGLES = 16 equiareal ONES.

PLATE 47: An EQUILATERAL TRIANGLE = a PENTAGON.

PLATES 11, 12: An EQUILATERAL TRIANGLE = a HEXAGON.

PLATE 13: An EQUILATERAL TRIANGLE = an OCTAGON.

PLATE 166: An EQUILATERAL TRIANGLE = a NONAGON.

PLATE 14:

1 EQUILATERAL TRIANGLE = 2 equiareal ONES.
1 EQUILATERAL TRIANGLE = 3 equiareal ONES.

PLATE 15:

1 EQUILATERAL TRIANGLE = 5 equiareal ONES.

PLATES 16, 17:

1 EQUILATERAL TRIANGLE = 7 equiareal ONES.

PLATES *16, 18*:

1 EQUILATERAL TRIANGLE = 13 equiareal ONES.

PLATE *19*:

1 EQUILATERAL TRIANGLE = 12 equiareal ONES.
3 equiareal EQUILATERAL TRIANGLES = 7 equiareal ONES.
An EQUILATERAL TRIANGLE = a GREEK CROSS.

PLATE *20*:

3 equiareal EQUILATERAL TRIANGLES = 4 equiareal ONES.
2 equiareal EQUILATERAL TRIANGLES = a SQUARE.

PLATE *21*:

1 EQUILATERAL TRIANGLE = 4 EQUILATERAL TRIANGLES with respective sides in the ratio of 1:2:3:4.
1 EQUILATERAL TRIANGLE = 6 EQUILATERAL TRIANGLES with respective sides in the ratio of 1:2:3:4:5:6.

PLATE *22*:

1 EQUILATERAL TRIANGLE = 9 EQUILATERAL TRIANGLES with respective sides in the ratio of 2:5:8:11:14:17:20:23:26.

PLATE *23*:

1 EQUILATERAL TRIANGLE = 2 equiareal HEXAGONS.
2 equiareal EQUILATERAL TRIANGLES = 3 equiareal HEXAGONS.

PLATE *24*:

5 equiareal EQUILATERAL TRIANGLES = a HEXAGON.

PLATE *25*:

20 equiareal EQUILATERAL TRIANGLES = a HEXAGON.

PLATE *26*:

3 EQUILATERAL TRIANGLES = a STELLATED NONAGON.

PLATE *47*:

An EQUILATERAL TRIANGLE = a PENTAGON.

PLATE *72*:

14 equiareal EQUILATERAL TRIANGLES = a HEXAGON.

PLATE *74*:

An EQUILATERAL TRIANGLE plus an equivalent SQUARE = a HEXAGON.

PLATE *75*:

8 equiareal EQUILATERAL TRIANGLES = 3 equiareal HEXAGONS.

PLATE *76*:

An EQUILATERAL TRIANGLE = 6 equiareal HEXAGONS.
3 equiareal EQUILATERAL TRIANGLES = 2 equiareal HEXAGONS.
8 equiareal EQUILATERAL TRIANGLES = 1 HEXAGON.

PLATE *79* (foot of plate):

A 5-piece dissection of an EQUILATERAL TRIANGLE and a SQUARE. (See PLATE 8 for a 4-piece dissection.)

PLATE *80*:

An EQUILATERAL TRIANGLE = 14 equiareal HEXAGONS.
An EQUILATERAL TRIANGLE = 3 HEXAGONS with respective sides in the ratio of 1:2:3.

PLATE 81:

2 equiareal EQUILATERAL TRIANGLES = a HEXAGON.
3 equiareal EQUILATERAL TRIANGLES = a HEXAGON.

PLATE 82:

3 EQUILATERAL TRIANGLES with areas in the ratio of 1:2:3 make a HEXAGON.

PLATE 83:

4 equiareal EQUILATERAL TRIANGLES = a HEXAGON.

PLATE 84:

12 equiareal EQUILATERAL TRIANGLES = a HEXAGON.

PLATE 85:

An EQUILATERAL TRIANGLE = a HEXAGRAM.

PLATE 104:

An EQUILATERAL TRIANGLE = a DECAGON.

PLATE 112:

An EQUILATERAL TRIANGLE = a DODECAGON.

PLATE 113:

2 equiareal EQUILATERAL TRIANGLES = a DODECAGON.

Transformations of the RECTANGLE

PLATE 33:

RECTANGLES SQUARED by one stepped cut; a special 2-piece dissection, and the simple formulas for finding all such related RECTANGLES and SQUARES.

PLATE 34:

RECTANGLES SQUARED by 3-piece dissection; the two general methods and the limiting cases of each.

PLATE 44:

RECTANGLES that may be SQUARED by INTEGRAL 3-piece dissection, and the simple formulas for finding all such related RECTANGLES and SQUARES in both of two possible series.... the first four of each series being given on this PLATE in tabulated form.

Transformations of the SQUARE

PLATE 8:

General method of SQUARING the ISOSCELES TRIANGLE, inclusive of the EQUILATERAL TRIANGLE (used as an example), by cutting it into 4 pieces and turning it "inside out." (See foot of PLATE 79 for a 5-piece SQUARING of the EQUILATERAL TRIANGLE.

PLATE 9:

2 equiareal SQUARES = an EQUILATERAL TRIANGLE.

PLATE 10:

3 equiareal SQUARES = an EQUILATERAL TRIANGLE.

PLATE 20:

1 SQUARE = 2 equiareal EQUILATERAL TRIANGLES.

PLATE 27:

Special dissections of SQUARES, each yielding an ISOSCELES TRIANGLE.
Special dissection of a SQUARE yielding 2 equiareal ISOSCELES TRIANGLES.

PLATE 28:

A SQUARE = either of 2 different ISOSCELES TRIANGLES.
A SQUARE = a DODECAGON SECTOR. (Also see foot of PLATE 8)
A SQUARE = a DODECAGON SEGMENT.

PLATE 29:

SQUARING the OCTAGRAM.
2 equiareal SQUARES = an OCTOGRAM.
A SQUARE = 2 equiareal GREEK CROSSES.
A SQUARE = 4 equiareal GREEK CROSSES.

PLATE 30:

SQUARING the GREEK CROSS.
2 equiareal SQUARES = a GREEK CROSS.
2 equiareal SQUARES = a STEPPED CROSS.

PLATES 30, 31:

SQUARING the STEPPED CROSS.... two examples, one rational.

PLATE 32:

SQUARING the QUADRATE CROSS.
SQUARING the SWASTIKA.

PLATE 33:

SQUARING the RECTANGLE by one stepped cut; a special
2-piece dissection, and the simple formulas for finding
all such related SQUARES and RECTANGLES.

PLATE 34:

SQUARING the RECTANGLE by 3-piece dissection; the
two general methods and the limiting cases of each.

PLATE 35:

SQUARING any 2 SQUARES by the general Grid Method.

PLATE 36:

SQUARING 3 equiareal SQUARES.
SQUARING 7 equiareal SQUARES.
SQUARING 10 equiareal SQUARES.

PLATE 37:

SQUARING 5 equiareal SQUARES.
SQUARING 8 equiareal SQUARES.
2 equiareal SQUARES = 5 equiareal SQUARES.
2 equiareal SQUARES = 9 equiareal SQUARES or 1 SQUARE.
2 equiareal SQUARES = 13 equiareal SQUARES.

PLATE 38:

1 SQUARE = 3 SQUARES with respective areas in the ratio of 2:3:4.

PLATE 39:

A _rational_ 5-piece dissection of a SQUARE = 3 unequal SQUARES.
A _rational_ 6-piece dissection of a SQUARE = 4 unequal SQUARES.

PLATE 40:

1 SQUARE = 6 SQUARES with areas in the ratio of 1:3:5:7:9:11,

PLATE 41:   or 3 in the ratio of 1:3:5, as shown.

1 SQUARE = 4 SQUARES with respective sides in the ratio of
the arithmetical progression, 13, 19, 25, 31; _a rational_ dissection.

PLATE 42:

1 SQUARE = 9 SQUARES with respective sides in the ratio of
the arithmetical progression, 2, 5, 8, 11, 14, 17, 20, 23, 26, which
yields an _all-rational_ dissection.

PLATE 43:

A _rational_ conversion of 4 equiareal SQUARES into another
4 SQUARES with respective sides in the ratio of the
arithmetical progression, 7, 15, 23, 31, with a common difference of 8.

PLATE 44:

The special cases of SQUARED RECTANGLES (yielded by
either of the 3-piece dissective methods of _PLATE 34_) for which
all dimensions become _INTEGRAL_, and the simple formulas
for finding all such related SQUARES and RECTANGLES,
the first four of each series being tabulated on this PLATE.

PLATE 45:

Transformation of a SQUARE FRAME, or HOLLOW SQUARE,
into a SOLID SQUARE.

PLATE 46:

Conversion of a HOLLOW SQUARE into a larger or smaller
ONE, and the numerical relations between such SQUARES.

PLATE 48:

SQUARING the PENTAGON.

PLATE 57:

A SQUARE = 2 identical PENTAGONS.

PLATE 58:

4 SQUARES with areas in the ratio of 1:2:3:4 = a PENTAGON.

PLATE 59:

SQUARING the PENTAGRAM.

PLATE 64:

SQUARING the HEXAGON.

PLATE 65:

2 equiareal SQUARES = a HEXAGON.

PLATE 66:

A SQUARE = 2 equiareal HEXAGONS.

PLATE 67:

3 IDENTICAL SQUARES = a HEXAGON.
A SQUARE = 3 equiareal HEXAGONS.

PLATE 74:

A SQUARE plus an EQUILATERAL TRIANGLE of the same area = a HEXAGON. (An example of composite dissection.)

PLATE 79:

2 SQUARES with areas in the ratio of 1:2 = a HEXAGON.

PLATE 85:

SQUARING the HEXAGRAM.

PLATES 89, 90:

SQUARING the OCTAGON.

PLATE 91:

2 equiareal SQUARES = an OCTAGON.

PLATE 92:

4 identical SQUARES = an OCTAGON.

PLATE 99:

SQUARING the NONAGON.

PLATE 105:

SQUARING the DECAGON.

PLATE 109:

2 SQUARES with an area-ratio of 2:3 = a DECAGON.

PLATE 110:

4 SQUARES with an area-ratio of 1:2:3:4 = a DECAGON.

PLATE 114:

3 identical SQUARES = a DODECAGON.

PLATES 114, 115:

SQUARING the DODECAGON.

PLATE 123:

A SQUARE = 2 identical DODECAGONS.
2 equiareal SQUARES = a DODECAGON.

PLATE 124:

2 SQUARES with an area-ratio of 1:2 = a DODECAGON.

PLATES 125, 126:

6 equiareal SQUARES = a DODECAGON.

PLATE 126:

SQUARING the DODECAGON DELTOID.

PLATE 127:

12 equiareal SQUARES = a DODECAGON.

PLATE 128:

A SQUARE = 3 equiareal DODECAGONS.
3 equiareal SQUARES = 2 equiareal DODECAGONS.

PLATE 129:

A SQUARE = 4 identical DODECAGONS.

PLATE 130:

A SQUARE = 6 equiareal DODECAGONS.

PLATE 134:

SQUARING the PENTADECAGON. (15 SIDES)

PLATE 136:

SQUARING the HEXADECAGON. (16 SIDES)

PLATE 138:

SQUARING the ICOSAGON. (20 SIDES)

PLATE 140:

SQUARING the ICOSITETRAGON. (24 SIDES)

Transformations of the
GREEK CROSS, LATIN CROSS, QUADRATE CROSS, STEPPED CROSS and SWASTIKA

PLATE 19:

A GREEK CROSS = an EQUILATERAL TRIANGLE.

PLATE 29:

2 identical GREEK CROSSES = a SQUARE.
4 equiareal GREEK CROSSES = a SQUARE.

PLATE 30:

SQUARING the GREEK CROSS.....two methods.
A GREEK CROSS = 2 identical SQUARES.
A STEPPED CROSS = 2 identical SQUARES.

PLATES 30, 31:

SQUARING the STEPPED CROSS.

PLATE 31:

A 2-IN-1 GREEK CROSS, by the "RING EXPANSION" method.
A 4-IN-1 GREEK CROSS, as per General Theorem on PLATE 69.

PLATE 32:

SQUARING the QUADRATE CROSS and the SWASTIKA.

PLATE 132:

A GREEK CROSS = a DODECAGON.
A LATIN CROSS = a DODECAGON.

Transformations of the
PENTAGON and PENTAGRAM

PLATE 4:

A PENTAGON = one SECTOR of a PENTAGON.

PLATE 5:

A PENTAGON = 2 similar ISOSCELES TRIANGLES with an area-ratio of 1:2.

PLATE 47:

A PENTAGON = an EQUILATERAL TRIANGLE.

PLATE 48:

SQUARING the PENTAGON.

PLATE 49:

A PENTAGON = a HEXAGON.

PLATE 50:

A 2-IN-1 PENTAGON..... illustrating the simple and general "RING EXPANSION" method of doubling any regular polygon whatsoever of more than 3 sides.

PLATE 51:

A 4-IN-1 PENTAGON.

PLATES 51, 52:

Some 5-IN-1 PENTAGONS, by two dissective methods.

PLATE 53:

A 9-IN-1 PENTAGON.

PLATE 54:

A 16-IN-1 PENTAGON.

PLATES 55, 56:

2 PENTAGONS, with respective sides in a given ratio, converted into 1 PENTAGON.... a general method of "RING EXPANSION" applicable to all regular polygons.

PLATE 57:

2 identical PENTAGONS = a SQUARE.

PLATE 58:

A PENTAGON = 4 SQUARES with areas in the ratio of 1:2:3:4.

PLATE 59:

A PENTAGRAM SQUARED.

PLATE 60:

A 4-IN-1 PENTAGRAM.
A 5-IN-1 PENTAGRAM.

PLATE 61:

2 identical PENTAGONS = a STELLATED DECAGON.

PLATE 62:

5 identical PENTAGONS = a PENTAGONAL "RING!"

PLATE 63:

4 identical PENTAGONS = a DECAGON RING.
10 identical PENTAGONS = a DECAGON RING.

Transformations of the
HEXAGON and HEXAGRAM

PLATES 11, 12:

A HEXAGON = an EQUILATERAL TRIANGLE.

PLATE 23:

2 equiareal HEXAGONS = an EQUILATERAL TRIANGLE.
3 identical HEXAGONS = 2 identical EQUILATERAL TRIANGLES.

PLATE 24:

A HEXAGON = 5 equiareal EQUILATERAL TRIANGLES.

PLATE 25:

A HEXAGON = 20 equiareal EQUILATERAL TRIANGLES.

PLATE 49:

A HEXAGON = a PENTAGON.

PLATE 64:

SQUARING the HEXAGON.

PLATE 65:

A HEXAGON = 2 equiareal SQUARES.

PLATE 66:

2 equiareal HEXAGONS = a SQUARE.

PLATE 67:

A HEXAGON = 3 identical SQUARES.
3 equiareal HEXAGONS = a SQUARE.

PLATE 68:

A 2-IN-1 HEXAGON.

PLATE 69:

A 3-IN-1 HEXAGON, and a 4-IN-1 HEXAGON. (Also see
note on PLATE 87)

PLATE 70:

A 6-IN-1 HEXAGON.

PLATE 71:

A 7-IN-1 HEXAGON.
4 equiareal HEXAGONS = 9 equiareal HEXAGONS.
(Also see PLATE 73)

PLATE 72:

A 12-IN-1 HEXAGON, and a 14-IN-1 HEXAGON.
3 equiareal HEXAGONS = 7 equiareal HEXAGONS.

PLATE 73:

A 13-IN-1 HEXAGON.
4 equiareal HEXAGONS = 9 equiareal HEXAGONS.
(Also see PLATE 71)

PLATE 74:

A HEXAGON = an EQUILATERAL TRIANGLE plus a SQUARE of
same area. (7 pieces)

PLATE 75:

3 identical HEXAGONS = 4 equiareal HEXAGONS.
3 equiareal HEXAGONS = 8 equiareal HEXAGONS.

PLATE 76:

A HEXAGON = 8 equiareal EQUILATERAL TRIANGLES.
A 9-IN-1 HEXAGON.
2 equiareal HEXAGONS = 3 identical EQUILATERAL TRIANGLES.
6 equiareal HEXAGONS = an EQUILATERAL TRIANGLE.
6 equiareal HEXAGONS = a HEXAGRAM.

PLATE 77:

1 HEXAGON = either 3 or 4 equiareal HEXAGONS of same pieces.

PLATE 78:

1 HEXAGON = 6 HEXAGONS with respective sides in the
ratio of 1:2:3:4:5:6.

PLATE **79**:

A HEXAGON = 2 SQUARES with an area ratio of **1:2**.

PLATE **80**:

3 HEXAGONS with respective sides in the ratio of **1:2:3** = an EQUILATERAL TRIANGLE.

14 equiareal HEXAGONS = an EQUILATERAL TRIANGLE.

PLATE **81**:

A HEXAGON = **3** identical EQUILATERAL TRIANGLES.

A HEXAGON = **2** equiareal EQUILATERAL TRIANGLES.

A HEXAGON = **2** HEXAGONS with an area-ratio of **1:3**.

PLATE **82**:

A HEXAGON = **3** EQUILATERAL TRIANGLES with an area-ratio of **1:2:3**.

PLATES **83,84**:

A HEXAGON = **4** equiareal EQUILATERAL TRIANGLES.

A HEXAGON = **12** equiareal EQUILATERAL TRIANGLES.

PLATE **85**:

A HEXAGRAM = an EQUILATERAL TRIANGLE.

A HEXAGRAM = **2** identical HEXAGONS.

SQUARING the HEXAGRAM.

PLATE **86**:

A **3**-IN-**1** HEXAGRAM.

A **4**-IN-**1** HEXAGRAM.

A HEXAGRAM = a DODECAGRAM.

PLATE **87**:

3 HEXAGONS = either of two HEXAGONAL RINGS.

(Also a **4**-IN-**1** HEXAGON, formed as noted on this <u>PLATE</u>.)

PLATE **88**:

A HEXAGON = a DODECAGON RING.

PLATE **116**:

A HEXAGON = a DODECAGON...a <u>**6**-piece</u> dissection.

PLATE **131**:

A HEXAGON = **6** component polygons of <u>same length of side</u>.

PLATE **133**:

A HEXAGON = a DODECAGRAM.

Transformations of the OCTAGON and OCTAGRAM

PLATE **4**:

An OCTAGON = one SECTOR of an OCTAGON.

An OCTAGON = one triangular SEGMENT of an OCTAGON.

PLATE **13**:

An OCTAGON = an EQUILATERAL TRIANGLE.

PLATE **29**:

An OCTAGRAM = **2** equiareal SQUARES.

PLATE 29:

SQUARING the OCTAGRAM.

PLATES 89,90:

SQUARING the OCTAGON.

PLATES 91,92:

An OCTAGON = 2 equiareal SQUARES.
An OCTAGON = 4 identical SQUARES. (Also see PLATE 153)

PLATE 93:

A 2-IN-1 OCTAGON .... two examples by different methods.

PLATE 94:

A 3-IN-1 OCTAGON.

PLATE 95:

A 4-IN-1 OCTAGON and an 8-IN-1 OCTAGON.

PLATES 96,97:

A 5-IN-1 OCTAGON and a 9-IN-1 OCTAGON.

PLATE 98:

An OCTAGON so dissected as to make either 2 or 4 equiareal OCTAGONS.

Transformations of the
NONAGON

PLATE 26:

A STELLATED NONAGON = 3 identical EQUILATERAL TRIANGLES.

PLATE 166: A REGULAR NONAGON = 1 EQUILATERAL TRIANGLE.

PLATE 99: SQUARING the NONAGON.

PLATE 100:

A 2-IN-1 NONAGON .... a typical example of the general "RING EXPANSION" method of doubling a regular polygon.

PLATE 101:

A 3-IN-1 NONAGON .... a typical example of the general "RING EXPANSION" method of tripling a regular polygon.

PLATE 102:

A 4-IN-1 NONAGON .... a typical example of quaduplication, as
PLATE 103:   per general theorem on PLATE 69.

A 5-IN-1 NONAGON .... a typical example of quintuplication by RING EXPANSION of a quadruple nucleus.

Transformations of the
DECAGON and DECAGRAM

PLATE 61:

A STELLATED DECAGON = 2 identical PENTAGONS.

PLATE 63:

A DECAGON RING = 4 identical PENTAGONS.
A DECAGON RING = 10 identical PENTAGONS.

PLATE 104:

A DECAGON = an EQUILATERAL TRIANGLE.

PLATE 105:

SQUARING the DECAGON.

PLATE 106:

A 2-IN-1 DECAGON.

PLATE 107:

A 3-IN-1 DECAGON.

PLATE 108:

A 4-IN-1 DECAGON and a 5-IN-1 DECAGON.

PLATE 109:

A DECAGON = 2 SQUARES with an area-ratio of 2:3.

PLATE 110:

A DECAGON = 4 SQUARES with an area-ratio of 1:2:3:4.

PLATE 111:

A 4-IN-1 DECAGRAM.

Transformations of the DODECAGON and DODECAGRAM

PLATE 3:

A DODECAGON SECTOR = an ISOSCELES RIGHT TRIANGLE.
A DODECAGON SEGMENT = an ISOSCELES RIGHT TRIANGLE.

PLATE 28:

A DODECAGON SECTOR = a SQUARE. (Also see foot of PLATE 8)
A DODECAGON SEGMENT = a SQUARE.

PLATE 86:

A DODECAGRAM = a HEXAGRAM.

PLATE 87:

A DODECAGON RING = a HEXAGON.

PLATE 112:

A DODECAGON = an EQUILATERAL TRIANGLE.

PLATE 113:

A DODECAGON = 2 equiareal EQUILATERAL TRIANGLES.

PLATE 114:

A DODECAGON = 3 identical SQUARES.

PLATES 114, 115:

SQUARING the DODECAGON .... two methods.

PLATE 116:

A DODECAGON = a HEXAGON .... a 6-piece dissection.

PLATES 117, 118:

A 2-IN-1 DODECAGON .... two examples by different methods.

PLATE 119:

1 DODECAGON = 2 DODECAGONS with an area-ratio of 1:2.

PLATE 120:

A 3-IN-1 DODECAGON.

PLATE 121:

A 4-IN-1 DODECAGON.

PLATE 122:

A 12-IN-1 DODECAGON.

PLATE 123:

A DODECAGON = 2 equiareal SQUARES.
2 identical DODECAGONS = a SQUARE.

PLATE 124:

A DODECAGON = 2 SQUARES with an area-ratio of 1:2.

PLATE 125:

A DODECAGON = an ISOSCELES RIGHT TRIANGLE.

PLATES 125, 126:

A DODECAGON = 6 equiareal SQUARES....two examples.

PLATE 126:

A DODECAGON DELTOID = a SQUARE.

PLATE 127:

A DODECAGON = 12 equiareal SQUARES.

PLATE 128:

3 equiareal DODECAGONS = a SQUARE.
2 identical DODECAGONS = 3 SQUARES.

PLATE 129:

4 identical DODECAGONS = a SQUARE.

PLATE 130:

6 equiareal DODECAGONS = a SQUARE.

PLATE 131:

4 equiareal DODECAGONS = 3 identical SQUARES.

PLATE 132:

A DODECAGON = a GREEK CROSS.
A DODECAGON = a LATIN CROSS.

PLATE 133:

A DODECAGRAM = a HEXAGON.

Transformations of the PENTADECAGON...15 SIDES.

PLATE 134:

SQUARING the PENTADECAGON.

PLATE 135:

A 2-IN-1 PENTADECAGON....a general method of doubling any regular polygon whatsoever of more than 3 sides, and herein named the "RING EXPANSION" method.

Transformations of the
HEXADECAGON....16 SIDES.

PLATE 136:

SQUARING the HEXADECAGON.

PLATE 137:

A 2-IN-1 HEXADECAGON.....a typical example of the "RHOMBIC" method of doubling any 4m-gon, that is, any regular polygon whose number of sides is a multiple of 4.

Transformations of the
ICOSAGON... 20 SIDES.

PLATE 138:

SQUARING the ICOSAGON.

PLATE 139:

A 2-IN-1 ICOSAGON.

Transformations of the
ICOSITETRAGON...24 SIDES.

PLATE 140:

SQUARING the ICOSITETRAGON.

PLATE 141:

A 2-IN-1 ICOSITETRAGON.

MISCELLANEOUS
TRANSFORMATIONS

PLATE 5:

The 30 SECTORS of the TRIACONTAGON re-assembled to form 4 similar ISOSCELES TRIANGLES with areas in the ratio of 1:2:3:4.

PLATE 6:

The 3-piece dissection of the GENERAL TRIANGLE to form a different one of same base and altitude.

PLATE 9:

General method of converting ANY 4 EQUAL TRIANGLES into 9 EQUAL TRIANGLES, all such being similar.

PLATE 33:

Simple formulas for finding all 2-piece INTEGRAL transformations of RECTANGLES into SQUARES.

PLATE 44:

Simple formulas for finding all 3-piece INTEGRAL transformations of RECTANGLES INTO SQUARES.

PLATE 50

The General "RING EXPANSION" method of converting 2 similar equiareal regular POLYGONS into ONE.

PLATE 142:

To cut from the center of a given EQUILATERAL TRIANGLE, SQUARE or HEXAGON, a similar polygon equal in area to $\frac{1}{13}$ th the area of the given polygon.

PLATE 143:

To dissect an EQUILATERAL TRIANGLE into a pair of unequal SCALENE TRIANGLES, each having INTEGRAL SIDES and a 60° ANGLE. (13 tabulated solutions). Dissection of an EQUILATERAL TRIANGLE into 8 SCALENE TRIANGLES, all with INTEGRAL sides.

PLATE 144:

To find an INTEGRAL RECTANGLE such that a 4-piece dissection will convert it into 2 INTEGRAL SQUARES of a given ratio. Formulated values for all cases.

PLATE 145:

A special 5-piece transformation of a RIGHT-ANGLED HEXAGON, or L-SHAPE, into a SQUARE.

PLATE 146:

A 20-IN-1 SQUARE, and a graphic demonstration of the generality of the PYTHAGOREAN THEOREM.

PLATE 147:

The "WHIRLING SQUARES," exemplifying a new concept in DISSECTIVE GEOMETRY — THE MITERED GRID.

PLATE 148:

The series of 5-piece conversions of 2 SQUARES into 1 SQUARE for which all values are RATIONAL and the SIDES and AREAS are INTEGRAL. (Formulated values and also a tabulation of the first 18 cases possible in this series). The one case for which two 4-piece conversions also exist.

PLATE 149:

SQUARING the 4 UNEQUAL SQUARES of the CHORDS of the OCTAGON; a typical example of the General "RING EXPANSION" method of SQUARING ANY NUMBER OF SQUARES, EQUAL OR NOT.

PLATE 150:

CONVERTIBLE SQUARES....from PERFORATED to SOLID, or vice versa.

PLATE 151:

Dissective transformations of CURVILINEAR ELEMENTS into SQUARES having INTEGRAL AREAS. Also see PLATE 152.

PLATE 152:

EQUIAREAL CURVILINEAR DISSECTIVE COMBINATIONS and EQUIVALENT SQUARES having INTEGRAL AREAS.

PLATE 153:

SQUARING a CUT-CORNER SQUARE by a direct 6-piece dissection, and application to the SQUARING of an OCTAGON QUADRANT, thus transforming the OCTAGON into 4 SQUARES, or into 1 SQUARE by combining the four.

PLATE 154:

Multiple transformations of a SQUARE FRAME, or HOLLOW SQUARE, either into 1 or 2 SOLID SQUARES or a GREEK CROSS.

PLATE 155:

EQUIVALENT FIGURES from the SAME 12 PIECES: the SWASTIKA, QUADRATE CROSS, SQUARE, and a ST. ANDREW'S CROSS plus a TETRASKELION.

PLATE 156:

Two ways to SQUARE a 41-SQUARE STEPPED CROSS.

PLATE 157:

SQUARING A CROSSLET.
SQUARING A TETRASKELION.

PLATE 158:

SQUARING AN EQUAL-ARMED CROSS... a General Method within stated limits which exclude the Greek cross. Formulated values for this series, & 6 RATIONAL transformations.

PLATE 159:

SQUARING the variable MALTESE CROSS... a General Method, the limiting case being the "VICTORIA" CROSS.

PLATE 160:

A General Method of dissecting a radially-symmetrical rectilinear figure to form a SQUARE, when such a SQUARE is the difference between the circumscribing SQUARE and the SQUARE of the 4 spandrels.

PLATE 161:

EXPANDING A PENTAGON by adding a PENTAGON RING formed from a DECAGON. (A special type of "RING EXPANSION" with other examples on PLATES 165 & 169.)

PLATE 162:

Cutting a PENTAGON into 5 equal areas radiating from 1 vertex.
Cutting a DECAGON into 10 equal areas radiating from 1 vertex.

PLATE 163:

A special TRIPLET of 49-IN-1 HEXAGONS in which all cuts are RATIONAL.

PLATE 164:

A 3-IN-1 CONCAVE HEXAGON = A 3-IN-1 CONVEX HEXAGON.

PLATE 165:

Expanding an OCTAGON by adding an OCTAGON "RING" formed from 4 SQUARES.

PLATE 166:

A 9-piece NONAGON makes an EQUILATERAL TRIANGLE.

PLATE 167:

Transformation of a NONAGON into 2 ISOSCELES TRIANGLES such that the vertex angle of one equals a base angle of the other. Also, a General 2-piece cut-up of any ISOSCELES TRIANGLE (having base angles of less than 60°) that also forms an ISOSCELES TRAPEZOID.

PLATE 168:

5 DECAGON assemblies, each from the same 20 pieces.

PLATE 169:

Expanding a DODECAGON by adding a DODECAGON "RING" formed from 8 EQUILATERAL TRIANGLES.

PLATE 170:

A DODECAGON = 3 equiareal ISOSCELES TRIANGLES.
A DODECAGON SECTOR = a SEGMENT = a SPANDREL.
The INSCRIBED TRIANGLE of a PENTAGON = the 4 SPANDRELS.
An OCTAGON "turns out" to be a TETRASKELION.

PLATE 171:

Transformation of REGULAR POLYGONS, each into its 2 COMPONENT POLYGRAMS.

PLATE 172:

Dissection of the DODECAGON, DECAGON & HEXAGON into different POLYGONS of SAME LENGTH OF SIDES.

PLATE 173:

RHOMBIC DISSECTION; 4 examples illustrating a theorem of DISSECTIVE GEOMETRY.

PLATE 174:

Transforming CONVEX POLYGONS into CONCAVE POLYGONS:
A REGULAR HEXAGON = a CONCAVE HEXAGON,
A SQUARE = a CONCAVE OCTAGON,
A SQUARE = a CONCAVE DODECAGON.

PLATE 175:

A REGULAR DODECAGON = 3 identical CONCAVE DODECAGONS.

PLATE 176:

SQUARING the CONCAVE HEXADECAGON.

PLATE 177:
5 DECAGONS, 2 of which are CONCAVE ones, transformed into 2 PENTAGONS.

PLATE 178:
10 PENTAGONS = 3 DECAGONS (2 of which are CONCAVE), every side of each piece and figure having the SAME LENGTH.

PLATE 179:
A COMPOSITE SQUARE formed by an EQUILATERAL TRIANGLE & a HEXAGON of equal areas. (8 pieces.)

PLATE 180:
A COMPOSITE SQUARE formed from an EQUILATERAL TRIANGLE & a DODECAGON of equal areas. (8 pieces.)

PLATE 181:
A COMPOSITE SQUARE formed from a PENTAGON & DODECAGON of equal areas. (10 pieces.)

PLATE 182:
A COMPOSITE SQUARE formed from a HEXAGON & DODECAGON of equal areas. (9 pieces.)

PLATE 183:
A COMPOSITE SQUARE formed from an OCTAGON & DODECAGON of equal areas. (11 pieces.)

PLATE 184: A COMPOSITE SQUARE formed from an EQUILATERAL TRIANGLE & a HEXAGON & a DODECAGON of equal areas. (17 pieces.)

PLATE 185: A COMPOSITE SQUARE formed from a PAIR of identical EQUILATERAL TRIANGLES & a PAIR of identical PENTAGONS, all of equal areas. (14 pieces)

PLATE 186: A COMPOSITE SQUARE formed from an EQUILATERAL TRIANGLE & a PENTAGON & HEXAGON & OCTAGON, all of equal areas. (19 pieces.)

PLATE 74: A COMPOSITE HEXAGON formed from an EQUILATERAL TRIANGLE & a SQUARE of equal areas. (7 pieces.)

PLATE 187:
A COMPOSITE HEXAGON formed from an EQUILATERAL TRIANGLE & a SQUARE & PENTAGON of equal areas. (13 pieces.)

PLATE 188:
A COMPOSITE DODECAGON formed from a SQUARE & HEXAGON & OCTAGON of equal areas. (17 pieces.)

PLATE 189: Multiple dissection of RIGHT TRIANGLES into ISOSCELES TRIANGLES, and formula for determining all possible cases.

PLATE 190: Multiple dissection of ISOSCELES TRIANGLES into ISOSCELES TRIANGLES, and the limiting conditions for 2-IN-1, 3-IN-1, 4-IN-1 and 5-IN-1 dissections of this type.

PLATE 191: Some INTEGRAL examples of the multiple dissection of ISOSCELES TRIANGLES into 3, 4 and 5 ISOSCELES TRIANGLES, and general forms for determining all primary INTEGRAL ISOSCELES TRIANGLES.

PLATE 192: An EQUILATERAL TRIANGLE formed from 13 OBLIQUE TRIANGLES, each having a 60° ANGLE and INTEGRAL SIDES, and the formation, by successive subtraction, of other EQUILATERAL TRIANGLES having the same properties as the original.

PLATE 193: Four "PROBLEMS" in Dissective Geometry and the "ANSWERS".

PLATE 194: A RATIONAL 4-piece transformation of an OBLIQUE TRIANGLE with INTEGRAL sides into a regular HEXAGON.

PLATE 195: A RATIONAL transformation of 10 equiareal regular HEXAGONS into an OBLIQUE SCALENE TRIANGLE with INTEGRAL sides.

PLATE 196: A UNIQUE ALL-INTEGRAL transformation of 10 equiareal OBLIQUE SCALENE TRIANGLES, having INTEGRAL sides and a 60° ANGLE, into a single SCALENE TRIANGLE having INTEGRAL sides and a 120° ANGLE.

PLATE 197: 4 PAIRS of non-similar INTEGRAL equiareal RIGHT TRIANGLES, each PAIR dissected to form 2 different PAIRS of non-similar equiareal INTEGRAL RIGHT and OBLIQUE TRIANGLES.

PLATE 198: Composite dissection of INTEGRAL RIGHT TRIANGLES, wherein the dissective elements are 2 non-similar INTEGRAL RIGHT TRIANGLES and 1 INTEGRAL OBLIQUE TRIANGLE that has ONE ANGLE THAT IS AN INTEGRAL MULTIPLE OF ANOTHER.

PLATE 199: 6 different examples of INTEGRAL RIGHT TRIANGLES dissected into 3 non-similar unequal INTEGRAL RIGHT TRIANGLES.

PLATE 200: An OBLIQUE SCALENE TRIANGLE WITH INTEGRAL sides dissected into 5 ISOSCELES TRIANGLES with INTEGRAL sides and with base angles increasing in the arithmetical progression of 1, 2, 3, 4, 5.

Chapter 5

Isosceles Triangles (Freese's Plates 1–7)

In his first seven plates Freese set out to dissect isosceles triangles to various figures, the simplest of them being themselves triangles. In Plates 1 and 2 he described methods for dissecting certain isosceles triangles to a pair of isosceles triangles. In Plate 3 he dissected an isosceles right triangle to three different figures: an equilateral triangle, an isosceles triangle with apex angle of 90°, and an isosceles triangle with an apex angle of 150°. In Plate 4 he dissected each of a regular pentagon, a regular hexagon, and a regular octagon to some isosceles triangle. In Plate 5, Freese dissected a regular 30-sided polygon to similar isosceles triangles with sides in the ratio of $1:2:3:4$. In Plate 6 he dissected an arbitrary triangle to an isosceles triangle, any isosceles triangle to a pair of congruent isosceles triangles, and an equilateral triangle to two different-sized equilateral triangles. Finally, in Plate 7 he took several combinations of isosceles triangles to equilateral triangles.

Perhaps the most memorable of these dissections are the hingeable dissections, namely that of an isosceles right triangle to an equilateral triangle in Plate 3, and Henry Taylor's dissection of an arbitrary triangle to a certain isosceles triangle in Plate 6.

Plate 1: In this plate Freese considered an isosceles triangle whose apex angle is 30°. He gave dissections to four different pairs of equal-area isosceles triangles, with apex angles of 15°, 60°, 90°, and 120°. Without proof, he claimed as a theorem that the isosceles triangle of apex angle 30° is unique in having four such dissections. However there are problems. First, there are an infinite number of such dissections for that triangle. Second, nothing unique comes from an apex of 30°. There is an infinite set of isosceles triangles for which there will be related dissections to four pairs of different congruent isosceles triangles. There are counter-examples readily available in each case. And there is not much reason to display them, since they are not particularly remarkable or even attractive.

Nonetheless, it is worth noting that the fourth dissection in this plate is hingeable, as we see in Figure I1.

Plate 2: Freese was the first to explore ways to dissect an isosceles triangle into six pieces that form either of two different pairs of congruent isosceles triangles. He drew two different such triangles, but did not give a precise description in terms of angles. The first triangle should have an apex angle of $\arccos(3/4) \approx 41.4096°$. (The pair of triangles numbered 5 and 6 together form an isosceles triangle. Dropping a perpendicular to the base of that isosceles triangle produces a right triangle with long leg of length 3/4 of the length of the hypotenuse.) The second triangle should have an apex angle of $\arccos(1/4) \approx 75.5225°$. (Again, the triangles numbered 5 and 6 together form an isosceles triangle. Dropping a perpendicular to the base of that isosceles triangle produces a right triangle with short leg 1/4 of the length of the hypotenuse.) Freese claimed, without proof, that these are the only two such isosceles triangles.

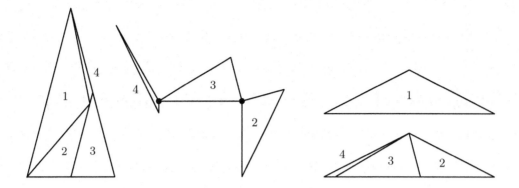

I1: Hinging the fourth dissection in Plate 1 [GNF]

Plate 3: The top dissection, of an equilateral triangle to a right triangle, appears to have been created by the same technique as the 4-piece dissection of an equilateral triangle and a square. This dissection is hingeable, as we see in Figures I3a and I3b.

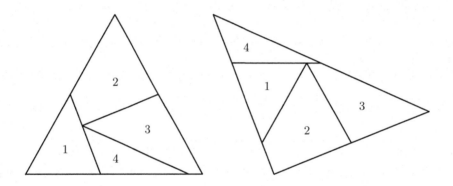

I3a: Hingeable equilateral triangle to isosceles right triangle

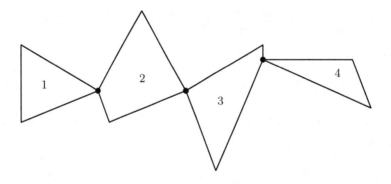

I3b: Hinged pieces for equilateral triangle to isosceles right triangle

The second dissection is of a sector of a dodecagon to a right triangle. (A *sector of a dodecagon* is an isosceles triangle that is essentially a pie slice from the center of the dodecagon.) Freese used a technique by Henry Taylor (1905) that dissects a triangle to another of the same height. The bottom dissection in the plate is of a segment of a dodecagon to a right triangle. (A *segment of a dodecagon* is an isosceles triangle that is bounded by two sides of the dodecagon and a chord of the dodecagon.) Freese used a strip technique to get his 4-piece dissection.

Plate 4: Freese was the first to dissect regular polygons to form isosceles triangles. In this plate he cut a pentagon into five pieces, a hexagon into four pieces, and an octagon into five pieces. He claimed without proof that he accomplished each of these in the fewest possible number of pieces. His dissection that turns a regular hexagon "inside out" to form an isosceles triangle, is hingeable, as we see in the middle of Figure I4a.

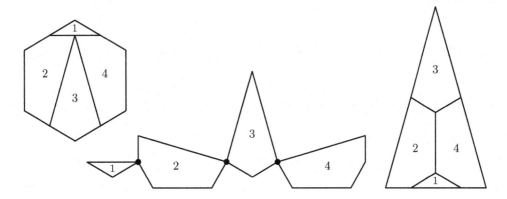

I4a: Hinging of regular hexagon to isosceles triangle [GNF]

Gavin Theobald found a way to dissect the regular pentagon into just four pieces that form the isosceles triangle on the right in Figure I4b. Gavin found this dissection by segmenting a pentagon strip designed by Harry Lindgren, as we see in Figure 4b.

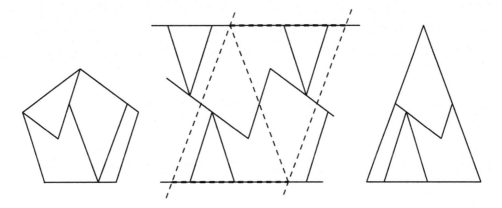

I4b: Regular pentagon to an isosceles triangle [Gavin Theobald]

Plate 5: Freese appears to have been the first to specifically dissect a 30-sided regular polygon into four similar isosceles triangles whose bases are 1, 2, 3, and 4 times the side length of the 30-gon. By the way, the official name of the 30-gon is indeed a triacontagon, as Antreas Hatzipolakis and John Conway have confirmed.

Plate 6: Henry Taylor (1905) described the general method that Freese sketched out at the top of the plate. The second dissection in the plate is a simple generalization of the second dissection in Plate 1. It is hingeable, with the triangle on the right accommodating two hinges along its base. Swinging the left and right pieces down creates a pocket into which we can slide the middle triangle.

The third dissection is a special case described by Harry Hart (1877). Hart gave ring expansions for two similar polygons that can be either circumscribed or inscribed. As I described in my 2002 book, we can hinge the pieces in the larger of the two polygons to produce a ring of pieces. The total number of pieces will be one more than twice the number of sides. We can see an application of Hart's dissection applied to regular pentagons in Plate 55, and the hinged pieces for the larger pentagon in Figure I55.

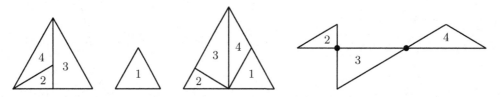

I6: Improved and hinged triangles for $(\sin \frac{\pi}{3})^2 + (\cos \frac{\pi}{3})^2 = 1$ [GNF]

Freese showed the limiting case for the third dissection in the lower righthand corner of Plate 6. The ratio of side lengths illustrates the identity $(\sin \frac{\pi}{3})^2 + (\cos \frac{\pi}{3})^2 = 1$. This dissection is hingeable, in a fashion analogous to regular pentagons as in Figure I56a. However, in the case of equilateral triangles, there are dissections that have just five pieces, by Harry Bradley (1930) and by Harry Lindgren (1956), with the latter being hingeable. Furthermore, there is a 4-piece hingeable dissection in Solution 5.1 in my 1997 book. I show a new 4-piece dissection in Figure I6, which is hingeable in three different ways. (Can you identify the two other ways?)

Plate 7: The first two dissections in this plate clearly use a minimal number of pieces, and both are hingeable. We can improve the 9-piece dissection at the bottom of the plate to an 8-piece dissection by not splitting one isosceles triangle into the two right triangles.

This chapter on isosceles triangles has been a quick warm-up to a more serious undertaking, namely the next chapter, focused on equilateral triangles, in which we will see a much greater variety of triangular dissections.

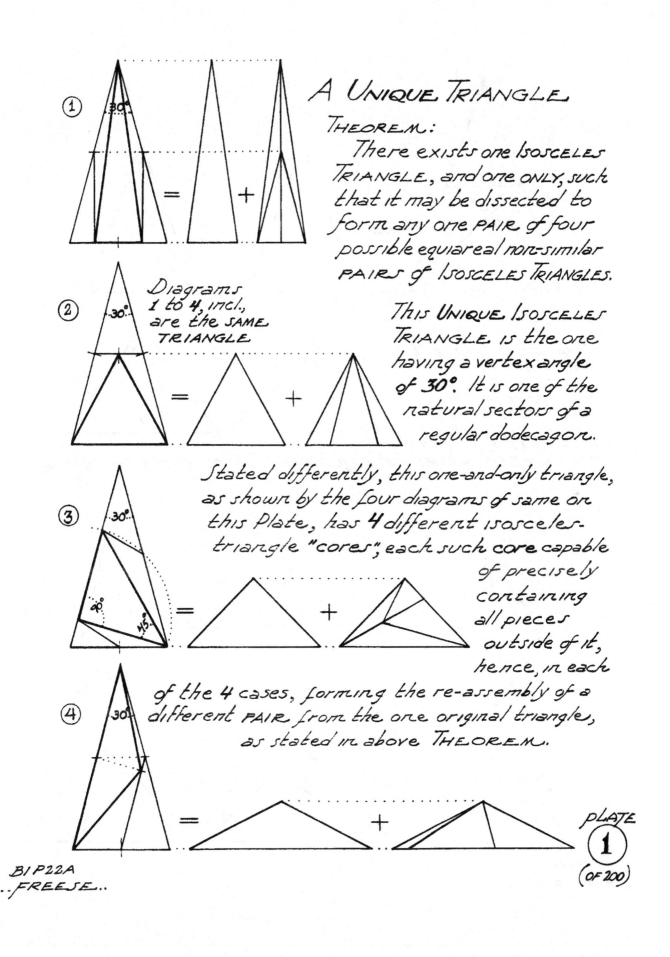

A Unique Triangle

Theorem:

There exists one Isosceles Triangle, and one ONLY, such that it may be dissected to form any one PAIR of four possible equiareal non-similar PAIRS of Isosceles Triangles.

This UNIQUE Isosceles Triangle is the one having a vertex angle of 30°. It is one of the natural sectors of a regular dodecagon.

Diagrams 1 to 4, incl., are the SAME TRIANGLE

Stated differently, this one-and-only triangle, as shown by the four diagrams of same on this Plate, has **4** different isosceles-triangle "cores", each such **core** capable of precisely containing all pieces outside of it, hence, in each of the 4 cases, forming the re-assembly of a different PAIR from the one original triangle, as stated in above THEOREM.

BIP22A
..FREESE..

PLATE 1 (OF 200)

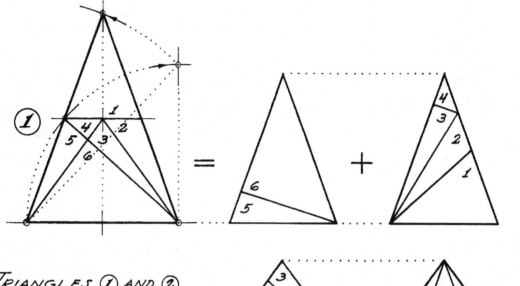

TRIANGLES ① AND ②
ARE THE ONLY TWO
ISOSCELES TRIANGLES
FOR EACH OF WHICH
THE PIECES OF A 6-PART
DISSECTION WILL ALSO
FORM EITHER OF TWO
DIFFERENT PAIRS OF EQUIAREAL ISOSCELES TRIANGLES.

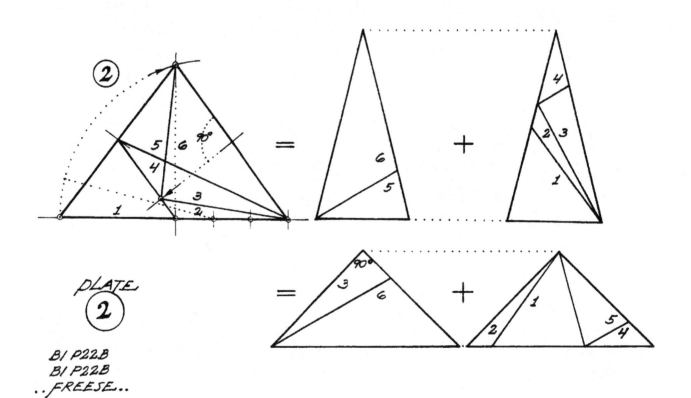

PLATE
②

BI P22B
BI P22B
..FREESE..

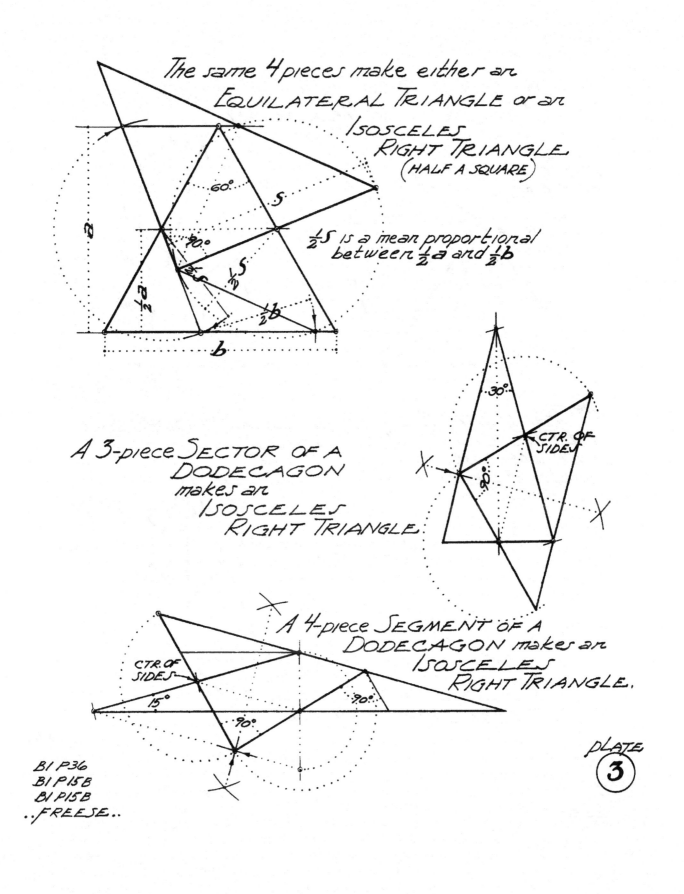

The same 4 pieces make either an
EQUILATERAL TRIANGLE or an
ISOSCELES
RIGHT TRIANGLE
(HALF A SQUARE)

60°

S

90°

$\frac{1}{2}$S is a mean proportional
between $\frac{1}{2}$a and $\frac{1}{2}$b

$\frac{1}{2}$S

a

$\frac{1}{2}$a

$\frac{1}{2}$b

b

A 3-piece SECTOR OF A
DODECAGON
makes an
ISOSCELES
RIGHT TRIANGLE

30°

CTR. OF
SIDES

90°

X

X

A 4-piece SEGMENT OF A
DODECAGON makes an
ISOSCELES
RIGHT TRIANGLE.

CTR. OF
SIDES

15°

90°

90°

BI P36
BI P15B
BI P15B
..FREESE..

PLATE
3

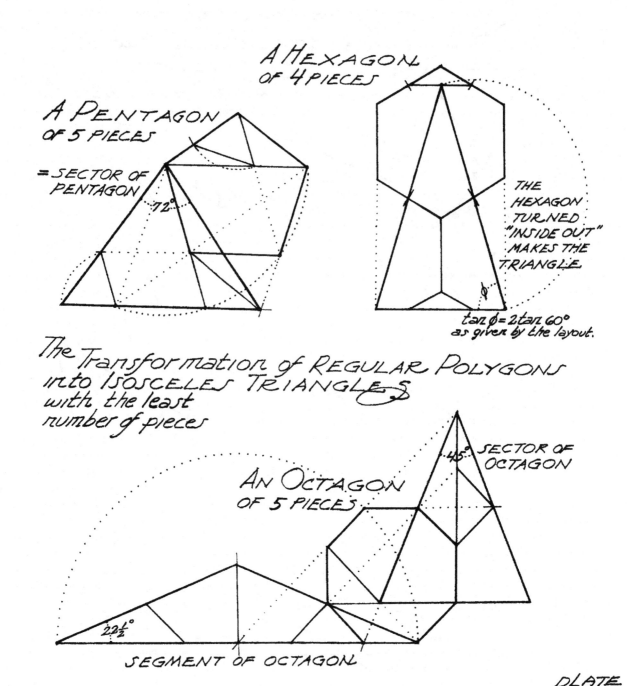

A HEXAGON
OF 4 PIECES

A PENTAGON
OF 5 PIECES

= SECTOR OF
PENTAGON

72°

THE
HEXAGON
TURNED
"INSIDE OUT"
MAKES THE
TRIANGLE

ϕ

tan ϕ = 2 tan 60°
as given by the layout.

The Transformation of REGULAR POLYGONS
into ISOSCELES TRIANGLES
with the least
number of pieces

AN OCTAGON
OF 5 PIECES

45° SECTOR OF
OCTAGON

22½°

SEGMENT OF OCTAGON

PLATE
④

BI P36AI
BI P36AI
BI P36B
..FREESE..

THE 30 SECTORS OF
THE TRIACONTAGON
RE-ASSEMBLED INTO
4 SIMILAR ISOSCELES
TRIANGLES
WITH CORRESPONDING
SIDES IN THE RATIO OF
1 : 2 : 3 : 4

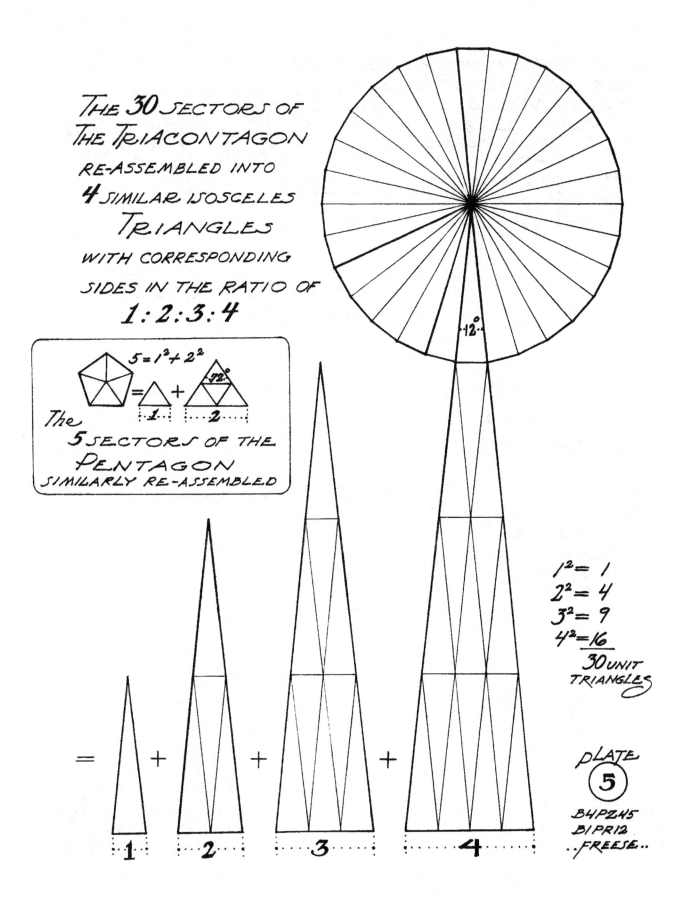

$5 = 1^2 + 2^2$

The
5 SECTORS OF THE
PENTAGON
SIMILARLY RE-ASSEMBLED

$1^2 = 1$
$2^2 = 4$
$3^2 = 9$
$4^2 = 16$
30 UNIT
TRIANGLES

PLATE
⑤

B4PZ45
B1PR12
..FREESE..

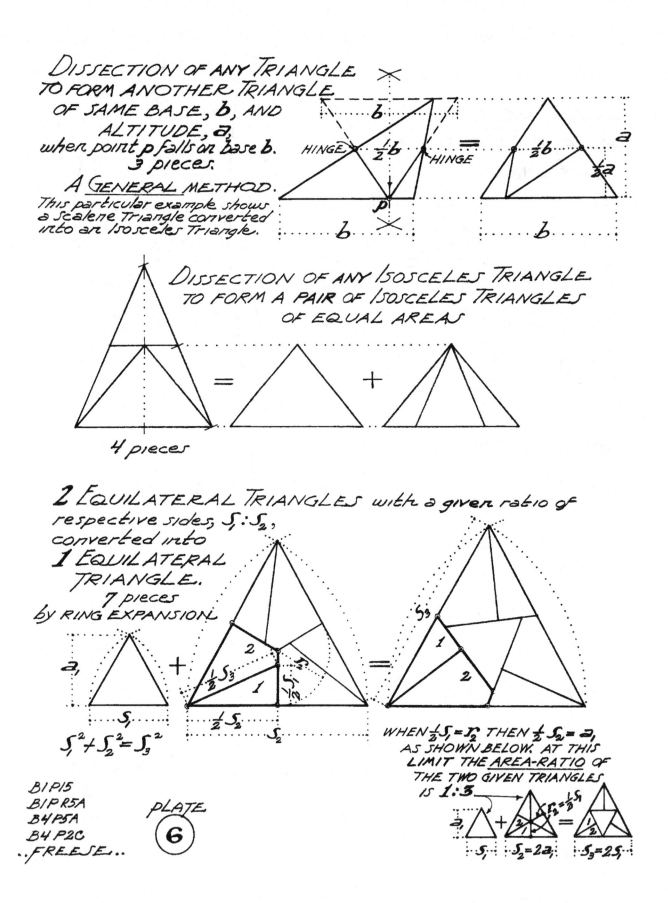

DISSECTION OF ANY TRIANGLE TO FORM ANOTHER TRIANGLE OF SAME BASE, b, AND ALTITUDE, a, when point p falls on base b. 3 pieces.

A GENERAL METHOD.
This particular example shows a Scalene Triangle converted into an Isosceles Triangle.

HINGE $\frac{1}{2}b$ HINGE p b

$\frac{1}{2}b$ a $\frac{1}{2}a$ b

DISSECTION OF ANY ISOSCELES TRIANGLE TO FORM A PAIR OF ISOSCELES TRIANGLES OF EQUAL AREAS

= +

4 pieces

2 EQUILATERAL TRIANGLES with a given ratio of respective sides, $S_1 : S_2$, converted into
1 EQUILATERAL TRIANGLE.
7 pieces
by RING EXPANSION

a_1 S_1 + $\frac{1}{2}S_3$ $\frac{1}{2}S_1$ 2 1 $\frac{1}{2}S_2$ S_2 = S_3 1 2

$S_1^2 + S_2^2 = S_3^2$

WHEN $\frac{1}{2}S_1 = I_2$ THEN $\frac{1}{2}S_2 = a_1$, AS SHOWN BELOW. AT THIS LIMIT THE AREA-RATIO OF THE TWO GIVEN TRIANGLES IS **1:3**

a_1 + $I_2 = \frac{1}{2}S_1$ 2 1 = $\frac{1}{2}$

S_1 | $S_2 = 2a_1$ | $S_3 = 2S_1$

B1 P15
B1 P R5A
B4 P5A
B4 P2C
..FREESE..

PLATE
6

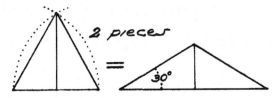

2 pieces

An EQUILATERAL TRIANGLE MAKES AN ISOSCELES TRIANGLE WITH BASE ANGLES OF 30°.

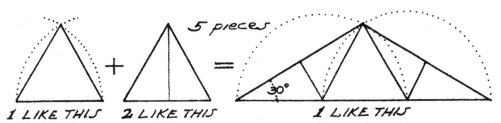

5 pieces

1 LIKE THIS   2 LIKE THIS   1 LIKE THIS

3 EQUILATERAL TRIANGLES MAKE AN ISOSCELES TRIANGLE WITH BASE ANGLES OF 30°.

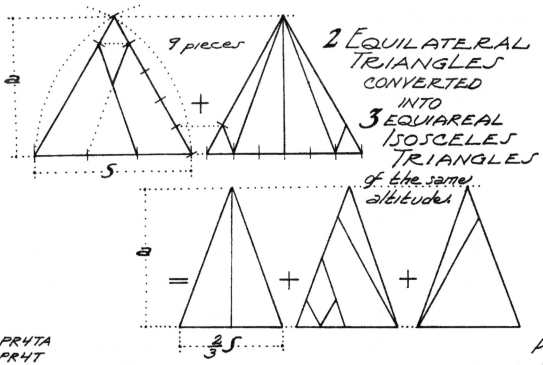

9 pieces

2 EQUILATERAL TRIANGLES CONVERTED INTO **3** EQUIAREAL ISOSCELES TRIANGLES of the same altitude.

PLATE **7**

Chapter 6

Equilateral Triangles   (Plates 8–26)

In this chapter we will feast on scrumptious dissections of equilateral triangles. Plates 8-13 present dissections of an equilateral triangle to squares, to a regular hexagon, and to a regular octagon. Freese also included in this group 4 congruent equilateral triangles to 9 congruent equilateral triangles, as well as 9 equilateral triangles to 16. Plates 14-20 give an equilateral triangle to various numbers of congruent equilateral triangles. Plates 21 and 22 serve up dissections of an equilateral triangle to different-sized equilateral triangles, demonstrating certain number identities. Plates 23-25 display dissections of congruent equilateral triangles to one or more regular hexagons. Plate 26 shows off a singular dissection of three congruent equilateral triangles to a "stellated nonagon."

A threesome of splendidly hingeable dissections, in Plates 8, 12, and 26, elevates the cavalcade of triangular dissections in this chapter to breath-taking heights.

Plate 8: On April 6, 1902, in his column in the *Weekly Dispatch*, Henry Dudeney posed the problem of dissecting an equilateral triangle to a square. Two weeks later he identified Charles W. McElroy as his only correspondent who had found a 4-piece solution, which he published after a delay of an additional two weeks. This is a most remarkable dissection, one that Dudeney may well not have anticipated, as I discussed in my second book. Subsequently, in his 1907 book, Dudeney identified the dissection as hingeable. Freese pointed out that the same technique works for any isosceles triangle within a certain range, but failed to note that it works on a much larger class of triangles that need not be isosceles. Although it may well be true, Freese's claim, namely that no other 4-piece squaring of the equilateral triangle is possible, is yet to be proved.

Near the bottom of this plate, Freese pointed out an interesting generalization of the dissection of an equilateral triangle to a square. The method also works for isosceles triangles over a wide range of base angles. While Freese did not explicitly state that these 4-piece dissections are hingeable, he clearly recognized that they were, since he indicated with circled dots the locations of the hinges. We see the hinged dissections in Figures I8a and I8b.

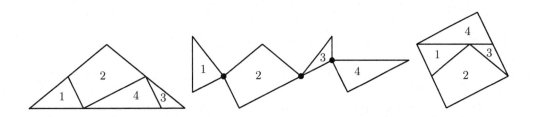

I8a: Hinging Freese's dissection of an isosceles triangle to a square

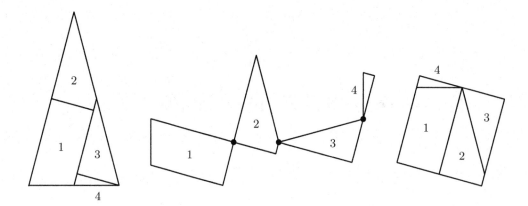

I8b: Hinging Freese's dissection of a second isosceles triangle to square

Plate 9: Freese was the first to dissect an equilateral triangle to two congruent squares. It would seem to be difficult to beat his 6-piece dissection. He apparently did not notice that his dissection (Figure I9a) is hingeable—a bit of a surprise—as we see in Figure 9b.

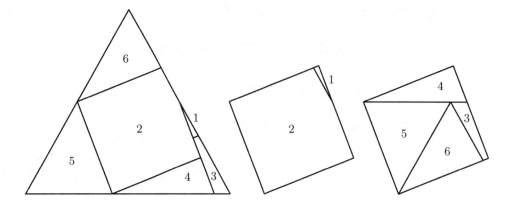

I9a: Freese's hingeable dissection of an equilateral triangle to two squares

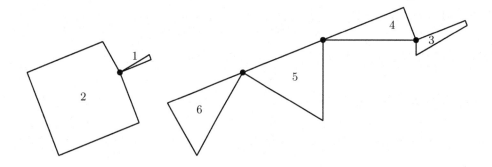

I9b: Hinged pieces for Freese's equilateral triangle to two squares [GNF]

Freese was also the first to dissect four congruent triangles to nine congruent triangles similar to the four. However, when the similar triangles are equilateral triangles, his 16-piece dissection

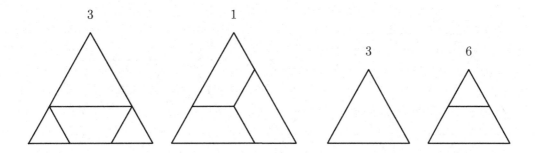

I9c: Improved four equilateral triangles to nine such triangles [GNF]

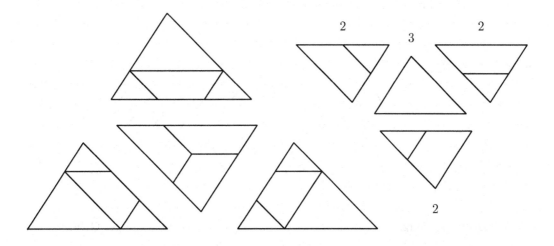

I9d: Four congruent triangles to nine similar congruent triangles [GNF]

has at least one piece more than is necessary, as we see from my 15-piece dissection in Figure I9c. Freese claimed that his technique works for any triangle, not just one that is equilateral. Indeed, my improvement will also work for any triangle, not just one that is equilateral. To get it to work, however, we need the variation that I show in Figure 9d. For the three larger triangles that have four pieces, place the trapezoid parallel to a different side in each. Similarly, for the six smaller triangles that have two pieces, place the trapezoid parallel to a given side in two each of them.

Plate 10: Freese was the first to dissect an equilateral triangle to three squares. It would seem difficult to beat his 8-piece dissection. He was also the first to dissect nine triangles to sixteen triangles. However, his 36-piece dissection has at least five pieces more than necessary, as we see with my 31-piece dissection in Figure I10a. Although Freese did not note it, a variation of his technique will work for any triangle, not just an equilateral one. And a suitable variation of my dissection in Figure I10a will also work, as in Figure 10b.

Plate 11: Freese based his 6-piece dissection of an equilateral triangle to a regular hexagon on his dissection of two triangles to one in Plate 14, recognizing that it is easy to convert two congruent equilateral triangles to a regular hexagon. In labeling the number of pieces as minimal, he probably meant that the number of pieces was smaller than for the dissection in Plate 12. In any event, his dissection uses one more piece than the 1951 dissection by Harry Lindgren in Figure I11.

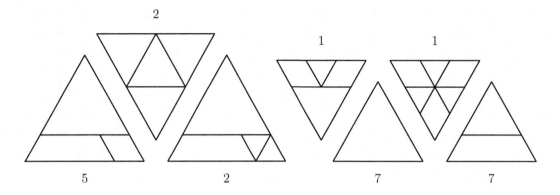

I10a: Improved nine equilateral triangles to sixteen such triangles [GNF]

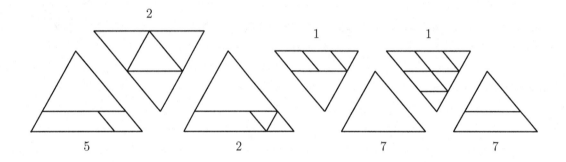

I10b: Nine congruent triangles to sixteen similar congruent triangles [GNF]

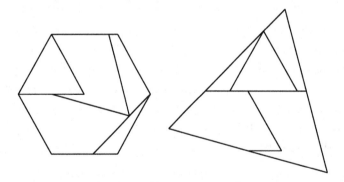

I11: Improved hexagon to an equilateral triangle [Harry Lindgren 1951]

Plate 12: Although using more pieces than Plate 11, the 9-piece dissection of an equilateral triangle to a hexagon has a handsome symmetry. The dissection derives directly from a dissection in the fourteenth century Persian manuscript, *Interlocks of Similar or Complementary Figures.* Freese underscored the 3-fold symmetry by emboldening the edges of the pieces in one third of the pieces, in both the equilateral triangles and the hexagon.

An interesting feature of the dissection is that it is fully hingeable. Freese showed how to swing two of the small right triangles in the equilateral triangle around to fit against a piece in the hexagon, and he did claim that he had turned the triangle "inside out." However it seems unlikely

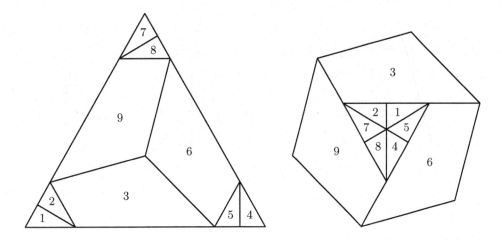

I12a: Dissection of an equilateral triangle to a regular hexagon [*Interlocks*]

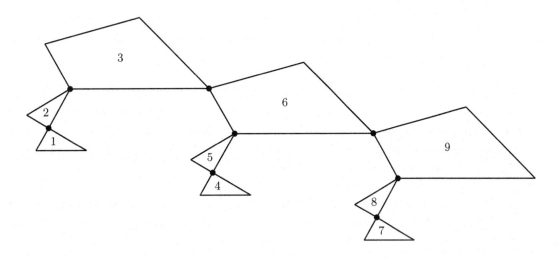

I12b: Hinging of equilateral triangle to a regular hexagon [GNF]

that he meant that the dissection is fully hingeable. Figure I12a shows the dissection with the pieces labeled, and Figure I12b shows a hinging of the pieces.

Unfurling the pieces from the equilateral triangle is easy. Just rotate pieces 1 and 2 away from piece 3, rotate pieces 4 and 5 away from piece 6, and rotate pieces 7 and 8 away from piece 9. Then rotate pieces 1-3 away from piece 6, and rotate pieces 7-9 away from piece 6.

Unfurling the pieces from the regular hexagon isn't much harder. Rotate pieces 7-9 as a unit part-way around the hinge between pieces 6 and 9. Then rotate pieces 1-3 as a unit part-way around the hinge between pieces 3 and 6. Finally, swing out pieces 7 and 8 away from piece 9, swing pieces 4 and 5 from piece 6, and swing pieces 1 and 2 from piece 3.

For my (2002) book I designed a hinged dissection of an equilateral triangle to a regular hexagon that uses just six pieces, but it lacks the wonderful symmetry and impressive pedigree of the foregoing dissection from over 700 years ago.

Plate 13: Harry Lindgren (1964) found the same 8-piece dissection of an equilateral triangle to a regular octagon as Freese. However, Gavin Theobald discovered an improvement (Figure I13)

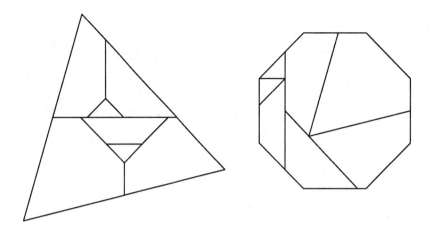

I13: Improved regular octagon to an equilateral triangle [Gavin Theobald]

to this dissection in the 1990's. Theobald's 7-piece dissection results from the crossposition of two T-strips, as I demonstrated in my 1997 book.

Plate 14: Freese's 5-piece dissection of two congruent triangles to one matches the number of pieces of the 1930 dissection by Harry Bradley. However, whereas Bradley's dissection will work for two triangles whose side lengths are in any ratio, Freese's works only for congruent triangles.

The dissection of three triangles to one is identical to the one that Plato described in his *Timaeus*. It is hingeable, as I demonstrated in my second book.

Plate 15: To dissect an equilateral triangle to five congruent ones, Freese first cut one of the five triangles off the apex of the large triangle. He then cut a small equilateral triangle off the righthand corner and swung it around, resulting in a parallelogram. Finally, he applied a slide technique to convert the parallelogram to a new parallelogram of height equal to the congruent triangles, and sliced the new parallelogram into four of the congruent triangles. Alfred Varsady (1989) also found a 9-piece dissection, but one that turns over two pieces.

Plate 16: To dissect an equilateral triangle to seven congruent triangles, Freese first cut one of the seven triangles off the apex of the large triangle. He then cut a small equilateral triangle off the righthand corner and swung it around, resulting in a parallelogram. Finally, he applied a slide technique to convert the parallelogram to a new parallelogram of height equal to the congruent triangles, and sliced the new parallelogram into six of the congruent triangles. David Collison matched this with a different 12-piece dissection, which he sent me in 1981. In labeling the number of pieces as "minimal, " Freese probably just meant that the number of pieces was smaller than for the dissection in Plate 17.

To dissect an equilateral triangle to 13 congruent ones, Freese followed the same general approach as in his dissection of an equilateral triangle to 7 congruent ones, except that he cut four of the thirteen triangles from the apex. He then cut a small equilateral triangle off the righthand corner and swung it around, resulting in a parallelogram. Once again, he applied a slide technique to convert the parallelogram to a new parallelogram of height equal to the congruent triangles, and sliced the new parallelogram into nine of the congruent triangles. The only additional trick was to convert the nine triangles into a parallelogram so that the slide technique applies. Again, in labeling the number of pieces as minimal, Freese probably meant that the number of pieces was smaller than for the dissection in Plate 18.

Plate 17: Hugo Steinhaus (1960) identified this second dissection of seven triangles to one. It has greater symmetry but also more pieces than the one in the previous plate. It is fully hingeable, in the sense that we can hinge together each pair of pieces that come from a small triangle.

Plate 18: Like the dissection in the previous plate, this second dissection of thirteen triangles to one derives from tessellations but uses more pieces than the dissection in Plate 16. We can modify this dissection to be fully hingeable, at the expense of three more pieces.

Plate 19: Freese's dissection of a Greek Cross to an equilateral triangle is identical to that of Harry Lindgren (1951). Freese's dissection of twelve triangles to one is a direct adaptation of the dissection of three triangles to one in Plate 14.

Freese was apparently the first to dissect seven triangles to three. David Paterson (1989) produced the same 21-piece dissection.

Plate 20: On page 295 of the *Strand Magazine* (volume 69, 1925), Henry Dudeney gave a 6-piece dissection of two equilateral triangles to a square. Not only did Freese beat Dudeney by one piece, but his dissection is translational.

Freese's 12-piece dissection of three triangles to four triangles is what you get when you take the 6-piece dissection of three triangles to one in Plate 14 and split the one triangle into four triangles in the obvious way. And the result is translational.

I21a: Improved equilateral triangles for $1^2 + 3^2 + 5^2 + 7^2 = (\sqrt{84})^2$ [GNF]

Plate 21: Freese gave a 9-piece dissection of four equilateral triangles with side ratios of $1:3:5:7$ to an equilateral triangle of side $\sqrt{84}$. This latter side length is the length of the third side of a (nonequilateral) triangle that has an angle of $120°$ sandwiched between sides of lengths 2 and 8. The triangles fit well together, with the 3-triangle and 7-triangle sitting side by side so that their opposing vertices are precisely $\sqrt{84}$ apart, and a slice of the 5-triangle is half this length. Freese made just two more cuts on the 5-triangle to get pieces that form the large triangle. However, we need just one more piece rather than two, if we cut a larger piece out of the 5-triangle, so that the

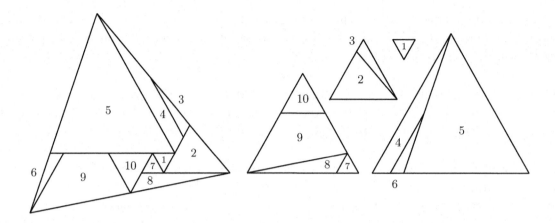

I21b: Hingeable dissection for triangles for $1^2 + 3^2 + 5^2 + 7^2 = (\sqrt{84})^2$ [GNF]

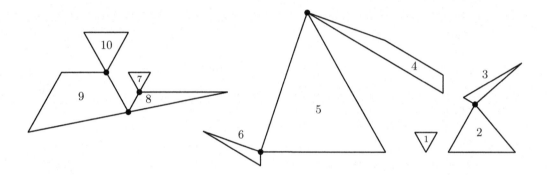

I21c: Hinged pieces for triangles for $1^2 + 3^2 + 5^2 + 7^2 = (\sqrt{84})^2$ [GNF]

cavity that we create accommodates the apex, once we rotate the piece by 60°. The result is the 8-piece dissection in Figure I21a.

Freese's original dissection at the top of Plate 21 is almost hingeable, with only the two pieces from the 3-triangle not hinged. If in the 5-triangle we cut a piece congruent to the thin triangle in the 3-triangle, we then get two congruent pieces 3 and 6 in Figure I21b, whose interchangeability makes the dissection hingeable. We see the resulting hinged pieces in Figure I21c.

At the bottom of the same plate, Freese gave an 11-piece dissection of six equilateral triangles with side ratios of $1\!:\!2\!:\!3\!:\!4\!:\!5\!:\!6$ to an equilateral triangle of side $\sqrt{91}$. This latter side length is the length of the third side of a (nonequilateral) triangle that has an angle of 120° sandwiched between sides of lengths 5 and 6. Again, I have improved on this dissection, taking advantage of the fact that $\sqrt{91}$ is also the length of the third side of a (nonequilateral) triangle that has an angle of 120° sandwiched between sides of lengths 1 and 9. The final trick is to cut a trapezoid out of the bottom of the 6-triangle, making room for the 5-triangle in the large triangle and also providing the piece that fills in between the 4-triangle and the 5-triangle along the bottom edge of the large triangle, as we see in Figure I21d.

Plate 22: Freese was the first to pose the dissection of triangles of side lengths 2, 5, 8, 11, 14, 17, 20, 23, and 26 to a triangle of side length 48. It seems difficult to improve on his 14-piece dissection. Apparently Freese did not consider the possibility of integer identities that are not based

I21d: Improved triangles for $1^2 + 2^2 + 3^2 + 4^2 + 5^2 + 6^2 = (\sqrt{91})^2$ [GNF]

on arithmetic progressions. I explored that possibility for equilateral triangles in my first (1997) book.

Plate 23: Freese seems to have been the first to dissect two congruent regular hexagons to an equilateral triangle. Lindgren (1964) later gave a different 6-piece dissection of a triangle to two hexagons, which is hingeable, although Lindgren did not mention the hingeability. Freese's dissection is also hingeable, and he also did not mention that fact. See Freese's dissection with pieces numbered in Figure I23a and my hinging of it in Figure I23b.

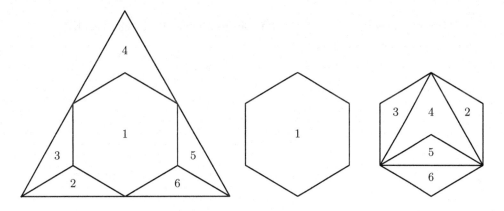

I23a: Freese's equilateral triangle to two regular hexagons

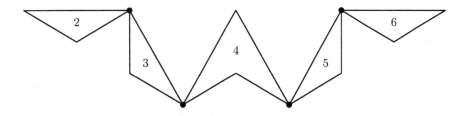

I23b: Hinging for Freese's equilateral triangle to two hexagons [GNF]

With regard to the dissection in the lower half of this plate, C. Stuart Elliott (1982–1983) later gave the same 6-piece dissection of three hexagons to two triangles. When we start with a different orientation of the figures, the dissection is translational.

Plate 24: Freese seems to have been the first to dissect five equilateral triangles to a regular hexagon. However he should have swapped the contents of Plates 24 and 25. Once you discover the twenty triangles to a hexagon in Plate 25, it is easy to group the triangles into five groups of four and produce the corresponding 17-piece dissection of five triangles to one. There is then only one problem—that dissection is not close to being minimal!

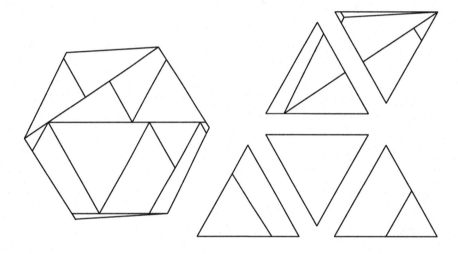

I24a: Improved five equilateral triangles to a hexagon [Gavin Theobald]

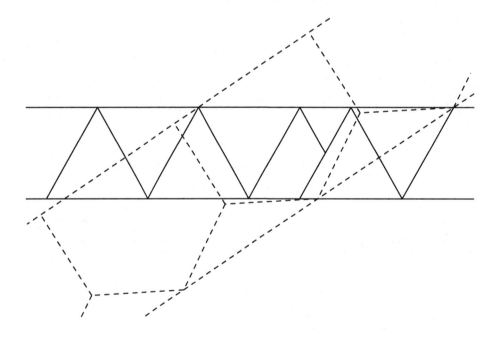

I24b: Crossposing five equilateral triangles to a hexagon

First, I found a 16-piece dissection by swapping parts of some pieces with parts of other pieces. Then I found a 14-piece dissection by using a T-strip. Finally, Gavin Theobald found a 13-piece dissection (Figure I24a) using a crossposition of two P-strips (Figure I24b)!

Plate 25: Freese was the first to dissect twenty triangles to a hexagon. He based his 28-piece dissection on a tessellation, in which two triangles form a small central hexagon, and 18 more triangles surround the small hexagon. It is hard to imagine improving on his dissection. The dissection is hingeable, and when we orient the triangles suitably, it is also translational.

Plate 26: The "stellated nonagon" in this plate is similar in appearance to a star, except that each pair of edges in the boundary that should match up don't lie together on the same straight line. Instead, the star-like figure is what I have termed a *pseudo-star* in my 2002 book and have defined fully in my 2015 article. The pseudo-star $\{p/q\}$ has p points, where q may be any positive real number in the range $1 \leq q < p$. Using the Law of Sines and the Law of Cosines, we can show that the angle of each point in Freese's stellated nonagon is approximately 92.69°, and thus the figure is approximately a $\{9/2.183\}$-pseudo-star.

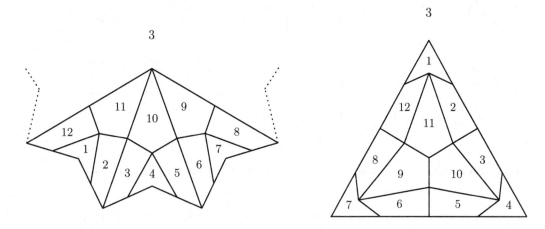

I26a: Hingeable three equilateral triangles to a "stellated nonagon"

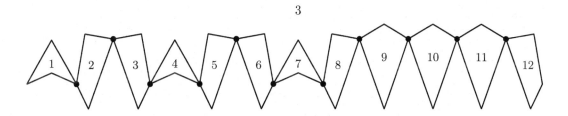

I26b: Hinged three equilateral triangles to a "stellated nonagon" [GNF]

We can hinge Freese's dissection, as I show in Figures I26a and I26b. To unfurl the equilateral triangle, swing piece 1 around the hinge between pieces 1 and 2, swing pieces 1-4 as a unit around the hinge between pieces 4 and 5, then swing pieces 1-7 as a unit around the hinge between pieces 7 and 8, then swing pieces 1-9 around the hinge between pieces 9 and 10. The rest is easy. To unfurl each third of the stellated nonagon, swing pieces 1-7 as a unit around the hinge between pieces 7 and 8, then the rest is easy. This dissection makes for an eye-catching animation, as my audience from the 2015 MOVES conference in New York City could verify.

We can reduce the number of pieces in this dissection if we merge together the two pieces that contribute to form each point of the stellated nonagon. To fit such a piece in the triangle, attach together pieces 4, 1, and 2 in Freese's diagram of the equilateral triangle of Plate 26, and then remove a merged piece from the right ride of the attached pieces. This eliminates nine pieces, producing the 27-piece dissection in Figure I26c. Too bad the resulting dissection is not hingeable!

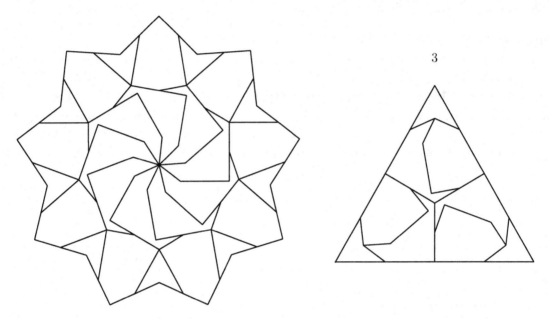

I26c: Improved three equilateral triangles to a "stellated nonagon" [GNF]

What a variety of dissection types that we have enjoyed in this chapter! From dissections of one equilateral triangle to certain regular polygons, to dissections of one or more congruent equilateral triangles to several congruent equilateral triangles, to an equilateral triangle to one or more squares or to one or more regular regular hexagons, to dissections expressing numeric identities to that last amazing dissection of a stellated figure. Can we expect any new types of dissections in our next chapter? Just wait and see ...

✱ SQUARING the EQUILATERAL TRIANGLE by turning it "INSIDE OUT"

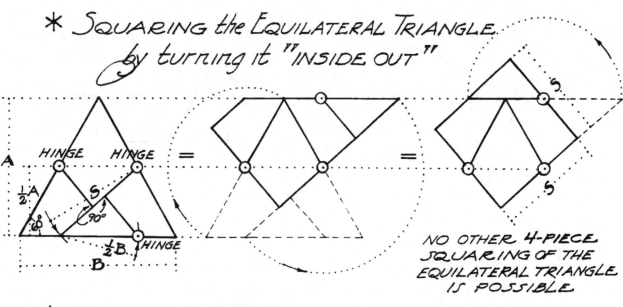

A HINGE HINGE = =

½A S 90° ½B HINGE B

NO OTHER 4-PIECE SQUARING OF THE EQUILATERAL TRIANGLE IS POSSIBLE

✱ This method of 4-piece SQUARING applies to any <u>ISOSCELES TRIANGLE</u> in which the altitude, A, is not less than $\frac{2}{5}$ the base, B, and the base angles are not greater than 75°. In all cases, S is a mean proportional between B $\frac{1}{2}$A, graphically found as follows.

The limiting cases of the above method of 4-piece SQUARING are shown below, the EQUILATERAL TRIANGLE thus coming within this range.

B 90° ½A S

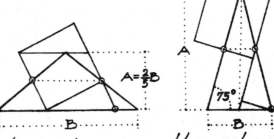

$A = \frac{2}{5}B$

LOWER LIMIT

A 75° B

UPPER LIMIT (SECTOR OF DODECAGON)

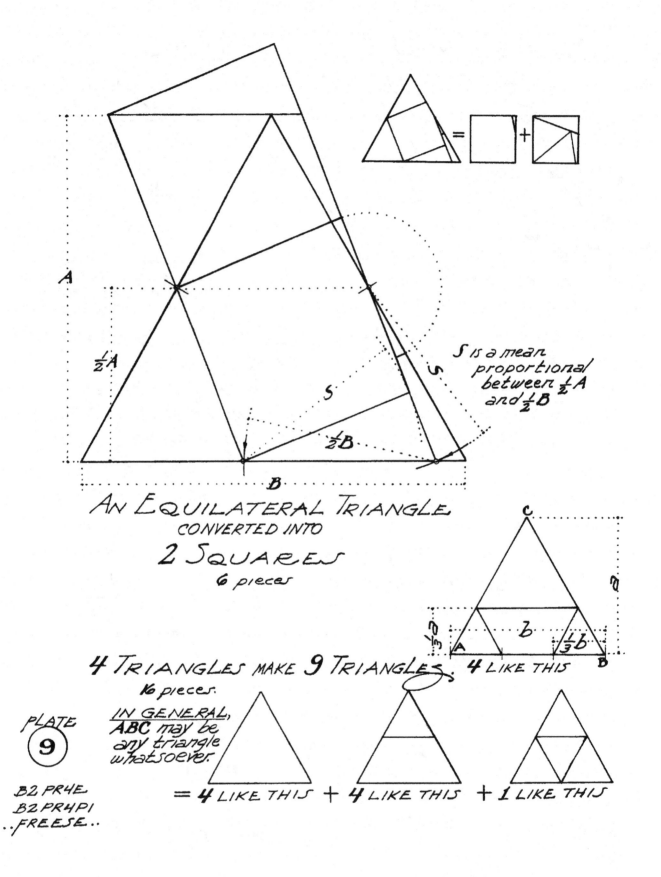

= S is a mean proportional between $\frac{1}{2}A$ and $\frac{1}{2}B$

AN EQUILATERAL TRIANGLE
CONVERTED INTO
2 SQUARES
6 pieces

4 TRIANGLES MAKE 9 TRIANGLES. 4 LIKE THIS

16 pieces.

IN GENERAL,
ABC may be
any triangle
whatsoever.

= 4 LIKE THIS + 4 LIKE THIS + 1 LIKE THIS

PLATE
9

B2 PR4E
B2 PR4PI
..FREESE..

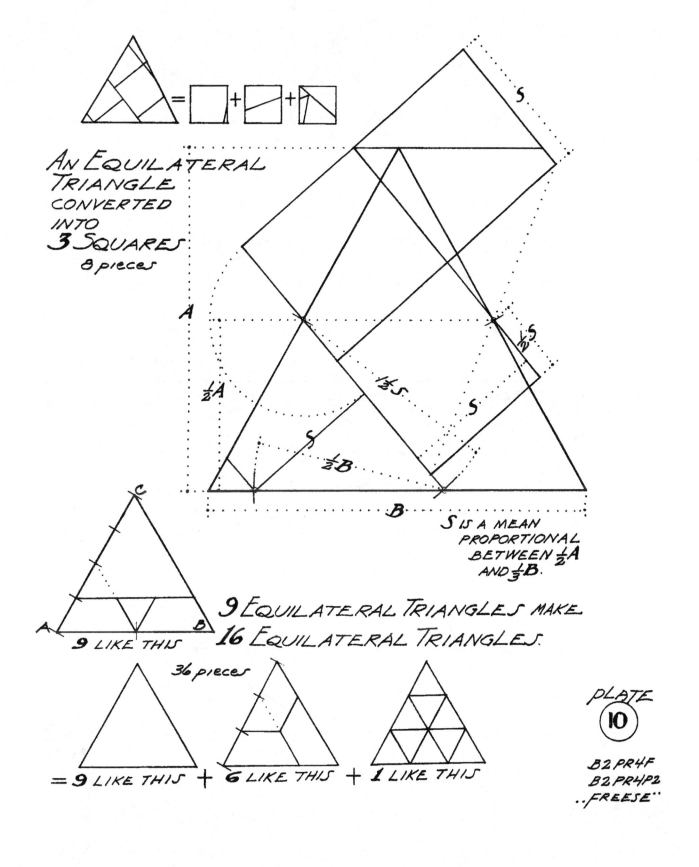

AN EQUILATERAL TRIANGLE CONVERTED INTO 3 SQUARES

8 pieces

S IS A MEAN PROPORTIONAL BETWEEN ½A AND ⅓B.

9 EQUILATERAL TRIANGLES MAKE 16 EQUILATERAL TRIANGLES.

9 LIKE THIS

36 pieces

= 9 LIKE THIS + 6 LIKE THIS + 1 LIKE THIS

PLATE 10

B2 PR4F
B2 PR4P2
"FREESE"

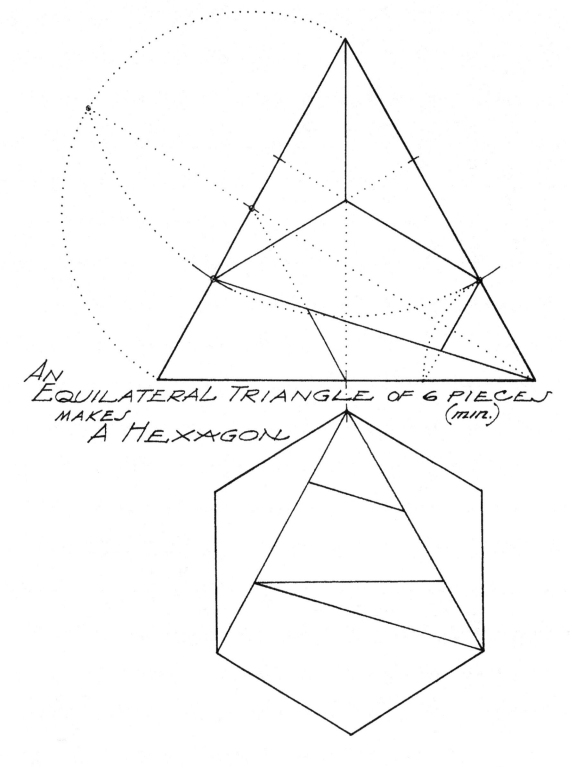

AN
EQUILATERAL TRIANGLE OF 6 PIECES
MAKES
A HEXAGON
(min.)

BIP3IB
..FREESE..

PLATE
11

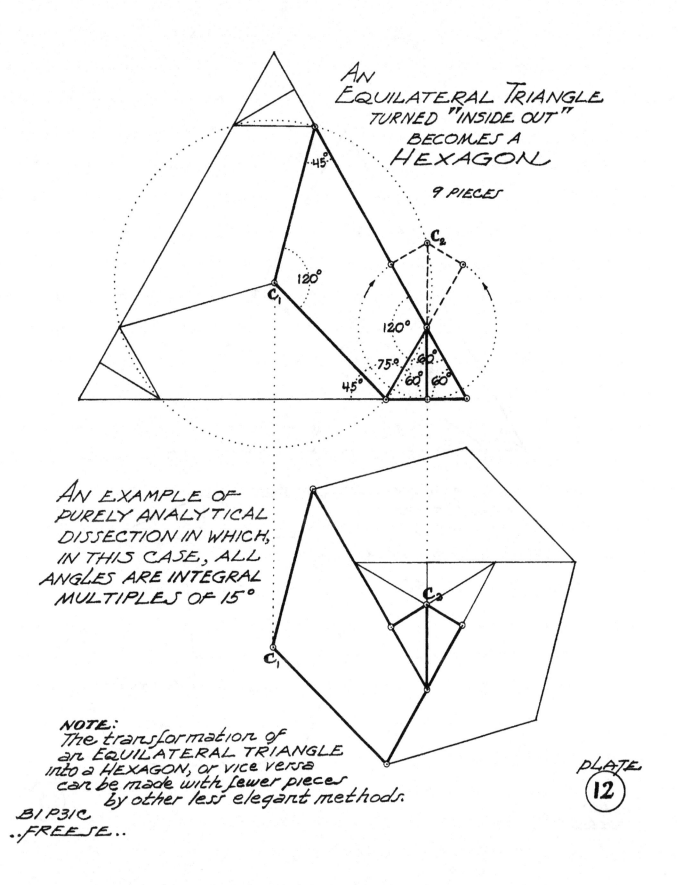

AN
EQUILATERAL TRIANGLE
TURNED "INSIDE OUT"
BECOMES A
HEXAGON

9 PIECES

45°

C₂

120°

C₁

120°

75°   60°

45°   60° 60°

AN EXAMPLE OF
PURELY ANALYTICAL
DISSECTION IN WHICH,
IN THIS CASE, ALL
ANGLES ARE INTEGRAL
MULTIPLES OF 15°

C₃

C₁

NOTE:
The transformation of
an EQUILATERAL TRIANGLE
into a HEXAGON, or vice versa
can be made with fewer pieces
by other less elegant methods.

B1 P31C
..FREESE..

PLATE
12

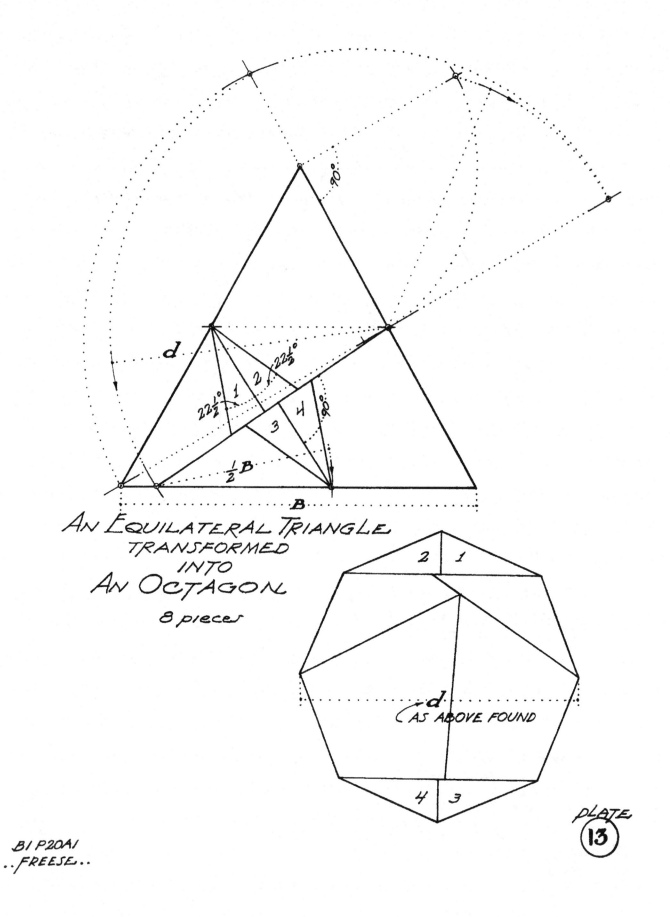

$22\frac{1}{2}°$ $22\frac{1}{2}°$ 90°

90°

d

1   2

3   4

$\frac{1}{2}B$

B

AN EQUILATERAL TRIANGLE
TRANSFORMED
INTO
AN OCTAGON

8 pieces

2   1

d
(AS ABOVE FOUND

4   3

PLATE
13

B1 P20A1
..FREESE..

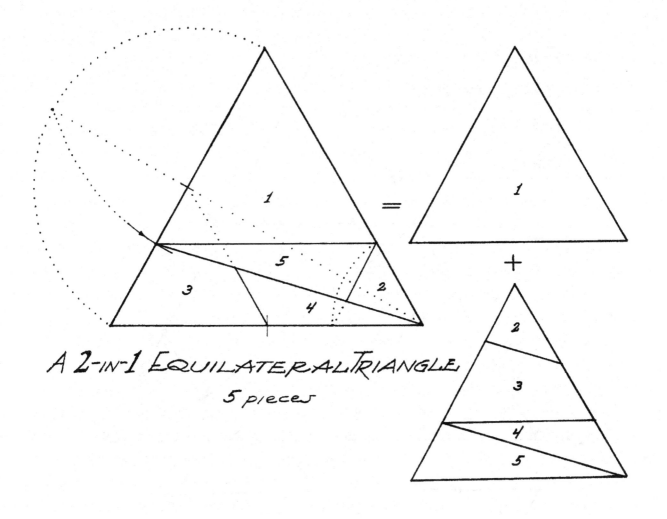

A 2-IN-1 EQUILATERAL TRIANGLE
5 pieces

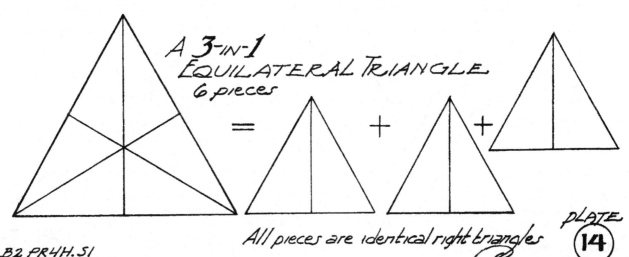

A 3-IN-1
EQUILATERAL TRIANGLE
6 pieces

All pieces are identical right triangles

PLATE
14

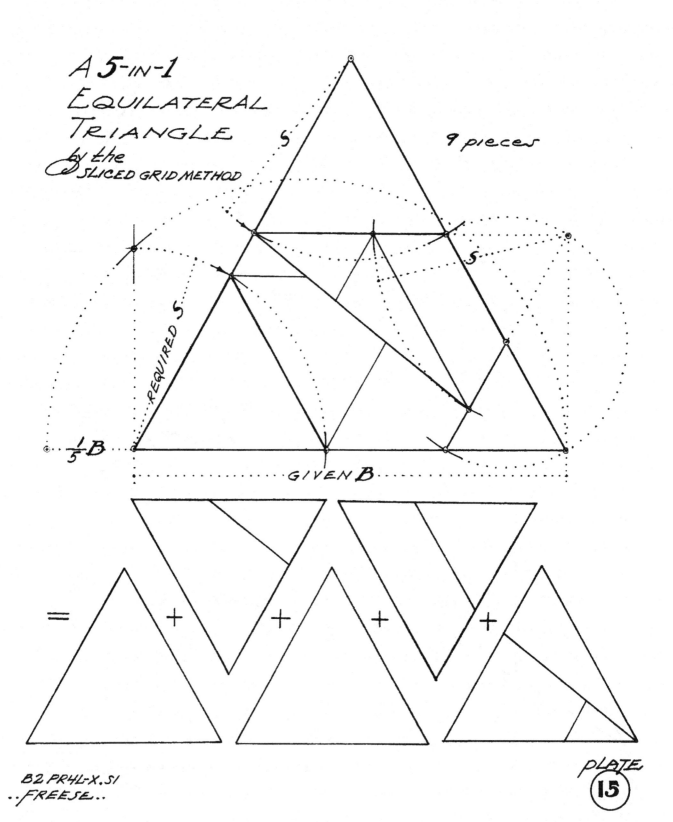

A 5-IN-1
EQUILATERAL
TRIANGLE
by the
SLICED GRID METHOD

9 pieces

REQUIRED S

$\frac{1}{5}$ B

GIVEN B

= + + + +

PLATE
15

B2 PR4L-X.SI
..FREESE..

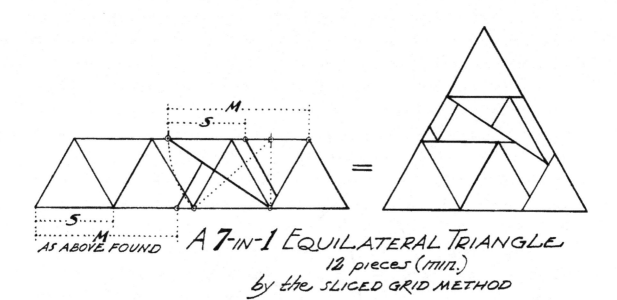

A 7-IN-1 EQUILATERAL TRIANGLE
12 pieces (min.)
by the SLICED GRID METHOD

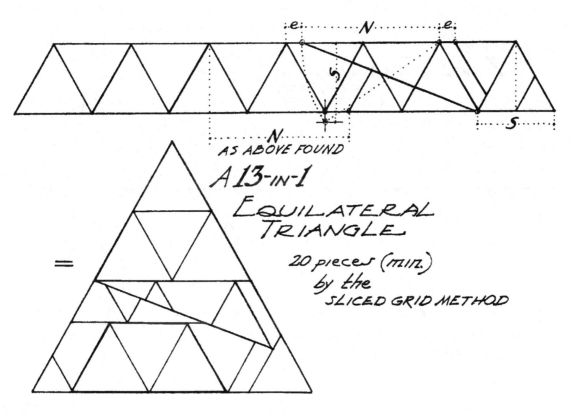

AS ABOVE FOUND

A 13-IN-1
EQUILATERAL
TRIANGLE

20 pieces (min.)
by the
SLICED GRID METHOD

PLATE
16

B4 PB7
B4 PB7
..FREESE..

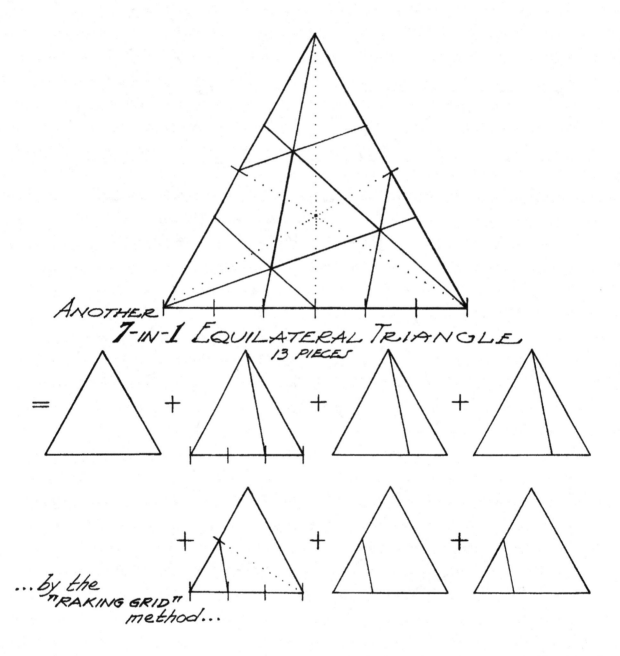

ANOTHER
7-IN-1 EQUILATERAL TRIANGLE
13 PIECES

...by the
"RAKING GRID"
method...

B2 PR4Q
..FREESE..

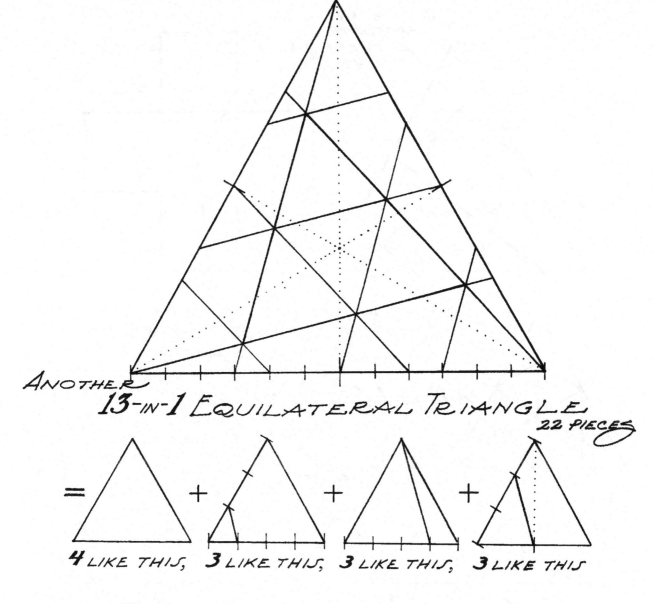

ANOTHER
13-IN-1 EQUILATERAL TRIANGLE
22 PIECES

= △ + △ + △ + △

4 LIKE THIS, 3 LIKE THIS, 3 LIKE THIS, 3 LIKE THIS

The 22 pieces are those that come within the
given triangle on the underlying RAKING GRID.

PLATE
18

B2 PR4R2
..FREESE..

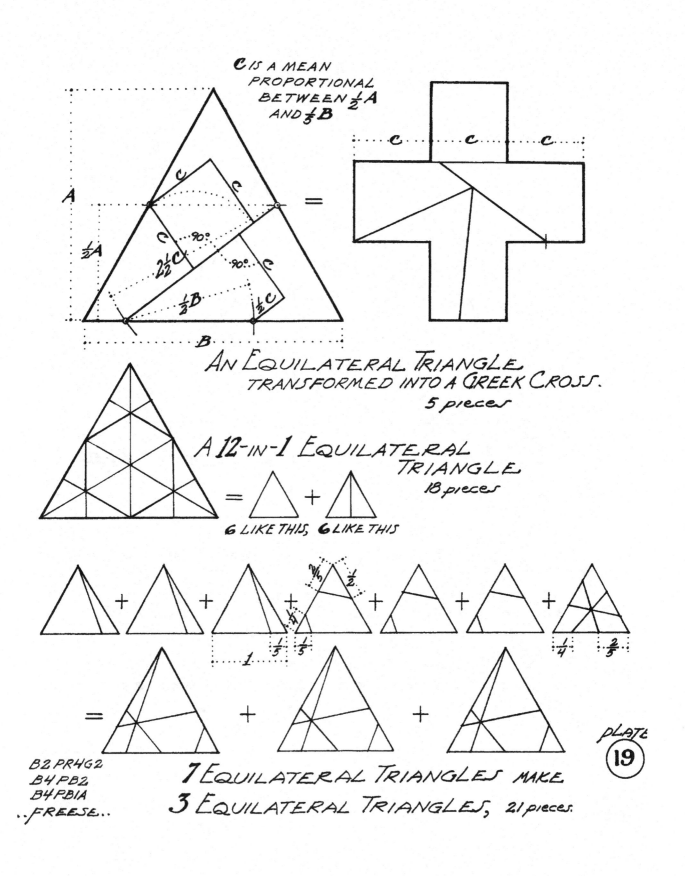

C IS A MEAN PROPORTIONAL BETWEEN $\frac{1}{2}A$ AND $\frac{1}{2}B$

AN EQUILATERAL TRIANGLE TRANSFORMED INTO A GREEK CROSS.
5 pieces

A 12-IN-1 EQUILATERAL TRIANGLE
18 pieces

= △ + △

6 LIKE THIS, 6 LIKE THIS

7 EQUILATERAL TRIANGLES MAKE 3 EQUILATERAL TRIANGLES, 21 pieces.

B2 PR4G2
B4 PB2
B4 PB1A
..FREESE..

PLATE
19

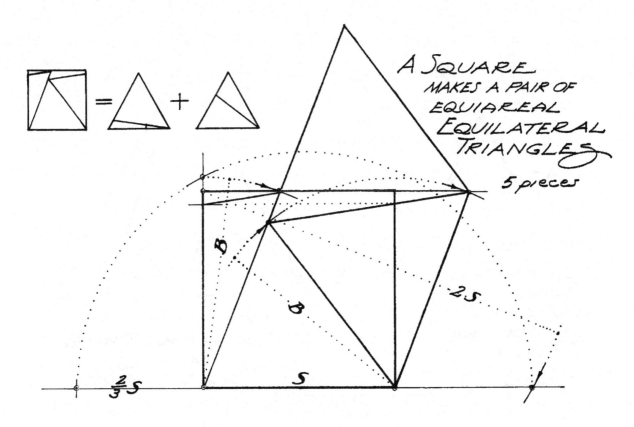

A SQUARE MAKES A PAIR OF EQUIAREAL EQUILATERAL TRIANGLES

5 pieces

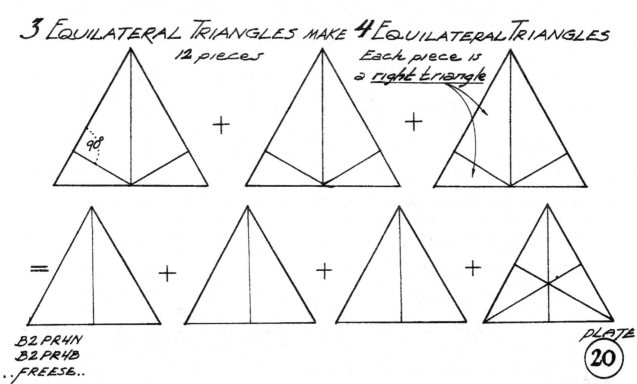

3 EQUILATERAL TRIANGLES MAKE 4 EQUILATERAL TRIANGLES

12 pieces

Each piece is a *right triangle*

90°

B2 PR4N
B2 PR4B
..FREESE..

PLATE 20

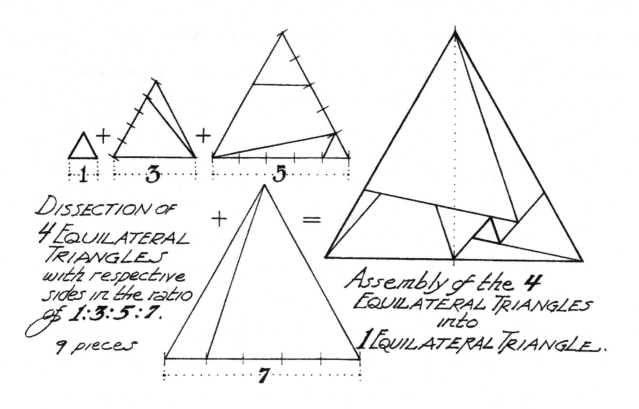

DISSECTION OF **4 EQUILATERAL TRIANGLES** with respective sides in the ratio of **1:3:5:7**.

9 pieces

Assembly of the **4 EQUILATERAL TRIANGLES** into **1 EQUILATERAL TRIANGLE**.

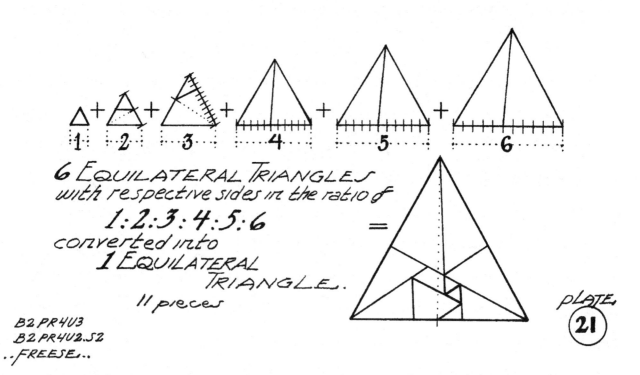

6 EQUILATERAL TRIANGLES with respective sides in the ratio of **1:2:3:4:5:6** converted into **1 EQUILATERAL TRIANGLE**.

11 pieces

PLATE **21**

9 EQUILATERAL TRIANGLES *with respective sides in the* RATIO *of* **2 : 5 : 8 : 11 : 14 : 17 : 20 : 23 : 26** *converted into*

1 EQUILATERAL TRIANGLE

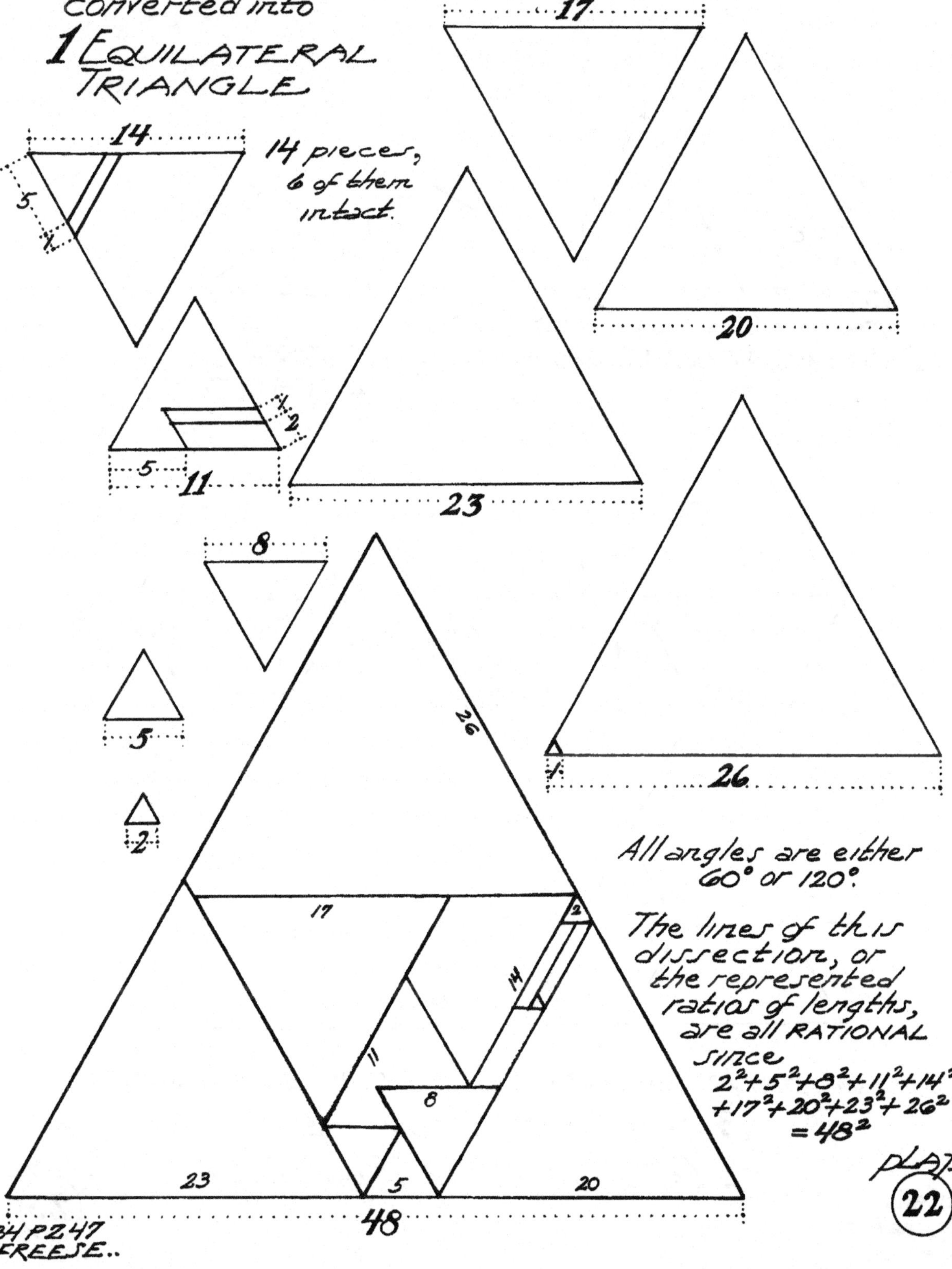

14 pieces, 6 of them intact.

14

5
4

17

20

11
5
4
2

23

8

5

2

26

26

48

26
17
14
11
8
5
23
5
20
48

2

All angles are either 60° or 120°.

The lines of this dissection, or the represented ratios of lengths, are all RATIONAL *since*

$$2^2 + 5^2 + 8^2 + 11^2 + 14^2 + 17^2 + 20^2 + 23^2 + 26^2 = 48^2$$

PLATE
22

B4 PZ 47
..FREESE..

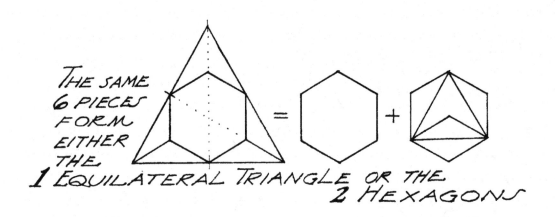

THE SAME 6 PIECES FORM EITHER THE 1 EQUILATERAL TRIANGLE OR THE 2 HEXAGONS

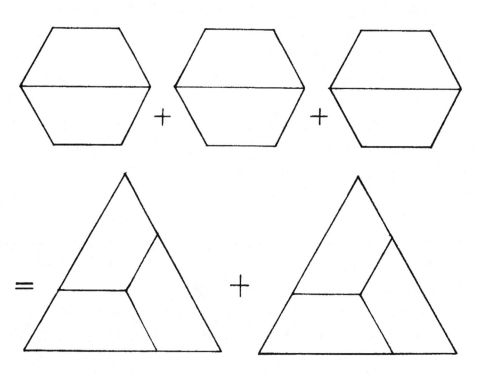

THE 6 IDENTICAL PIECES FORM EITHER THE
3 HEXAGONS OR THE
2 EQUILATERAL TRIANGLES.

B2 PRGL.51
B2 PRGM.51
..FREESE..

PLATE
23

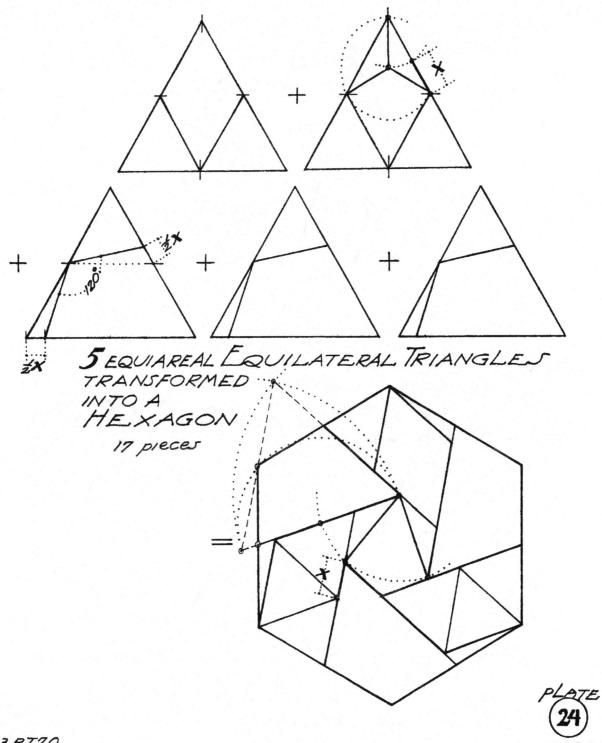

5 EQUIAREAL EQUILATERAL TRIANGLES
TRANSFORMED
INTO A
HEXAGON
17 pieces

B3 PT7.0
..FREESE..

PLATE
24

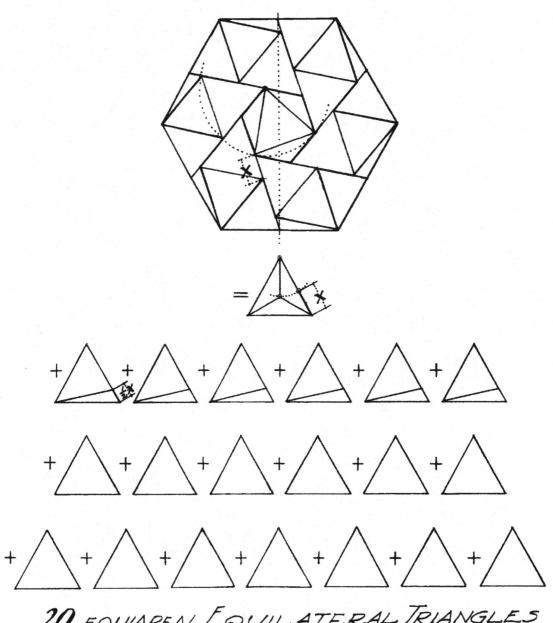

20 EQUIAREAL EQUILATERAL TRIANGLES
ASSEMBLED INTO A
HEXAGON
28 pieces

PLATE
25

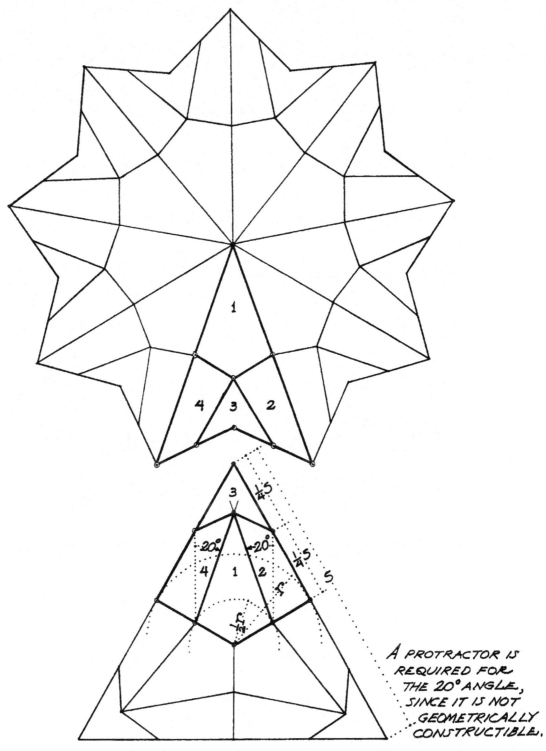

A PROTRACTOR IS
REQUIRED FOR
THE 20° ANGLE,
SINCE IT IS NOT
GEOMETRICALLY
CONSTRUCTIBLE.

3 OF THESE *12-PIECE* EQUILATERAL TRIANGLES
MAKE THE STELLATED NONAGON
36 PIECES IN ALL

PLATE
26

Chapter 7

Squares, Crosses, Rectangles (Plates 27–46)

Geometric figures with right angles, such as rectangles and squares, are perhaps the most cooperative of figures when it comes to dissections. This is not surprising since they fit together so easily to make a strip for a strip dissection or to tessellate the plane, and they convert to each other easily by using a P-slide or a step dissection. We see examples of such dissections in Plates 28, 33, and 34, respectively. While Freese did display dissections of regular polygons to squares in other chapters, he corralled many superb dissections in this chapter. In Plates 27 and 28 he dissected various isosceles triangles to squares. In Plates 29-32, he created squares from crosses and cross-related figures, including the octagram, Greek Cross, stepped cross, quadrate cross, and even the swastika. After dissecting rectangles to squares in Plates 33 and 34, Freese dissected sets of squares to other squares, especially congruent squares, in Plates 35-37. His most inventive square dissections are of squares of different areas to larger ones, as in Plates 38-43, climaxing with squares whose sidelengths are in arithmetic progression. Finally, Freese homed in on his so-called "hollow squares," which he dissected to normal squares and even to a second hollow square.

Especially satisfying are his symmetrical dissection of an octagram to a square in Plate 29 and his (hingeable!) dissection of a hollow square to another hollow square in Plate 46. Considering the fundamental nature of some of the dissection problems in this chapter, it is surprising that people were later able to make do with fewer pieces than Freese. Among the many dissections that have found improvement are those in Plates 29, 32, 39, 40, 41, 43, and 45. And too bad that Freese never saw that the 3-way dissection in Plate 28 is hingeable!

Plate 27: In this plate Freese gave seven simple dissections of a square to an isosceles triangle. One uses two pieces, another uses four pieces, and five use three pieces. An eighth dissection converts a square to congruent isosceles triangles in three pieces. Four of the dissections are hingeable.

Plate 28: Freese showed how to cut a square into five pieces that form either an isosceles triangle whose apex is 30° or an isosceles triangle whose apex is $2\arctan(1/4)$, which is approximately 28.072°. A pleasant surprise is that this 3-way dissection is hingeable, as we see in Figures I28a and I28b.

Freese also gave a 3-piece dissection of a square to a dodecagon sector and a 4-piece dissection of a square to a dodecagon segment. (See Plate 3 for definitions of a sector and a segment.)

Plate 29: Freese was the first to dissect the octagram to two squares. His 9-piece dissection is translational. Subsequently, C. Stuart Elliott (in 1985–1986) improved on it with the 7-piece dissection in Figure 29a. Too bad that Elliott's dissection is not translational.

Approximately 700 years ago the anonymous Persian manuscript *Interlocks of Similar or Complementary Figures* included the same 8-piece dissection of an octagram to a square that Freese

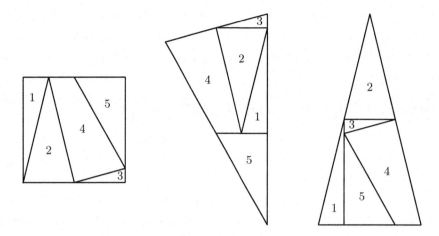

I28a: Freese's dissection of square to two different isosceles triangles

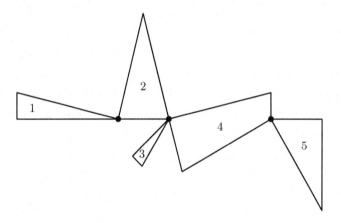

I28b: Hinged pieces for square to two different isosceles triangles [GNF]

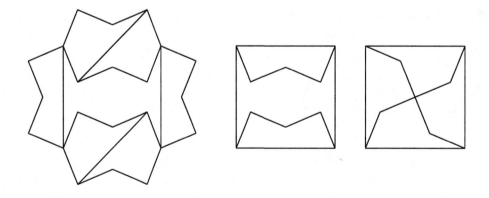

I29a: Improved two squares to an octagram [Stuart Elliott]

identified. When suitably arranged, it is translational. Harry Lindgren (1964) also matched the 8-piece dissection, which I improved (in 1972b) to the 7-piece dissection in Figure I29b.

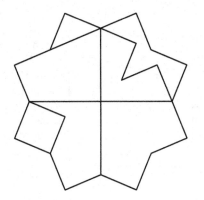 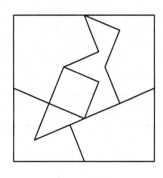

I29b: Improved octagram to a square [GNF 1972]

Sam Loyd first gave the dissection of two Greek Crosses to a square in *Tit-Bits* on May 15, 1897, on page 117. It is hingeable. The dissection of four Greek Crosses to a square seems to have been original and hard to beat. When suitably arranged, both of these two dissections are translational.

Plate 30: In this plate, Freese handled both the Greek Cross and the "Stepped Cross." Henry Dudeney (1931b) had presented the same dissection of the Stepped Cross to a square, which Dudeney had attributed to E. B. Escott in his column in the *Strand Magazine*, vol. 74 (1927), page 106. The dissection is both hingeable and translational.

Henry Dudeney also discussed the dissection of a Greek Cross to two squares in the second column on page 33 of his (1917) book. With appropriate orientations and labeling, that dissection is both hingeable and translational.

The Greek Cross to a square is a minor variation from a 3-piece Greek Cross to a (2×1)-rectangle that Henry Dudeney gave in (1917). Freese gave the 4-piece dissection of a Greek Cross to a square that was described by Sam Loyd in *Tit-bits* on April 24, 1897, on page 59. That dissection is both hingeable and translational, with appropriate labeling. Freese also gave, with dashed edges, a 4-piece translational dissection very close to that in Don Lemon's (1890) book.

Finally in this plate, Freese displayed a dissection of a Greek Cross to two squares. That dissection is hingeable and, when suitably arranged, translational.

Plate 31: The dissection of a cross with one more step than the Stepped Cross in Plate 30 is very similar to the dissection in that previous plate, and thus is both hingeable and translational. C. Stuart Elliott gave the same 4-piece dissection in his (1985–1986) article.

In his column in the *Weekly Dispatch* on August 26, 1900, Henry Dudeney gave a 5-piece dissection of two Greek Crosses to one. That delightful dissection is both hingeable and translational. The dissection method might better be termed a tessellation method than "ring expansion."

The 8-piece dissection of four Greek Crosses to one is translational and surely minimal.

Plate 32: Freese appears to have been the first to propose dissecting a "quadrate cross" to a square. His 8-piece dissection has 4-fold rotational symmetry, and is both hingeable and translational. Yet there is a less symmetrical dissection (Figure I32a) that uses only seven pieces. Out of a set of related 7-piece dissections, this one is also translational.

Freese also appears to have been the first to dissect the swastika to a square. He considered two different dimensions for the swastika, the first composed of seventeen small squares, and the second

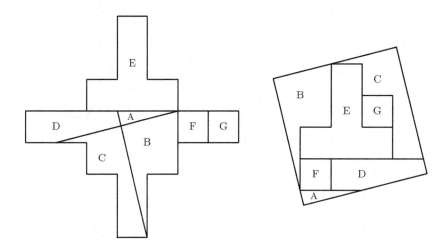

I32a: Translational improved quadrate cross to a square [GNF]

of twenty-five squares. The first he dissected into eight pieces, which is not as good as the 6-piece dissection (Figure I32b) of Lindgren (1964), although Freese's dissection is translational. On the other hand, Freese's 4-piece dissection of the second swastika to a square seems tough to beat.

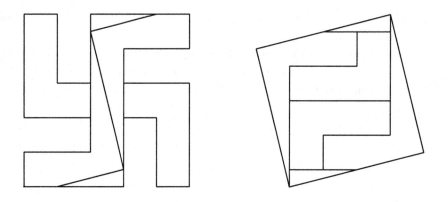

I32b: Improved swastika to a square [Harry Lindgren]

Plate 33: In this plate Freese showed four different applications of the *step technique*, with which we can dissect an $n^2 \times (n+1)^2$ rectangle to an $n(n+1)$-square using just two pieces. How to do this was known long before Freese finished his manuscript. Henry Dudeney (1926) described a general approach to this, identifying the first five cases, and noting that it would work as long as the ratios of the side lengths of the rectangle were as he described. Dudeney wrote that cutting a 9×16 rectangle to a square was a well-known puzzle, given in all the old books, but that no one had ever attempted the general rule before he did. As for having described a few earlier cases, that honor falls to Girolamo Cardano (1663), and also to Leonardo da Vinci. All of the examples on this plate are translational.

Plate 34: In this plate Freese reviewed two well-known techniques for converting a rectangle to a square using just 3 pieces. Method 1 employs the so-called strip method of converting one parallelogram to another. Paul-Jean Busschop (1876) made use of it, as did Henry Dudeney (1917). Method 2 is the so-called P-slide method of converting one parallelogram to another. Philip Kelland (1855) gave an early application of it and Dudeney (1917) found many applications of it. All dissections in this plate are translational, but the two produced by Method 1 need to have the square oriented differently. The two dissections at the bottom of the plate are hingeable.

Freese's characterization of the figure on the bottom right needs some explanation. The limit of Method 2 should really have the rightmost rectangle in the case $B = 4A$ be split into two right triangles by an edge from its upper left corner to its lower right corner, and the upper rectangle of the square should also have a similar split. Freese of course noticed that he could then improve the dissection by gluing the two right triangles together to get the 2-piece dissection.

Plate 35: For 5-piece dissections of two squares to one, Freese first presented the 5-piece dissection by Henry Perigal (1873) and then gave an uncentered version, illustrated earlier by Arthur Siddons (1932). Both versions are translational, and the first is hingeable.

Plate 36: The first two dissections in this plate are straightforward applications of the P-slide technique. Philip Kelland (1855) gave the first, which Henry Dudeney 1907) also later produced. The second is an elaboration of a solution given by Henry Dudeney (1917). Abū'l-Wafā described the third dissection, which Emile Fourrey (1907) discussed in detail. All three versions are translational.

Plate 37: The dissections of eight squares to one, and of five squares to one, follow the same pattern as described by Abū'l-Wafā and discussed in detail by Emile Fourrey (1907). The dissections of five squares to two, and of thirteen squares to two, use tessellation techniques and seem fairly straightforward. All of these dissections are translational, though certain squares in the dissections of 5 squares to 2 and 13 squares to 2 must be appropriately oriented before translation. The dissections of 8 squares to 1 and of 5 squares to 1 are hingeable.

Plate 38: The dissection of squares of areas 2, 3, and 4 to a square of area 9 may well have been original. It would seem difficult to do better than the seven pieces that Freese achieved. With an appropriate orientation, this dissection is translational.

Plate 39: Both dissections in this plate are translational. The first of the "unique examples" in this plate is certainly not unique. Geoffrey Mott-Smith (1946) gave a puzzle with a shape of linoleum created by the combining of squares of side 1, 4, and 8, in which the object was to cut this shape into a square of side 9. Treating the shape as unglued will give a 5-piece solution different from Freese's supposedly unique example. Furthermore, in (1997) I described four infinite classes of three noncongruent squares that have 5-piece dissections to a larger square, namely the Pythagoras-plus class, the PP-plus class, Cossali's class, and the square-sum class, plus a few incidental sets of three unequal squares. Included in the square-sum class is my dissection of squares for $8^2 + 9^2 + 12^2 = 17^2$, in Figure I39a.

Freese was apparently the first to give a dissection of squares for $2^2 + 4^2 + 5^2 + 6^2 = 9^2$. He gave no proof for his claim that his second example in Plate 39 is the only possible 6-piece dissection of four unequal squares to a larger square. Actually, his claim is ambiguous, since it is not clear whether he was concerned with an identity other than $2^2 + 4^2 + 5^2 + 6^2 = 9^2$ or a different dissection of squares for that identity. Clearly there is another 6-piece dissection realizing that identity, as we see in Figure I39b.

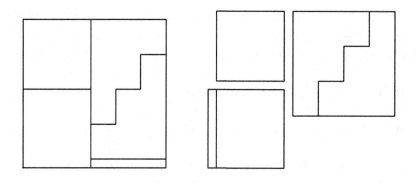

I39a: Squares for $8^2 + 9^2 + 12^2 = 17^2$ [GNF]

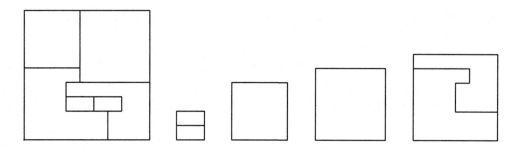

I39b: Another dissection for squares for $2^2 + 4^2 + 5^2 + 6^2 = 9^2$ [GNF]

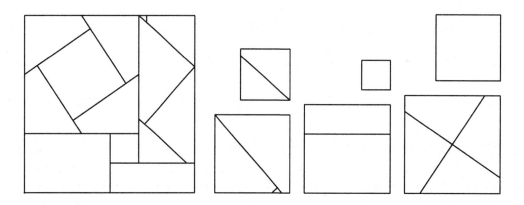

I40: Improved squares for $1^2 + (\sqrt{3})^2 + (\sqrt{5})^2 + (\sqrt{7})^2 + 3^2 + (\sqrt{11})^2 = 6^2$ [GNF]

Plate 40: The dissection of squares for $1^2 + (\sqrt{3})^2 + (\sqrt{5})^2 + (\sqrt{7})^2 + 3^2 + (\sqrt{11})^2 = 6^2$ is an intriguing puzzle that no one had previously posed. Observe that Freese implicitly grouped squares together to fill in the area of the 6-square. In particular, the 1- and $\sqrt{5}$-squares combine to form a (3×2)-rectangle. Freese then placed this rectangle on top of a (3×1)-rectangle which he produced from the $\sqrt{3}$-square. Similarly, he formed a (6×3)-rectangle by converting the $\sqrt{7}$- and $\sqrt{11}$-squares. With squares appropriately oriented, the dissection is translational.

Freese also stated a side constraint that the three smallest squares should incidentally be dissected to a square. This constraint seems opportunistic, and if we drop it and use several more tricks, we can improve on Plate 40 by one piece, as we see in Figure I40, and yet still maintain translationality. (Note that the $\sqrt{7}$-square and the $\sqrt{11}$-square must be appropriately oriented.) First, pair the $\sqrt{5}$- and $\sqrt{11}$-squares together to produce a 4-square, using a 5-piece dissection such as Perigal's (See Plate 35). Next, pair the $\sqrt{3}$- and $\sqrt{7}$-squares together to produce a (2×5)-rectangle which will pack against the 4-square inside the 6-square. Then, cut the 3-square to two rectangles which together with the 1-square will fill in the remaining area in the 6-square. If we use three pieces each for the $\sqrt{3}$- and $\sqrt{7}$-squares, then we don't have any improvement. Yet fortuitously, we can do this latter with just five pieces. What makes it possible is the fact that there is a right triangle with legs of length 2 and $\sqrt{3}$ and an hypotenuse of length $\sqrt{7}$. One of the right triangles that Freese drew in this plate illustrates that relationship! Indeed, the tiny right triangle in the improved dissection of Figure I40 has sides of length $2t$, $t\sqrt{3}$, and $t\sqrt{7}$, for $t = 1 - \sqrt{3}/2$.

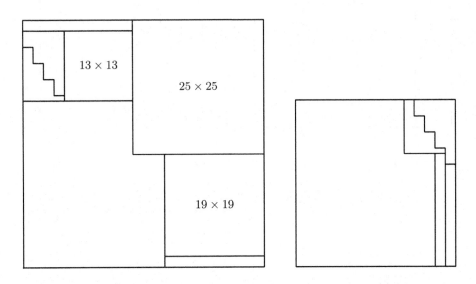

I41a: Improved squares for $13^2 + 19^2 + 25^2 + 31^2 = 46^2$ [GNF]

Plate 41: The dissection of squares for $13^2 + 19^2 + 25^2 + 31^2 = 46^2$ is a unique dissection puzzle that no one had previously posed. But look out! Freese made a mistake in drawing this figure. The horizontally oriented rectangle sitting beneath the 19-square should be 15 units long, rather than 19. The piece in the lower left-hand corner of the 46-square should extend 4 units further to the right, filling in to the left of the now shorter rectangle.

In any event, I have improved on Freese's dissection by one piece, as we see in Figure I41a. I cut only the 31-square, as Freese did, and leave the other squares positioned exactly the same in the 46-square. I then use a step dissection to convert an (8×13)-rectangle in the 46-square to a (10×11)-rectangle with a 4-square attached, which neatly fills a crucial area in the 31-square.

At the bottom of this plate, Freese claimed that there is another 9-piece dissection for a similar identity, namely $7^2 + 15^2 + 23^2 + 31^2 = 42^2$, but he didn't present that dissection. Challenged, I set out to find that dissection, and was rewarded when I discovered an 8-piece dissection (Figure I41b). The small squares seem to pack well into the large square if we trim rectangles off the larger two of the small squares and do a bit of "juggling." Can a reader improve on the 8-piece dissection?

Plate 42: The dissection of squares for $2^2 + 5^2 + 8^2 + 11^2 + 14^2 + 17^2 + 20^2 + 23^2 + 26^2 = 48^2$ appears to have been an original dissection puzzle. Freese's 13-piece dissection seems tough to improve on.

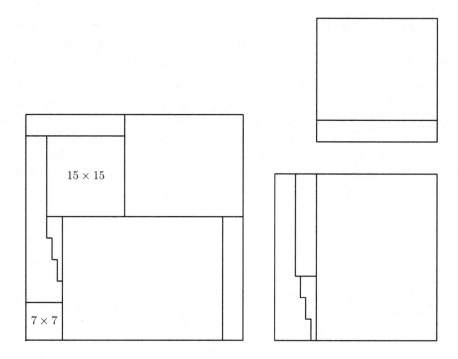

I41b: Improved squares for $7^2 + 15^2 + 23^2 + 31^2 = 42^2$ [GNF]

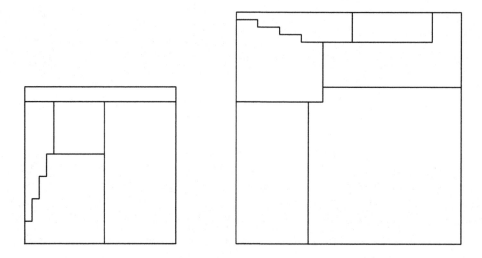

I43: 21- and 31-squares for improved $4 \times 21^2 = 7^2 + 15^2 + 23^2 + 31^2$ [GNF]

Plate 43: The dissection of squares for $4 \times 21^2 = 7^2 + 15^2 + 23^2 + 31^2$ also appears to have been original. However, I have found a way to improve Freese's 14-piece dissection, bringing it down to twelve pieces.

I cut three of Freese's 21-squares as he did, and similarly with the 7-, 15-, and 23-squares. Thus I show only the changed (rightmost) 21-square and the 31-square in Figure I43. The main change is that I use a step approach to convert an (11×12)-rectangle with a (4×7)-rectangle attached in the 21-square to a 12-square with a 4-square attached in the 31-square. The step in the 21-square has width 1 and riser height 3.

Plate 44: This plate illustrates two common ways to dissect a rectangle to a square. The first employs the P-strip method, used commonly since the latter part of the nineteenth century. The second method utilizes a P-slide. Neither of these methods was original to Freese. He did identify cases of these 3-piece dissections in which all edge lengths are integral, using Pythagorean triples. As discussed in the comments for Plate 34, these dissections are translational. In Plate 33, Freese identified 2-piece step-cut dissections of rectangles to squares which allow integral edge lengths.

I45a: Improved hollow square to square [GNF]

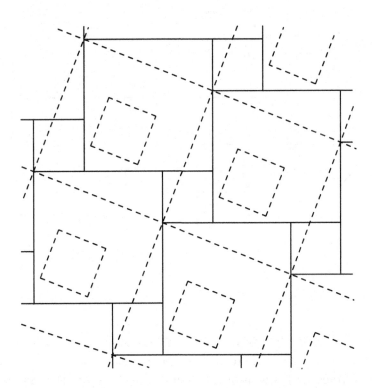

I45b: Tessellation for improved hollow square to square [GNF]

Plate 45: The dissection of a hollow square to a square seems to have been original with Freese. For the dissection at the top of the plate, Freese noted that if he had allowed turning over, then he could have achieved an 8-piece rather than a 12-piece dissection. Yet for a sufficiently small hole, I have been able to improve the dissection to five pieces. (See Figure I45a.) To qualify for this dissection, the ratio of the side of the hole to the side of the hollow square must be less than the real root of $2z^3 - 2z^2 + 3z - 1 = 0$, which is approximately .3966 . I use as a basis the dissection of two squares to one by the ninth century Islamic mathematician Thābit ibn Qurra. We see the tessellation for this dissection in Figure I45b.

This size range includes the case in which the hole has side length exactly one third of the side length of the square. For that case Bert Baetens (1997) found a special 5-piece dissection that is hingeable, as we see in Figures I45c and I45d.

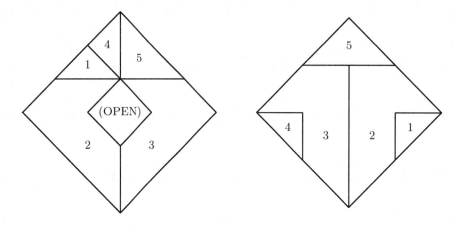

I45c: 3-Square with 1-square hole to square [Baetens]

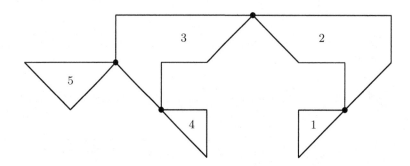

I45d: Hinged pieces for 3-square with 1-square hole to square [Baetens]

In the case that the hole is larger, and we respect Freese's restriction of not turning over pieces, Gavin Theobald has found the lovely 8-piece dissection (with curved cuts!) in Figure I45e. It works whenever the ratio of the side of the hole to the side of the hollow square is less than $1/\sqrt{2} \approx .707$. This dissection derives from the tessellation in Figure I45f, which is the tessellation for Perigal's dissection that we have already seen in Figure 1.5. The circular cut allows us to rotate the square hole and the four adjacent pieces bounded by the circle and the sides of the square.

At the bottom of the plate, Freese gave a limiting case, for which he noted that he did not need to split the right triangle into two isosceles triangles, because the right triangle is an isosceles right triangle. Thus his approach gives an 8-piece dissection. However, for this case I have found the

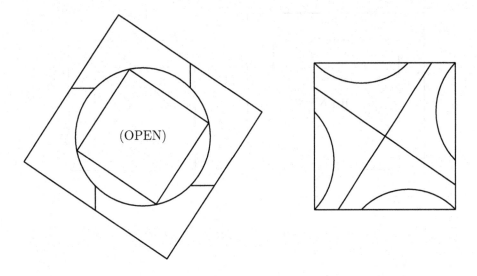

I45e: Improved hollow square with larger hole to square [Gavin Theobald]

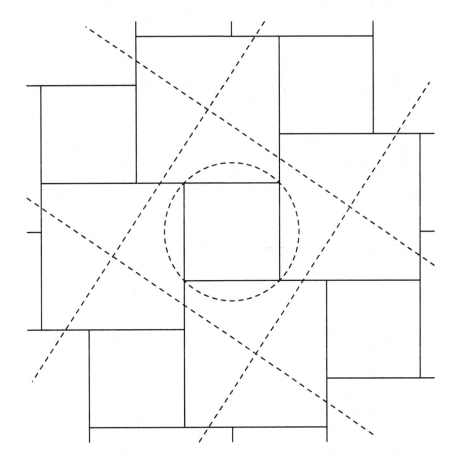

I45f: Tessellation for improved hollow square with larger hole to square [Gavin Theobald]

6-piece dissection in Figure I45g.

We survey two more special cases: Anton Hanegraaf found a 4-piece dissection of a 5-square with

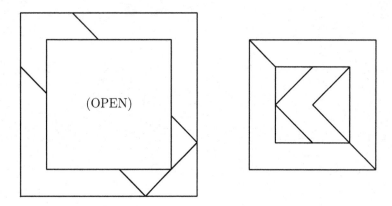

I45g: Improved minimum-width hollow square to square [GNF]

a 3-square hole to a 4-square (Figure I45h). Figure I45i shows a 6-piece dissection of a 10-square with an 8-square hole to a 6-square.

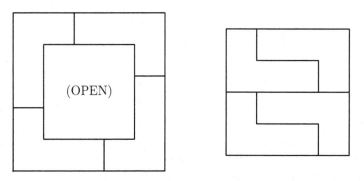

I45h: A 5-square with a 3-square hole to a 4-square [Anton Hanegraaf]

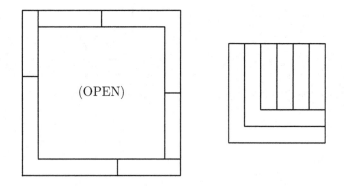

I45i: A 10-square with an 8-square hole to a 6-square [GNF]

Plate 46: This dissection puzzle of one hollow square to another hollow square is comparable to the dissection of two squares to two other squares. The method involves a double application of Perigal's dissection of two squares to one. Freese's appears to have been the earliest example of this dissection. Before I saw his manuscript, I had rediscovered this dissection and included it in my 1997 book. This dissection is translational.

This brings us to the end of our chapter on dissections of squares. We have enjoyed dissections of Greek Crosses to squares and dissections of congruent squares to larger squares, which Freese had well under control. Yet he also introduced new dissection problems, and for those we have found a variety of ways to improve his dissections. Those include his dissections of octagrams to squares, of a quadrate cross to a square, and a swastika to a square. Freese's dissection problems of squares with either areas in arithmetic progression or with side lengths in arithmetic progression were also novel, as was his dissection problem of a hollow square. It has been exciting to discover dissections that use fewer pieces than Freese's dissections. So I enthusiastically acknowledge the fertile imagination from which these new problems sprang!

At the same time, we should note that while Freese presented dissections of squares for integer identities in Plates 39 and 41–43, he missed out on an abundance of integer identities. Aside from Plate 39, he focused on integer identities that had integers in arithmetic progression, missing out on many possibilities. I have covered them in some detail in a chapter of my first (1997) book.

It is disappointing that Freese did not consider dissections of cubes and rectangular solids. H. S. M. Coxeter mentioned early (and clever) dissections of a $2 \times 1 \times 1$ rectangular block to a cube by William Cheney (1933) and Albert Wheeler (1935) in his (1939) revision of Rouse Ball's book, only two pages further on from where Freese had supplied his own handwritten comments. Cheney and Wheeler had used the same strip and P-slide techniques that Freese described in his Plate 44. It is ironic that an architect, for whom designing 3-dimensional structures such as houses was second nature and who wrote the book *Perspective Projection* for drawing 3D views of houses, had missed this opportunity.

More understandable is the omission of dissections of cubes for the identity $3^3 + 4^3 + 5^3 = 6^3$ by Herbert Richmond (1944) and by John Leech (anonymously, 1951), both of which appeared in a somewhat obscure British periodical.

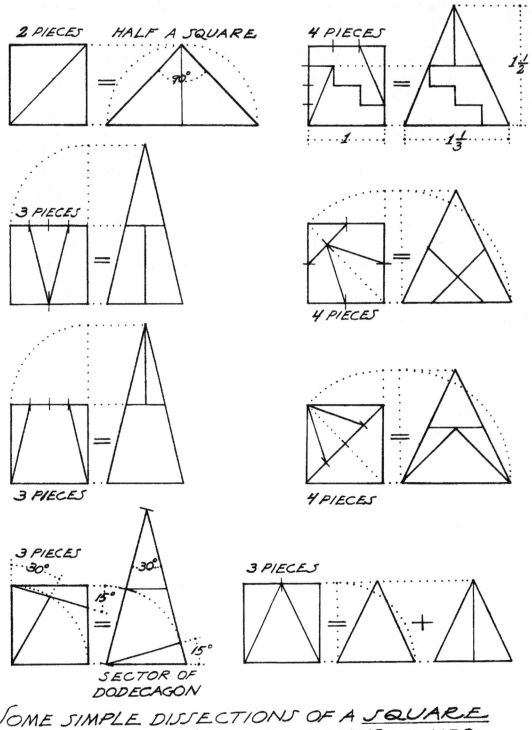

2 PIECES HALF A SQUARE 4 PIECES

3 PIECES

3 PIECES

4 PIECES

4 PIECES

3 PIECES
30° 30° 15° 15°
SECTOR OF
DODECAGON

3 PIECES

SOME SIMPLE DISSECTIONS OF A SQUARE
RE-ASSEMBLED INTO VARIOUS DERIVED
ISOSCELES TRIANGLES
(BUT NOT THE EQUILATERAL)

PLATE
27

BI PIGAI
BI PIGB
..FREESE..

THE 5 PIECES OF THE SQUARE MAKE **EITHER** OF THE TWO DIFFERENT ISOSCELES TRIANGLES

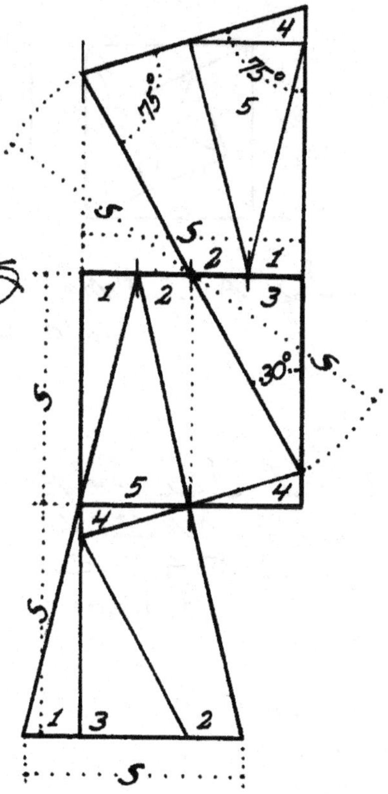

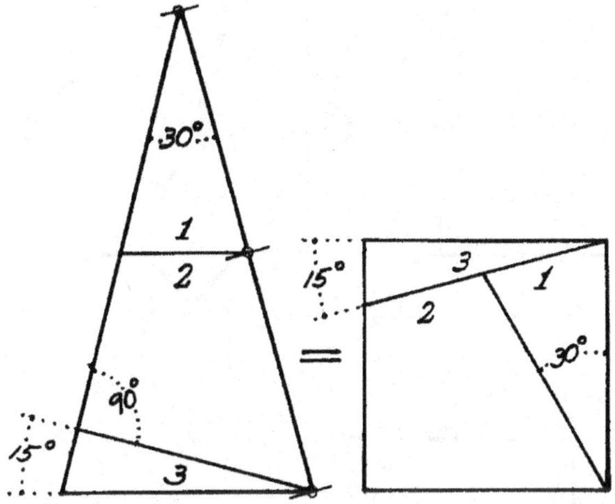

A 3-PIECE SQUARE MAKES A DODECAGON SECTOR

A 4-PIECE SQUARE MAKES A DODECAGON SEGMENT

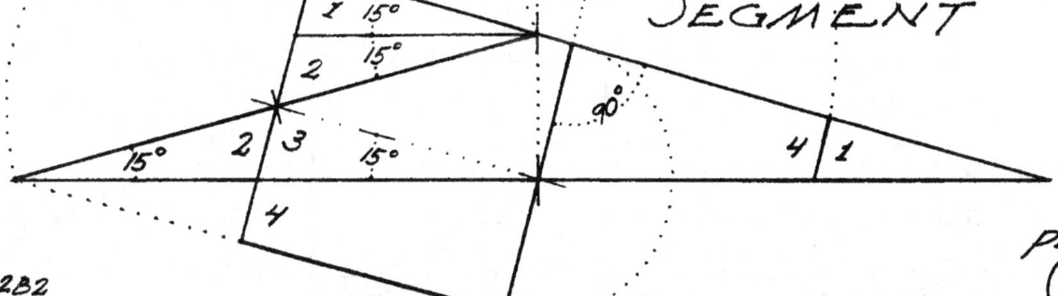

B2 PR2B2
B1 P22
B1 P22
..FREESE..

PLATE 28

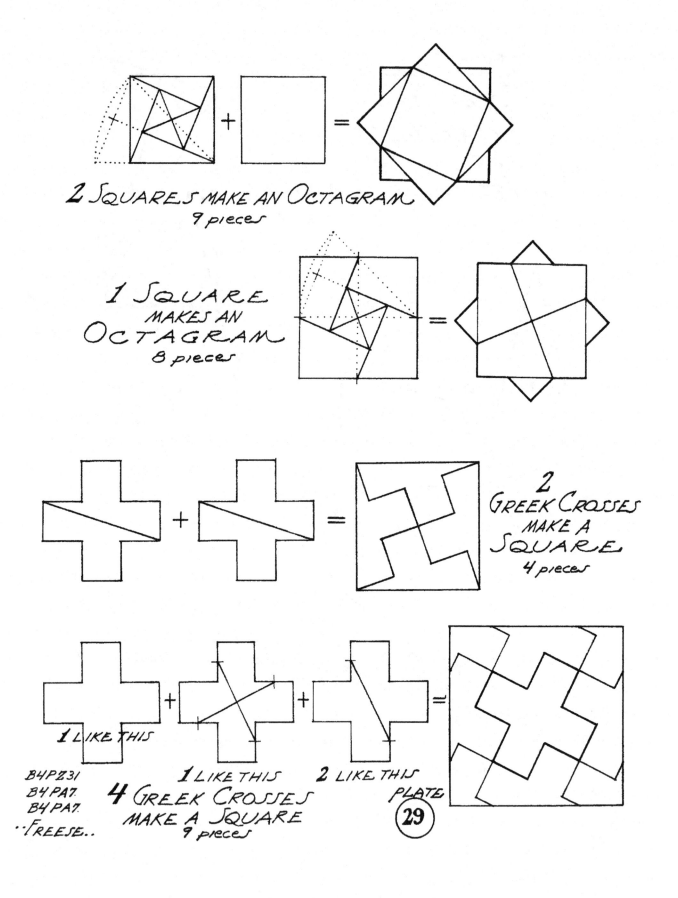

2 SQUARES MAKE AN OCTAGRAM
9 pieces

1 SQUARE MAKES AN OCTAGRAM
8 pieces

2 GREEK CROSSES MAKE A SQUARE
4 pieces

1 LIKE THIS + *1 LIKE THIS* + *2 LIKE THIS* =

4 GREEK CROSSES MAKE A SQUARE
9 pieces

B4 PZ 31
B4 PAT.
B4 PAT.
··FREESE··

PLATE
(29)

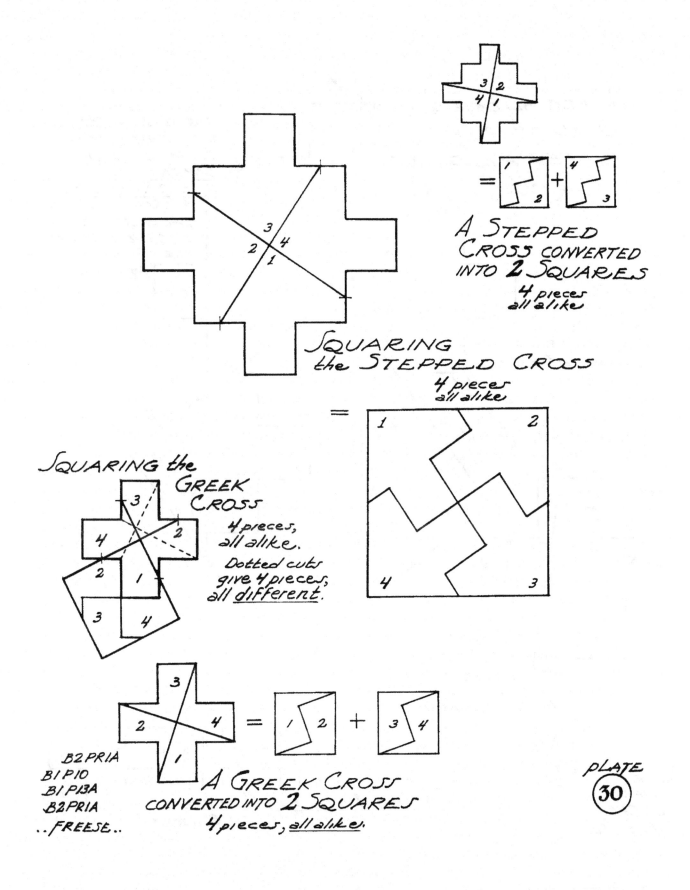

A STEPPED CROSS CONVERTED INTO **2** SQUARES
4 pieces all alike

SQUARING the STEPPED CROSS
4 pieces all alike

SQUARING the GREEK CROSS
4 pieces, all alike.
Dotted cuts give 4 pieces, all <u>different</u>.

A GREEK CROSS CONVERTED INTO **2** SQUARES
4 pieces, all alike.

B2 PR1A
B1 P10
B1 P13A
B2 PR1A
..FREESE..

PLATE
30

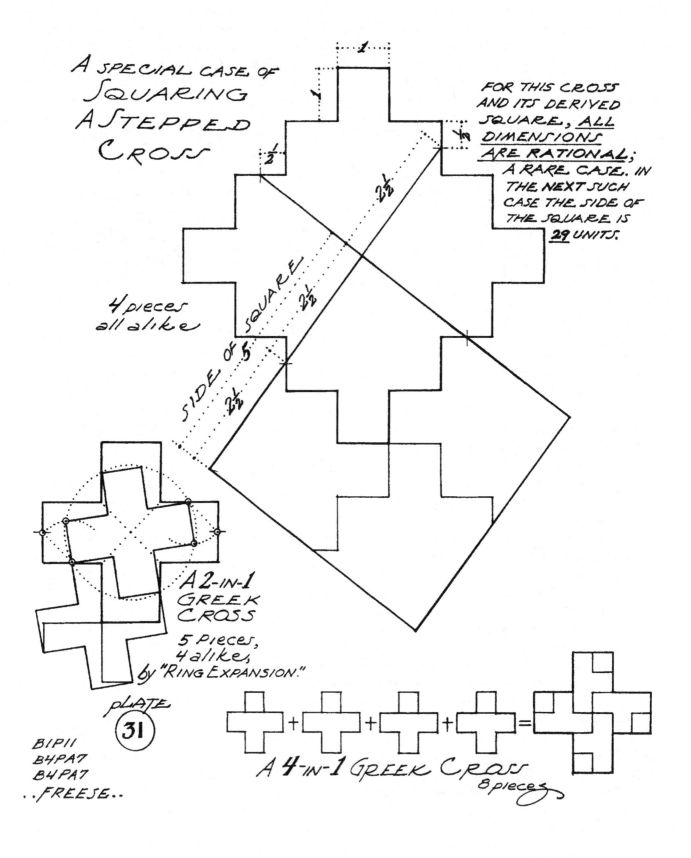

A SPECIAL CASE OF
SQUARING
A STEPPED
CROSS

FOR THIS CROSS
AND ITS DERIVED
SQUARE, *ALL
DIMENSIONS
ARE RATIONAL*;
A RARE CASE. IN
THE NEXT SUCH
CASE THE SIDE OF
THE SQUARE IS
29 UNITS.

4 pieces
all alike

SIDE OF SQUARE

A 2-IN-1
GREEK
CROSS

5 pieces,
4 alike;
by "RING EXPANSION."

PLATE
31

BIP11
B4PA7
B4PA7
..FREESE..

A 4-IN-1 GREEK CROSS
8 pieces

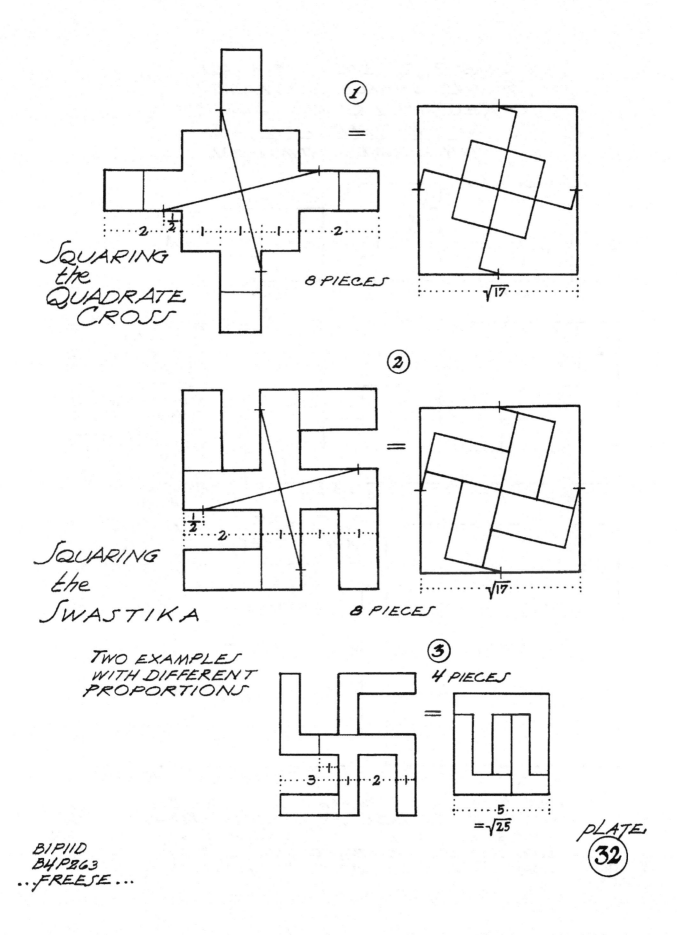

① =

SQUARING
the
QUADRATE
CROSS

8 PIECES

2   ½   2   √17

②

SQUARING
the
SWASTIKA

=

8 PIECES

½   2   √17

TWO EXAMPLES
WITH DIFFERENT
PROPORTIONS

③ 4 PIECES

=

3   2   5
=√25

PLATE
32

BIP11D
B4 P263
...FREESE...

Some Particular Cases of the **2-Piece** Conversion of Rectangles into Squares *by One Stepped Cut*, and with all dimensions integral.

2 pieces, alike.

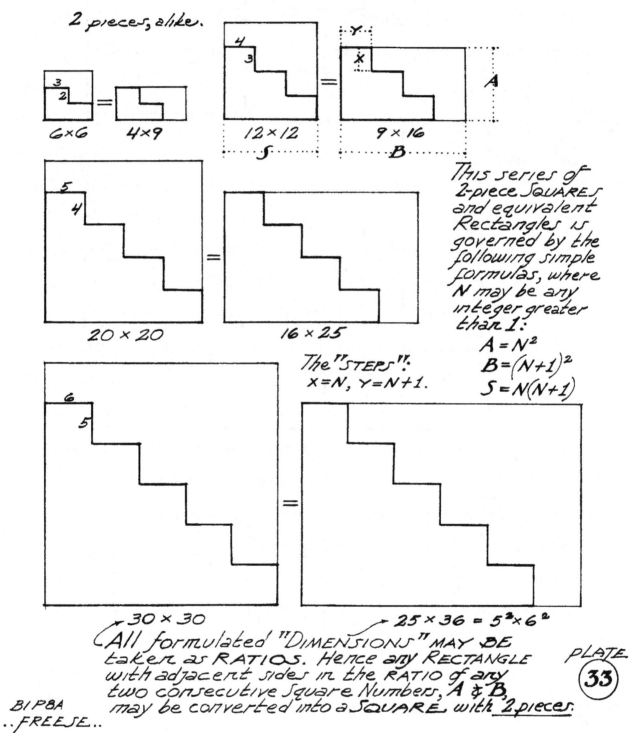

6×6   4×9

12×12   S   9×16   B

20 × 20   16 × 25

The "STEPS":
$x = N$, $y = N+1$.

This series of 2-piece SQUARES and equivalent Rectangles is governed by the following simple formulas, where N may be any integer greater than 1:

$A = N^2$
$B = (N+1)^2$
$S = N(N+1)$

30 × 30   25 × 36 = $5^2 × 6^2$

All formulated "DIMENSIONS" may be taken as RATIOS. Hence any RECTANGLE with adjacent sides in the RATIO of any two consecutive Square Numbers, A & B, may be converted into a SQUARE with 2 pieces.

PLATE
(33)

BIP8A
..FREESE..

The two 3-PIECE methods of converting RECTANGLES into SQUARES

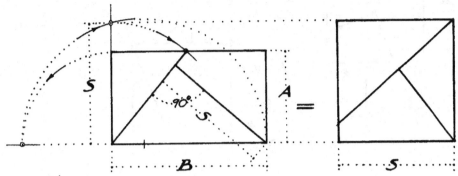

METHOD 1:
B NOT GREATER THAN 2A.

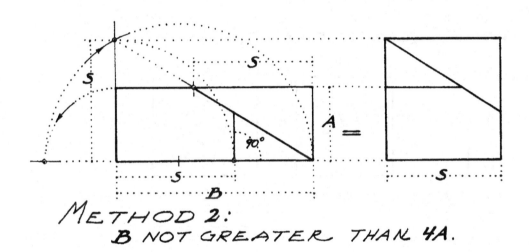

METHOD 2:
B NOT GREATER THAN 4A.

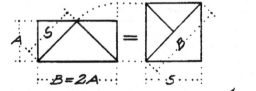

THE LIMIT OF METHOD 1.

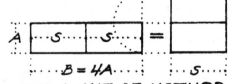

THE LIMIT OF METHOD 2.
(Reduces to 2 pieces)

If B, respectively, exceeds these limits, then a 3-piece dissection is impossible except under Method 2 when B equals 6A.

B1 P2B
..FREESE..

PLATE
34

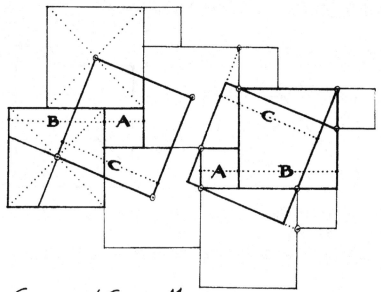

The General GRID METHOD of converting any 2 SQUARES into 1 SQUARE.

5 pieces

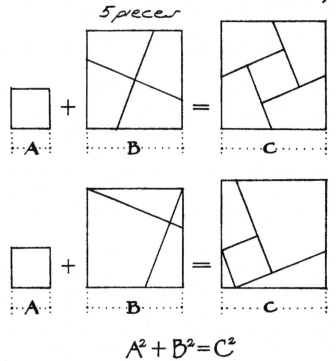

$$A^2 + B^2 = C^2$$

PLATE
35

B3 PS 34.0
..FREESE..

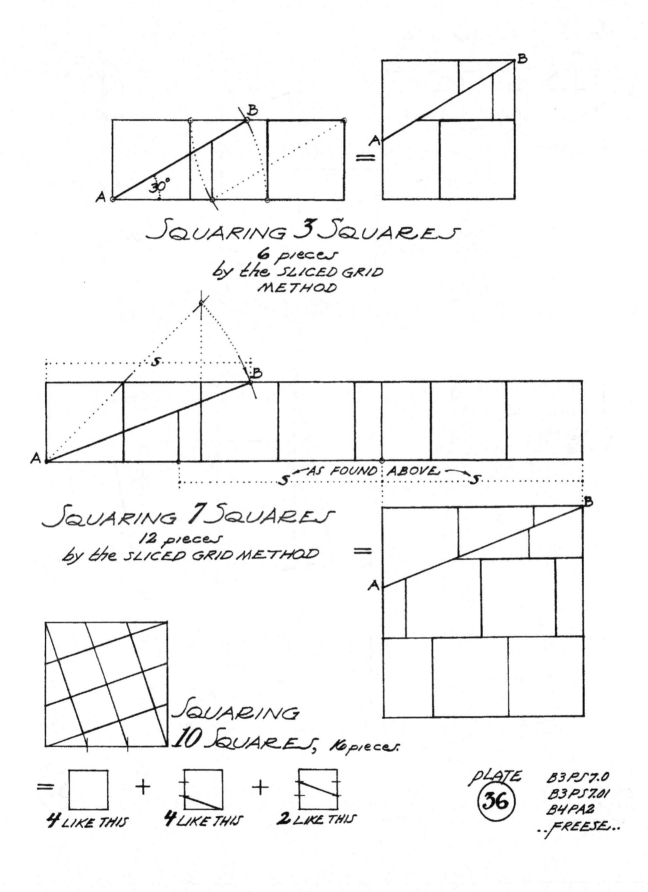

SQUARING 3 SQUARES

6 pieces
by the SLICED GRID
METHOD

30°

SQUARING 7 SQUARES
12 pieces
by the SLICED GRID METHOD =

S AS FOUND ABOVE S

SQUARING
10 SQUARES, 16 pieces.

= □ + □ + □

4 LIKE THIS   **4** LIKE THIS   **2** LIKE THIS

PLATE
36

B3 PS7.0
B3 PS7.01
B4 PA2
..FREESE..

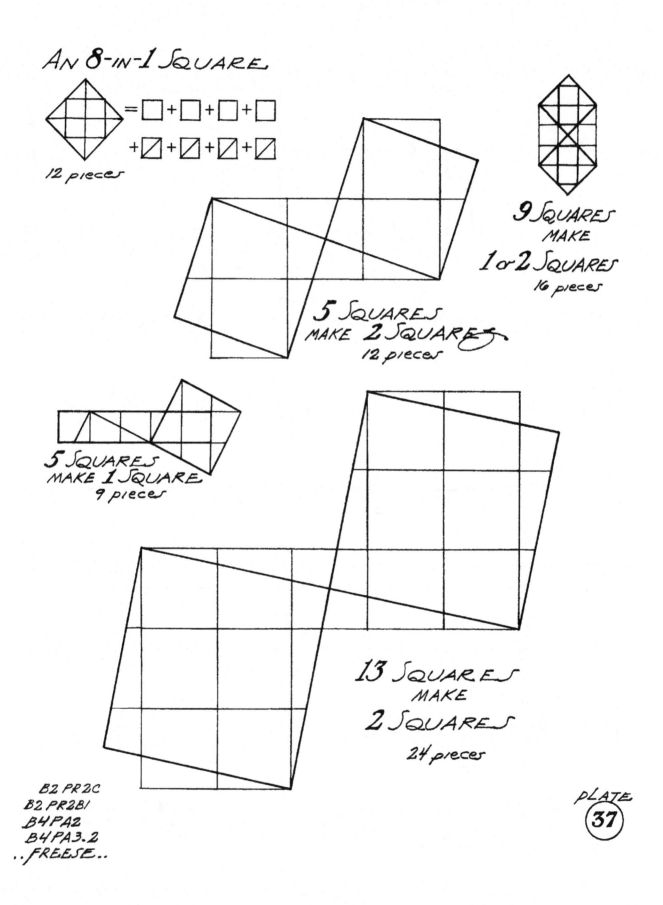

AN 8-IN-1 SQUARE

12 pieces

9 SQUARES
MAKE
1 or 2 SQUARES
16 pieces

5 SQUARES
MAKE 2 SQUARES
12 pieces

5 SQUARES
MAKE 1 SQUARE
9 pieces

13 SQUARES
MAKE
2 SQUARES
24 pieces

B2 PR2C
B2 PR2B1
B4PA2
B4PA3.2
..FREESE..

PLATE
37

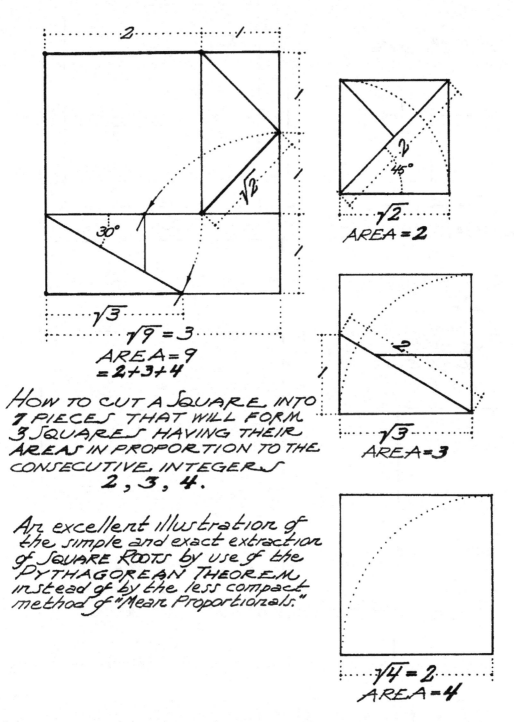

2　　1

√2

30°

√3

√9 = 3
AREA = 9
= 2 + 3 + 4

2
45°

√2
AREA = 2

1

2

√3
AREA = 3

√4 = 2
AREA = 4

How to cut a Square into 7 pieces that will form 3 Squares having their areas in proportion to the consecutive integers 2, 3, 4.

An excellent illustration of the simple and exact extraction of Square Roots by use of the Pythagorean Theorem instead of by the less compact method of "Mean Proportionals."

PLATE
38

....TWO UNIQUE EXAMPLES....

$$9^2 + 12^2 + 20^2 = 25^2$$

3 UNEQUAL SQUARES
MAKE
1 SQUARE
5 pieces
This is the only possible 5-piece
SQUARING OF ANY 3 UNEQUAL
SQUARES

=

4 UNEQUAL SQUARES MAKE
1 SQUARE

$$2^2 + 4^2 + 5^2 + 6^2 = 9^2$$

6 pieces
This is the only possible 6-piece
SQUARING OF ANY 4 UNEQUAL
SQUARES

PLATE
39

B4PA3.5
B4PA3.6
..FREESE..

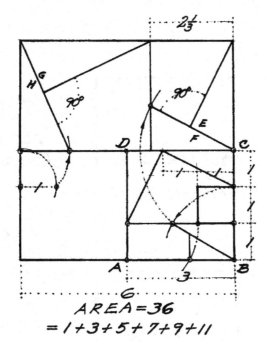

2⅖

6

AREA = 36
= 1 + 3 + 5 + 7 + 9 + 11

HOW TO CUT A SQUARE INTO 14 PIECES THAT MAY BE RE-ASSEMBLED INTO 6 SQUARES HAVING THEIR AREAS PROPORTIONATE TO THE FIRST SIX ODD INTEGERS

1, 3, 5, 7, 9, 11;

the SQUARE, ABCD, in itself being dissected to form, with its 7 pieces, 3 SQUARES with the ratio of AREAS

1 : 3 : 5.

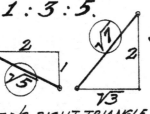

The simple RIGHT-TRIANGLE METHOD of extracting the irrational Square Roots, as incorporated in above dissection.

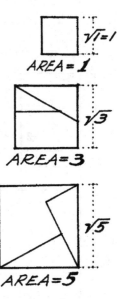

$\sqrt{1} = 1$

AREA = 1

$\sqrt{3}$

AREA = 3

$\sqrt{5}$

AREA = 5

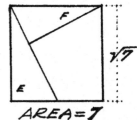

$\sqrt{7}$

AREA = 7

$\sqrt{9} = 3$

AREA = 9

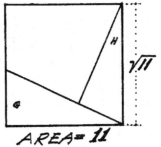

$\sqrt{11}$

AREA = 11

B4 PZ49
..FREESE..

PLATE
40

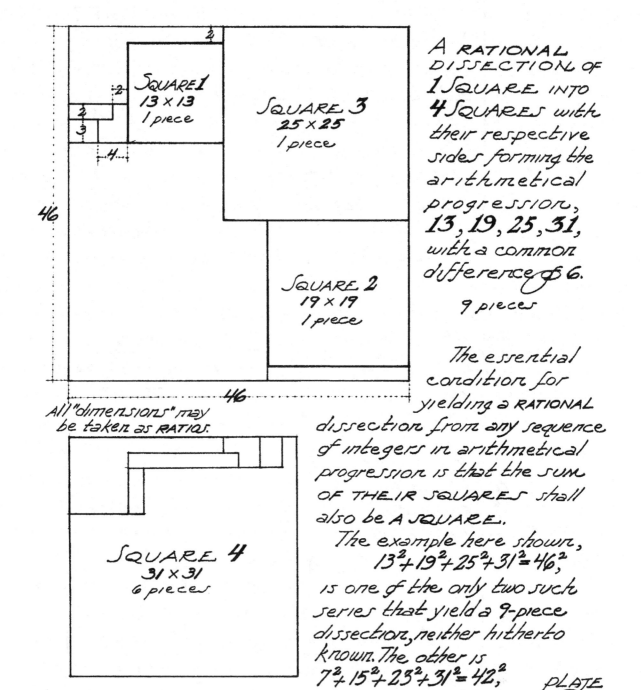

Square 1
13 × 13
1 piece

Square 3
25 × 25
1 piece

Square 2
19 × 19
1 piece

46

46

2

2

2

3

4

All "dimensions" may
be taken as RATIOS.

Square 4
31 × 31
6 pieces

A RATIONAL
DISSECTION OF
1 SQUARE INTO
4 SQUARES with
their respective
sides forming the
arithmetical
progression,
13, 19, 25, 31,
with a common
difference of 6.

9 pieces

The essential
condition for
yielding a RATIONAL
dissection from any sequence
of integers in arithmetical
progression is that the sum
OF THEIR SQUARES shall
also be A SQUARE.

The example here shown,
$$13^2 + 19^2 + 25^2 + 31^2 = 46^2,$$
is one of the only two such
series that yield a 9-piece
dissection, neither hitherto
known. The other is
$$7^2 + 15^2 + 23^2 + 31^2 = 42^2,$$
with a common diff. of 8.

PLATE
41

B4 PZ46A
..FREESE..

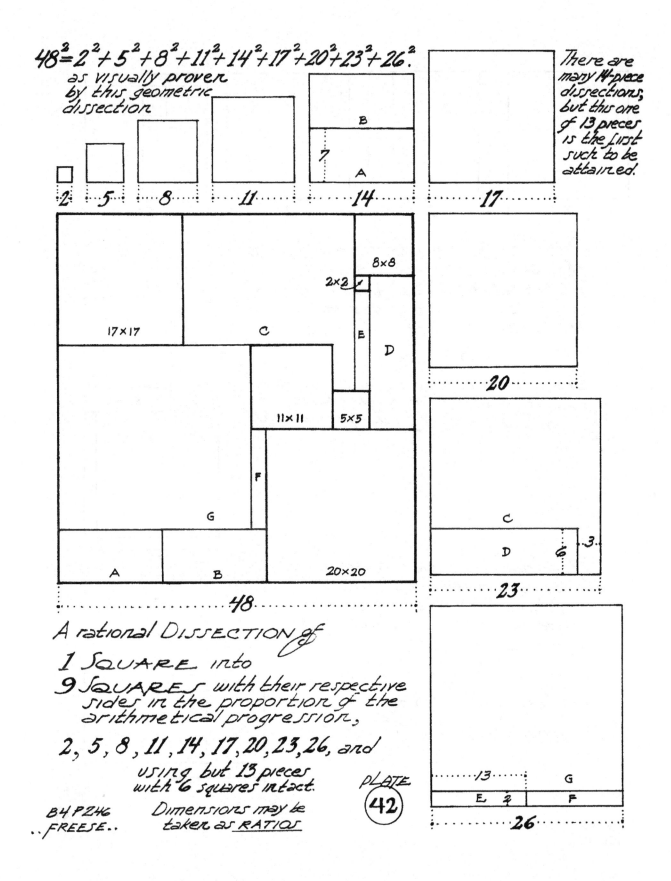

$$48^2 = 2^2 + 5^2 + 8^2 + 11^2 + 14^2 + 17^2 + 20^2 + 23^2 + 26^2$$

as visually proven by this geometric dissection

2   5   8   11   14   17

There are many 14-piece dissections, but this one of 13 pieces is the first such to be attained.

B
7
A

8×8

2×2
17×17   C   E   D

11×11   5×5

F

G

20

A   B   20×20

48

23

C
D   6   3

A rational DISSECTION of

1 SQUARE into

9 SQUARES with their respective sides in the proportion of the arithmetical progression,

2, 5, 8, 11, 14, 17, 20, 23, 26, and

using but 13 pieces with 6 squares intact.

PLATE
42

B4 P246
..FREESE..

Dimensions may be taken as RATIOS

13   G
E   2   F
26

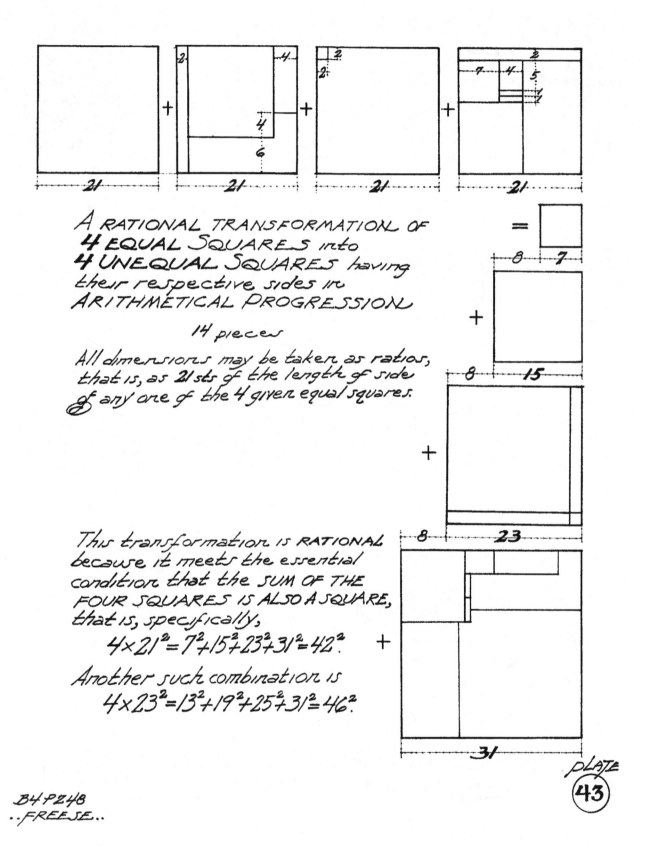

A RATIONAL TRANSFORMATION OF
4 EQUAL SQUARES into
4 UNEQUAL SQUARES having
their respective sides in
ARITHMETICAL PROGRESSION

14 pieces

All dimensions may be taken as ratios,
that is, as 21 sts of the length of side
of any one of the 4 given equal squares.

This transformation is RATIONAL
because it meets the essential
condition that the SUM OF THE
FOUR SQUARES IS ALSO A SQUARE,
that is, specifically,

$$4 \times 21^2 = 7^2 + 15^2 + 23^2 + 31^2 = 42^2.$$

Another such combination is
$$4 \times 23^2 = 13^2 + 19^2 + 25^2 + 31^2 = 46^2.$$

PLATE
43

TWO PARTICULAR SERIES OF 3-PIECE TRANSFORMATIONS OF RECTANGLES INTO SQUARES
IN WHICH
ALL DIMENSIONS ARE INTEGRAL

TABULATION OF THE FIRST FOUR SUCH TRANSFORMATIONS IN SERIES 1.

X	Y	Z	$A=Y^2$	$B=Z^2$	$S=ZY$
3	4	5	16	25	20
5	12	13	144	169	156
8	15	17	64	289	255
7	24	25	576	625	600

In both series, X, Y, Z, are the sides of any integral right triangle. Take Y as the longer leg in all formulas, with the additional restriction, in Series 2, only, that Y shall be less than 2X, since a 3-piece rational dissection is impossible in Series 2 if Y exceeds 2X in these formulas, that is, if B exceeds 4A.

① Diagram labels: $B=Z^2$, XY, Z^2-XY, $A=Y^2$, $S=ZY$, $Z(Y-X)$, ZX, $S=ZY$

② Diagram labels: $Y(Y-X)$, $S=XY$, $Z(2X-Y)$, ZX, $X(Y-X)$, $Z(Y-X)$, $A=X^2$, $S=XY$, $Y(Y-X)$, $S=XY$, $B=Y^2$

In both Series, dimensional & formulated values may be taken as RATIOS.

No RECTANGLE exists that belongs to BOTH series, since no right triangle exists in which X = Z.

TABULATION OF THE FIRST FOUR SUCH TRANSFORMATIONS IN SERIES 2.

X	Y	Z	$A=X^2$	$B=Y^2$	$S=XY$
3	4	5	9	16	12
8	15	17	64	225	120
20	21	29	400	441	420
28	45	54	784	2025	1260

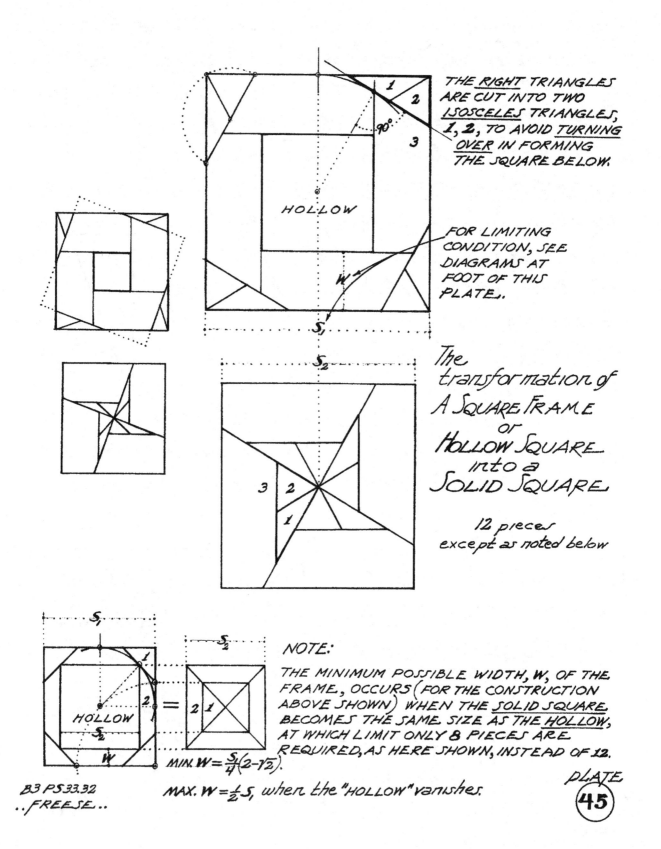

THE <u>RIGHT</u> TRIANGLES
ARE CUT INTO TWO
<u>ISOSCELES</u> TRIANGLES,
<u>1, 2,</u> TO AVOID <u>TURNING</u>
<u>OVER</u> IN FORMING
THE SQUARE BELOW.

90°

1
2
3

HOLLOW

FOR LIMITING
CONDITION, SEE
DIAGRAMS AT
FOOT OF THIS
PLATE.

W

S_1

S_2

3 2

1

The
transformation of
A SQUARE FRAME
or
HOLLOW SQUARE
into a
SOLID SQUARE

12 pieces
except as noted below

S_1

S_2

1
2

HOLLOW

S_2

W

2 1

NOTE:

THE MINIMUM POSSIBLE WIDTH, <u>W,</u> OF THE
FRAME, OCCURS (FOR THE CONSTRUCTION
ABOVE SHOWN) WHEN THE <u>SOLID SQUARE</u>
BECOMES THE SAME SIZE AS THE <u>HOLLOW,</u>
AT WHICH LIMIT ONLY <u>8</u> PIECES ARE
REQUIRED, AS HERE SHOWN, INSTEAD OF <u>12.</u>

MIN. $W = \frac{S_1}{4}\left(2 - \sqrt{2}\right)$.

MAX. $W = \frac{1}{2}S$, when the "HOLLOW" vanishes.

B3 PS33.32
..FREESE..

PLATE
45

The direct conversion of
A HOLLOW SQUARE
into another larger or smaller
HOLLOW SQUARE

with 4 pieces,
all alike

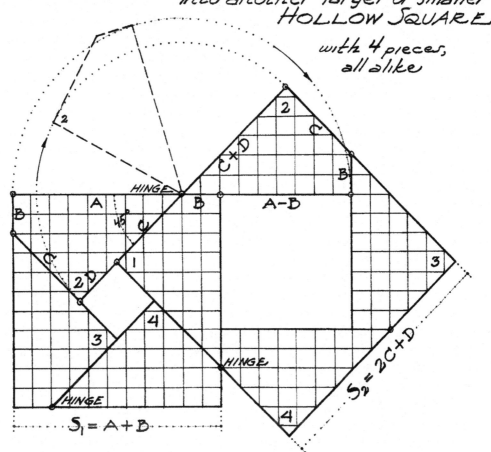

A = B   B = O

The LIMITS of this
process.
One square solidifies.

A geometric process of
turning one figure "INSIDE OUT"
to form the other.

Numerical relations:
Given; $S_1 = A + B$, where B is less
    than A, but is not zero.
Then, $C = \frac{1}{2}(A-B)\sqrt{2}$. $D = B\sqrt{2}$.

$C + D = \frac{1}{2}(A+B)\sqrt{2}$. $\therefore S_2 = A\sqrt{2}$.

If both S_1 and S_2 are given, where
    S_2 must be less than $S_1\sqrt{2}$,
Then, $\frac{B}{A} = \frac{S_1\sqrt{2} - S_2}{S_2}$.

PLATE
46

Chapter 8

Pentagons and Pentagrams (Plates 47–63)

Following the success with squares and related figures in the previous chapter, Freese set his sights on geometric figures with 5-fold symmetry, specifically pentagons and pentagrams. In Plates 47–49 he dissected a regular pentagon to each of an equilateral triangle, a square, and a regular hexagon. In Plates 50–54, he dissected a number of congruent regular pentagons to a single large pentagon. He handled the cases of 2, 4, 5, 9, and 16 congruent pentagons, with the square numbers 4, 9, and 16 leading to the easiest of those dissections.

Plates 55 and 56 showcase dissections of noncongruent regular pentagons to a larger pentagon, while Plates 57 and 58 present a dissection of two congruent pentagons to a square, and four squares of relative areas $1:2:3:4$ to a regular pentagon. Plates 59 and 60 startle us with pentagrams, namely a pentagram to a square in the first plate, and five congruent pentagrams to a large pentagram and four pentagrams to a pentagram in the second. Lastly, we see in Plates 61–63 the conversion of several congruent regular pentagons to unusual figures such as a "stellated decagon" and three different types of rings.

While there have been relatively few improvements to the dissections of Plates 47–54 since Freese completed his manuscript, there have been many revelations for Plates 55–63. This may well reflect the fact that in the latter part of this chapter Freese was more inventive in choosing the ensemble of figures to dissect. Also, there seem to have been more interesting issues regarding hingeability in the latter portion of the chapter.

Two of my favorite dissections in this chapter appear in the latter group of plates. First, at the top of Plate 56, we see the dissection of pentagons whose sidelengths realize the identity $(\sin\frac{\pi}{5})^2 + (\cos\frac{\pi}{5})^2 = 1$. It's beautifully symmetric and wonderfully hingeable, with the added attraction that we can improve it while still maintaining hingeability. Second, at the top of Plate 60, Freese gave a dissection of five congruent pentagrams to a large pentagram. How enchanted I was when I first glimpsed its 5-fold symmetry in Lindgren's 1964 book, where he attributed it to that mysterious "Irving L. Freese"!

Plate 47: Michael Goldberg (1952) published a 6-piece dissection of a regular pentagon to an equilateral triangle. He crossposed a T-strip for the triangle with a P-strip for the pentagon. Freese used the same technique and the same strips, although he produced a different 6-piece dissection.

Plate 48: Robert Brodie (1891) appears to have been the first person to find a 6-piece dissection of a regular pentagon to a square. There are infinitely many variations of Brodie's dissection, consistent with Freese's assertion about the number of 6-piece dissections. Freese's dissection is notable in that there are no small pieces and his dissection appears to have been the first such T-strip dissection. Freese's claim that no 5-piece dissection is possible has not yet been proved.

Plate 49: Lindgren (1964) gave a similar 7-piece dissection of a regular pentagon to a regular hexagon. His and Freese's dissections use the same pair of P-strips, but cross them differently.

Plate 50: This dissection of two congruent regular pentagons to one is an incomparable variation of a specialization of the method of Harry Hart (1877). Whereas Hart's approach works on similar regular polygons, Freese's approach works on two congruent copies of a polygon. Perhaps not unanticipated, considering the generality of Hart's approach, Freese's 11-piece dissection uses more pieces than necessary.

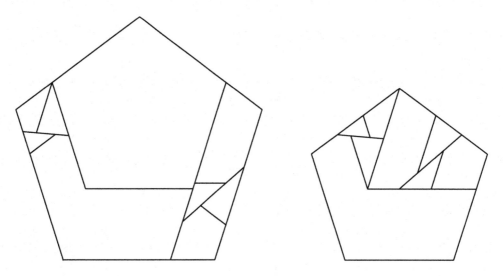

I50a: Improved two congruent regular pentagons to one [Gavin Theobald]

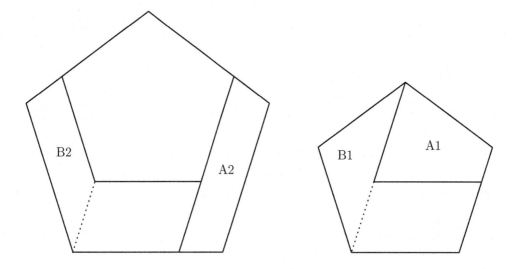

I50b: Trapezoids A and B for the Q-slides for two pentagons to one [Gavin Theobald]

Gavin Theobald later found the magnificent 9-piece dissection in Figure I50a. Gavin fit one of the small pentagons in the top corner of the large pentagon. As shown in Figure I50b, he cut trapezoid A1 from the other small pentagon and then applied Lindgren's Q-slide (see page 42 of (1997)) to A1, producing trapezoid A2. The Q-slide works because the area of A1 equals the area of A2, and they have the same set of angles. Theobald then applied the Q-slide to the triangular portion B1 of the remaining piece, producing trapezoid B2, suitably attached to the remaining part of that piece. Since the area of B1 equals the area of B2, the triangle is a trapezoid (with one of the sides being of length 0), and B1 and B2 have the same set of angles, the Q-slide again works.

Plate 51: The dissection of four congruent pentagons to one at the top of this plate is the same as the dissection in Lindgren (1964). The second dissection on this plate, of five congruent pentagons to one, is different from, but related to, the 12-piece dissection in Langford (1956). In labeling the number of pieces as minimal, Freese probably meant that the number of pieces is smaller than for any of the dissections that he showed in Plate 52.

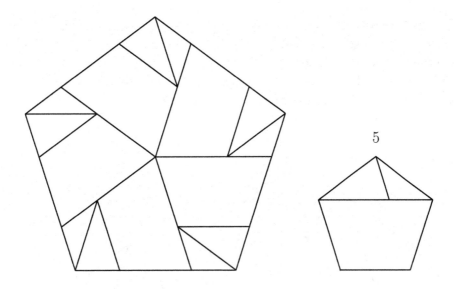

I52a: Fourth dissection of five pentagons to one [GNF]

Plate 52: In this plate Freese showed only some of the possible dissections of five regular pentagons each into the pieces 1, 2, and 3. He focused on dissecting the five regular pentagons identically, and then assembling the large regular pentagon to have 5-fold rotational symmetry. Thus he assembled each triple of pieces 1, 2, and 3 to an isosceles trapezoid with two 72°-angles and two 108°-angles. In particular, Freese packed the isosceles trapezoids within the large pentagon so that the shorter of the two parallel sides of the trapezoid follows the longer as we proceed clockwise around the large pentagon. Note that Freese packed the trapezoid with pieces 1, 2, and 3 in one of two ways, such as in his dissection 2 or dissection 3. He could have packed the trapezoids in a mirror image way of what he did in dissection 1, giving a fourth dissection (in Figure I52a), which he did not mention or present. Furthermore, he could have supplied four more versions, namely the mirror image of the four large pentagons already described, giving a total of eight different dissections, rather than three.

Of the eight different five-to-one dissections of pentagons for which Freese cut each small pentagon identically, versions 1 and 3, and their mirror image dissections, are hingeable, though Freese did not note the hingeability for them. We see the hinged pieces for each of dissection 1 and dissection 3 in Figure I52b. Of course, if we do not consider dissections in which a large pentagon is rotationally symmetric, then there are even more dissections, both hingeable and unhingeable.

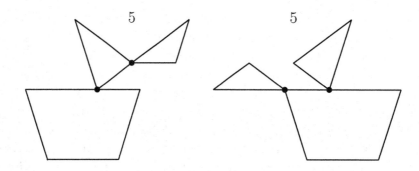

I52b: Hinged pieces for dissection 1 and for dissection 3 [GNF]

Plate 53: Freese seemed to be the first person to pose the puzzle of dissecting nine regular pentagons to one. Yet Freese's symmetrical solution is inferior to Robert Reid's later 14-piece solution, though Reid's has less symmetry. Rather than maximizing the number of uncut pentagons, as Freese did, Reid produced a 72°-rhombus for each of the five corners of the five corners of the large pentagon. I have modified Reid's solution so that the dissection once again has reflection symmetry, as we see in Figure I53.

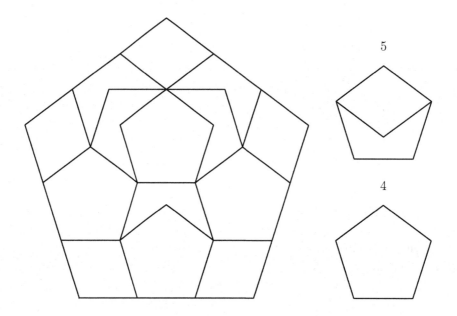

I53: Symmetric dissection of nine pentagons to one

Plate 54: Freese's 26-piece dissection of sixteen pentagons to one seems to have been original. Note that Freese could have arranged the pieces within the large pentagon to achieve 5-fold rotational symmetry.

Plate 55: The "ring expansion" that Freese gave for a dissection of two regular pentagons of different areas to a larger regular pentagon is a special case of similar polygons that can be inscribed in a circle or circumscribed about a circle. Harry Hart (1877) discovered these cases, which I illustrated in my (1997) book. As I discussed in my (2002) book, if the polygons are inscribable in a circle, we can hinge the pieces in the larger of the two polygons. Since Freese preferred regular over irregular polygons, I show the case for regular pentagons in Figure I55a.

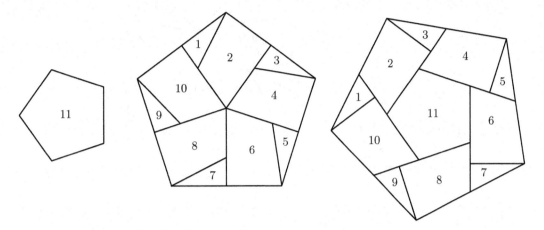

I55a: Hart's dissection applied to two pentagons of different areas to a pentagon

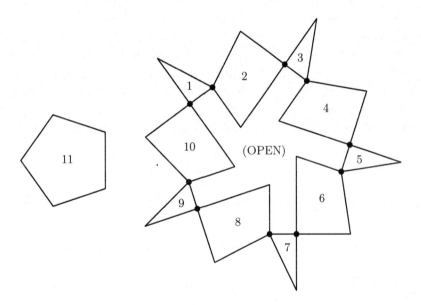

I55b: Hinged pieces for two pentagons of different areas to a pentagon [GNF]

The hinging in Figure I55b allows the ten hinged pieces to swing without obstruction from one configuration to the other. In the configuration on the right, we see that an edge of piece 10 is parallel to an edge of piece 2, and similarly for the pairs of pieces 2 and 4, 4 and 6, 6 and 8, and 8 and 10. If we simultaneously rotate pieces 1, 3, 5, 7, and 9 counterclockwise, those parallel edges stay parallel to each other, until we have completely "opened up" the hinged assemblage and can drop the small pentagon into the middle, forming the pentagon on the right in Figure I55a. Similarly, if we rotate pieces 1, 3, 5, 7, and 9 clockwise, the parallel edges stay parallel until we have closed up the assemblage, achieving the pentagon in the center of Figure I55a. With the cyclic hinging of pieces 1 through 10 and the coordinated motion of those pieces, this hinged dissection is truly a kinetic winner.

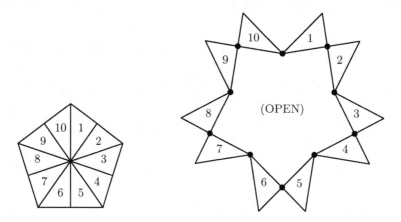

I56a: Hinged pieces for pentagons for $(\sin \frac{\pi}{5})^2 + (\cos \frac{\pi}{5})^2 = 1$ [GNF]

Plate 56: In the top portion of this plate, Freese identified one extreme in the dissection from Plate 55, with the uncut pentagon as large as possible. In particular, he drew the case for regular pentagons illustrating the identity $(\sin \frac{\pi}{5})^2 + (\cos \frac{\pi}{5})^2 = 1$. Although Freese did not notice it, his 11-piece dissection is fully hingeable. We see a captivating cyclic hinging in Figure I56a. To unfurl the ten pieces from the configuration on the left, simultaneously push each of the hinges at the midpoints of the sides of the pentagon away from the center at the same rate of speed, until the hole that opens up in the middle is in the shape of a regular decagon. Then simultaneously pull the other five hinges away from the center at the same rate of speed, until pairs of the short sides of the right triangles are flush against each other, so that we can drop the larger regular pentagon into the middle.

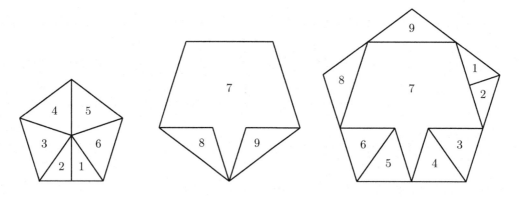

I56b: Improved pentagons for $(\sin \frac{\pi}{5})^2 + (\cos \frac{\pi}{5})^2 = 1$ [GNF]

Even more surprising, however, is my dissection in Figure I56b, which uses two fewer pieces. This 9-piece dissection is also fully hingeable, as we see in Figure I56c. Can you imagine the kinetic ballet created by an animation that I shared with the audience at the MOVES 2015 conference?

A similar dissection exists for every regular polygon with an odd number of sides greater than 5. For a regular polygon with k sides, the dissection, which has $(3k + 3)/2$ pieces, illustrates the

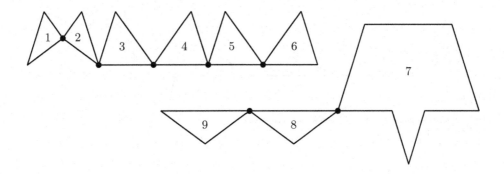

I56c: Hinged pieces: improved pentagons for $(\sin\frac{\pi}{5})^2 + (\cos\frac{\pi}{5})^2 = 1$ [GNF]

identity $(\sin\frac{\pi}{k})^2 + (\cos\frac{\pi}{k})^2 = 1$. We see a 12-piece dissection for regular heptagons in Figure I56d, with its hinged pieces in Figure I56e. There are similar dissections realizing $(\sin\frac{\pi}{k})^2 + (\cos\frac{\pi}{k})^2 = 1$ for regular polygons with an even number k of edges. In that case the number of pieces is $(3k+2)/2$. For $k = 6$ we can do much better, since Harry Lindgren identified a 6-piece hingeable dissection for hexagons in his 1964 book, which I reproduce in Figure I81c.

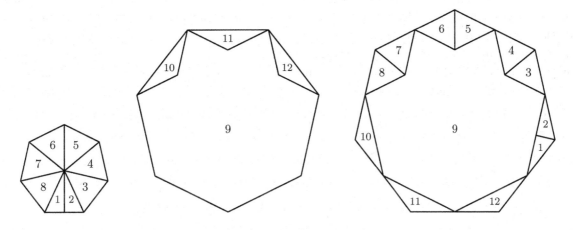

I56d: Improved heptagons for $(\sin\frac{\pi}{7})^2 + (\cos\frac{\pi}{7})^2 = 1$ [GNF]

In the bottom portion of Plate 56, Freese considered pairs of similar pentagons for which the larger of the two takes up an even larger proportion of the total area. He gave a method that leaves even more of the larger pentagon intact, at the expense of five additional pieces. This approach might be useful in a dissection in which the larger pentagon is already assembled from other figures and we wish to avoid having the cut lines from the earlier assembly cross the cut lines from Plate 55. This would happen if we applied the ring expansion of Plate 55 twice to assemble three congruent pentagons to one, requiring 31 pieces. Using the technique in the lower portion of Plate 56, along with a trick that avoids cut lines near the vertices of the larger pentagon, would give a 26-piece dissection. However, an approach such as strips would probably do better for three congruent regular pentagons.

Plate 57: The puzzle of dissecting two pentagons to a square seems to have been original. I know of no improvement to Freese's 8-piece dissection. Without discussion or any label, Harry Lindgren (page 16 of (1964)) gave a dissection with the same strips but with one strip reversed.

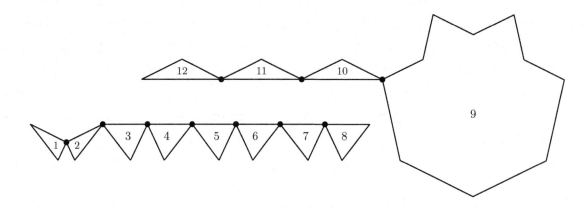

I56e: Hinged pieces: heptagons for $(\sin \frac{\pi}{7})^2 + (\cos \frac{\pi}{7})^2 = 1$ [GNF]

Plate 58: The dissection of four squares whose areas are in the ratio of $1:2:3:4$ to a pentagon also seems to have been an original dissection puzzle. Freese took advantage of the following fact: If you draw lines from the center of the pentagon to each of its five vertices, you split the pentagon into five isosceles triangles. Splitting each of these triangles in half, you obtain ten right triangles. Since the sum of the numbers from 1 to 4 is 10, you can accomplish this dissection by dissecting the smallest square to one of the right triangles, the next larger square to one of the isosceles triangles, the next square up to one of the right triangles plus one of the isosceles triangles, and the largest square to two of the isosceles triangles. Freese used P-strip and T-strip techniques to perform those four dissections in a total of 19 pieces.

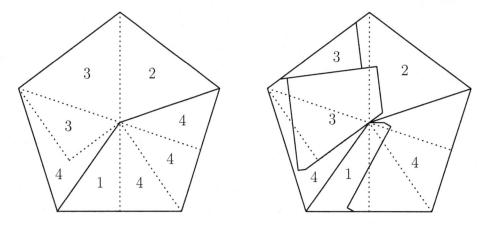

I58a: Partitioning the pentagon for squares of area ratio $1:2:3:4$, plus further refinement

I improved on Freese's dissection by two pieces, by partitioning the areas of the isosceles and right triangles more carefully. Then Gavin Theobald further improved my result by two more pieces. Looking at the pentagon on the left in Figure I58a, we split the isosceles and right triangles so that the piece labeled 1 has area equal to the smallest square, the piece labeled 2 has area equal to that of the next largest square, the pieces labeled with 3 have area equal to that of the next largest square, and the remaining pieces have the same area as the largest square.

Examining the pentagon on the right of Figure I58a, we see that piece 1 has gained and lost quadrilaterals of equal area, as had piece 2, and has the pieces labeled with 3, and similarly for the

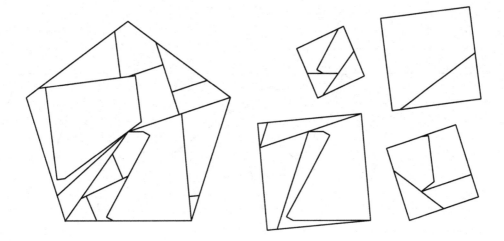

I58b: Improved squares of area ratio 1:2:3:4 to a pentagon [Gavin Theobald]

pieces labeled with 4. There are three different swaps that we performed. Can you identify them? The result of these swaps is to reduce the number of pieces from the second largest square, while not increasing the number of pieces from the other squares. Gavin's resulting 15-piece dissection is in Figure I58b.

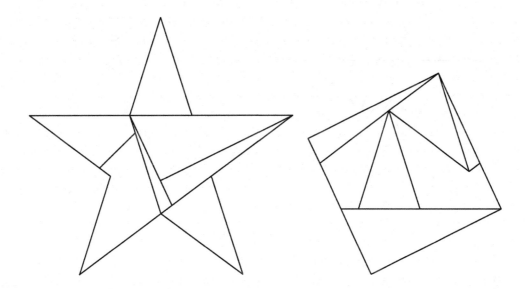

I59a: Improved pentagram to a square [Philip Tilson]

Plate 59: Freese was apparently the first to propose dissecting a pentagram to a square, but just by a whisker! A year after Freese died, Martin Gardner (1958) proposed this puzzle, requiring no more than the nine pieces that Freese had used. In response, Harry Lindgren (1958) gave an 8-piece dissection, using a piecemeal approach. (Freese's dissection uses a P-slide.) Two decades later, Philip Tilson (1978–1979) found a 7-piece dissection (Figure I59a) by using a P-strip method (Figure I59b).

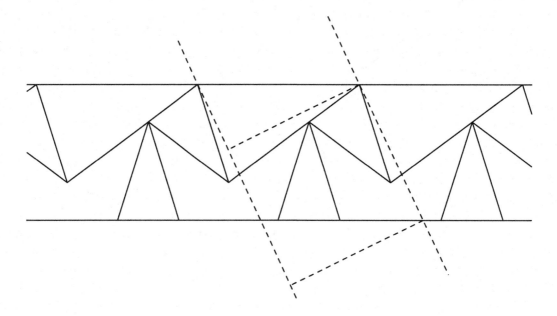

I59b: P-strip for improved pentagram to a square [Philip Tilson]

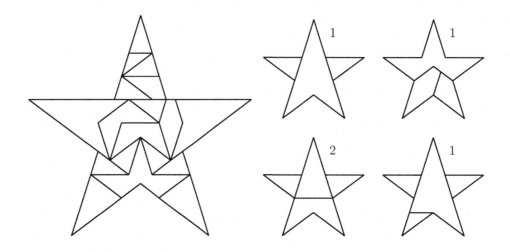

I60a: Improved five pentagrams to one [GNF 1974]

Plate 60: Freese was apparently the first to pose the puzzle of dissecting five pentagrams to one. In 1974 I improved his 20-piece dissection by two pieces (Figure I60a). I identified two pieces from the small pentagrams that each contain three points from the pentagrams. Each such piece allowed me to cut a small pentagram into three rather than four or more pieces. Then I used trial and error to fit the pieces together efficiently.

Freese was also the first to pose the puzzle of dissecting four pentagrams to one. Note that he mistakenly labeled his 14-piece dissection as having sixteen pieces. If we use the same dissection of the large pentagram, but rearrange the pieces within the four small pentagrams, we can arrive at the translational variant in Figure I60b. Stuart Elliott (1985–1986) improved Freese's dissection by two pieces, as we see in Figure I60c, though it doesn't seem possible to render a translational variant of Elliott's dissection.

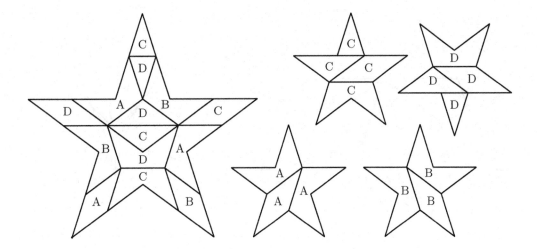

I60b: Translational variant of Freese's four pentagrams to one

I60c: Improved four pentagrams to one [Stuart Elliott]

Plate 61: This dissection, of two pentagons to a "stellated decagon," is similar to Freese's dissection of three equilateral triangles to a "stellated nonagon" in Plate 26. Thus we may also view his stellated decagon as a pseudo-star. I leave to the reader the task of determining the exact value of q, as I discussed with respect to Plate 26. In both dissections from Plates 26 and 61, Freese split the simpler figures into "deltoids" (kite-shaped pieces), each of which he dissected to an arrowhead-shaped figure. These arrowhead-shaped pieces assemble to form the pseudo-star.

Freese's dissection of a stellated decagon is hingeable, as I show in Figures I61a and I61b. (Pieces 4, 8, 12, 16, and 20 are too small to label in the pentagon; piece 3 points to unlabeled piece 4, piece 7 points to unlabeled piece 8, etc.) To unfurl a pentagon, first swing pieces 13-20 as a unit around the hinge between pieces 10 and 13, then swing pieces 17-20 as a unit around the hinge between pieces 14 and 17, then swing pieces 9-20 around the hinge between pieces 6 and 9, and next swing pieces 5-20 round the hinge between pieces 2 and 5. Then swing piece 20 out from piece 18, allowing us to then swing pieces 18, 19, and 17 into a linear sequence. Repeat this procedure

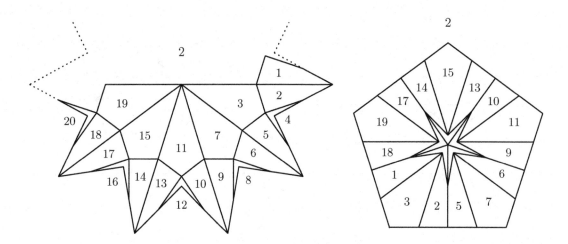

I61a: Hingeable two pentagons to stellated decagon

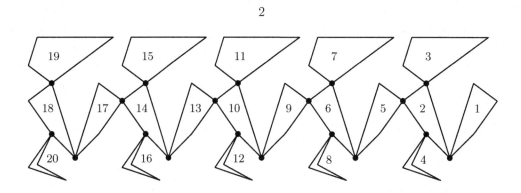

I61b: Hinged pieces for two pentagons to stellated decagon [GNF]

in turn for the groups of pieces 13-16, 9-12, 5-8, and finally 1-4. Try to picture the dance of these pieces, but don't hypnotize yourself!

Freese's dissection is also translational when suitably arranged. The arrangement requires that one of the two pentagons be rotated 180° with respect to the other. Yet there is a better dissection that is translational, as shall see momentarily.

We can also improve on Plate 61 if we focus on minimality rather than hingeability. Following the lead from Figure I26c, take a portion from one side of the kite-shaped piece, after having merged an identical-shaped piece onto the other side of the kite-shaped piece. Altogether this eliminates ten pieces, producing the 30-piece dissection in Figure I61c. This improved dissection is still translational, as we can now verify. The labeled pieces in the stellated decagon are translations of the pieces from one of the two pentagons. The remaining pieces are translations of the pieces from the other pentagon in the case that the other pentagon had already been rotated 36° around its center.

Wonderful! And yet we are still not done! In each pentagon, we have five V-shaped pieces. If we explore ways to merge these pieces and cut holes in other pieces to accommodate the merged pieces, we can arrive at the 25-piece dissection in Figure I61d. In the process of juggling V-shaped pieces and parts of them, I have changed all but one piece from Figure I61c!

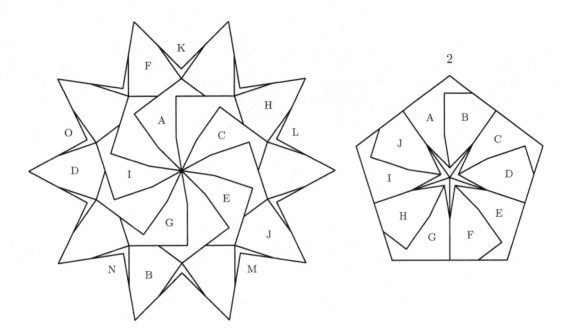

I61c: Translational improved two pentagons to stellated decagon [GNF]

Plate 62: A regular pentagon whose side length is three times that of a small version of the same will have nine times the area. Similarly, if the pentagon has twice the side length, it will have four times the area. Thus if we cut a hole in the shape of a 2-pentagon from within a 3-pentagon, the resulting area will be that of five times the area of a 1-pentagon. Freese gave a shrewd dissection that demonstrates this fact. He split each small pentagon into five identical isosceles triangles, and then merged two of the triangles back together.

Freese claimed that his example, in the case of $n = 5$, is a typical method. Yet his example is completely analogous only in the case that n is an odd number. The inner side length of the ring will be $\lfloor n/2 \rfloor$ and the outer side length of the ring will be $\lceil n/2 \rceil$. Then $(\lceil n/2 \rceil)^2 - (\lfloor n/2 \rfloor)^2 = n$. In this case, we cut each n-gon into $n - 1$ pieces, so that the total number of pieces is $n(n - 1)$.

Returning to $n = 5$, we can get a hingeable dissection as in Figure I62a if we are willing to tolerate using one more piece. I show the hinged pieces in Figure I62b. The hinging leads to a gentle promenade, which will carry over to the next several hingings. This hingeable dissection generalizes easily to an n-gon, which will contain $n^2 - n + 1$ pieces. You see the corresponding dissection for heptagons in Figure I62c. There are three identical hinged sequences of seven pieces, two mirror-image sequences of six pieces, and two mirror-image sequences of five pieces.

Returning once again to the problem of the pentagonal ring, it is possible to use fewer than Freese's twenty pieces for the unhingeable dissection. I have found a dissection (Figure I62d) with as few as fourteen pieces. The trick is to add and subtract a certain small obtuse triangle (with angles of 18°, 54°, and 108°) from and to various pieces. For example, the trapezoid in the middle of the bottom part of the ring is the combination of one of Freese's isosceles triangles with two of these small obtuse triangles. Also, each of the two small pentagons that I cut identically have an irregular pentagonal piece that combines an isosceles triangle with two obtuse triangles, but now into that irregular pentagonal shape.

Freese's statement in the lower lefthand corner of the plate, about dissecting n n-gons to $n-1$, is correct under an appropriate interpretation. The side length on the outside of the ring must be

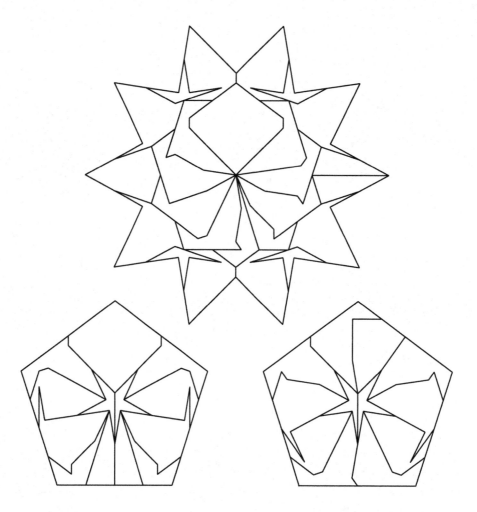

I61d: Further improved two pentagons to stellated decagon [GNF]

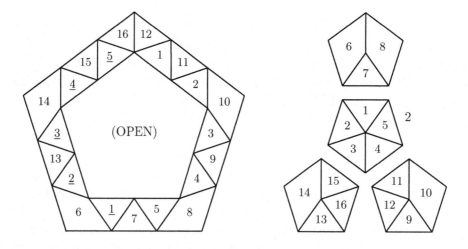

I62a: Hingeable five pentagons to a pentagonal "ring" [GNF]

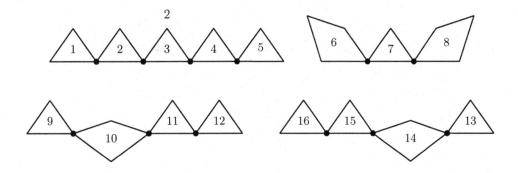

I62b: Hinged pieces for five pentagons to a pentagonal "ring" [GNF]

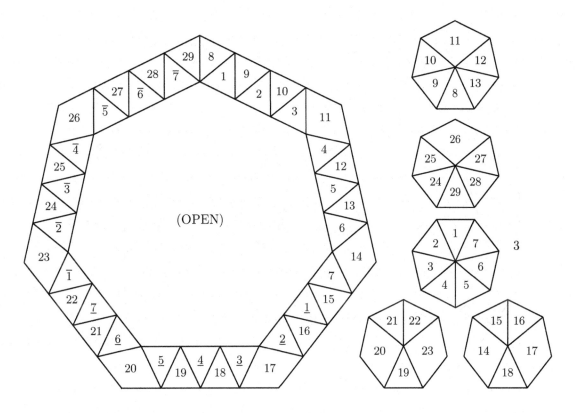

I62c: Hingeable seven heptagons to a heptagonal "ring" [GNF]

$(n+1)/2$ and the side length on the inside of the ring must be $(n-1)/2$. Squaring both expressions and taking the difference gives n. For example, when $n = 4$, the square ring will have an outer side of length 2.5 and an inner side of length of 1.5. In addition, whenever n is even, Freese's method needs to split one of the isosceles triangles from each n-gon, so that it will form a parallelogram.

For $n = 3$, Freese's 6-piece dissection would seem to be hard to beat. For other values of n, different strategies may do better. For example, for $n = 4$, there are several 8-piece dissections. For $n = 6$, there is a 15-piece dissection.

Plate 63: Freese discovered two different ways to dissect some number of pentagons to a decagon ring. In the first, he created a thin ring from four pentagons, with the inner vertices of the ring positioned angularly between consecutive pairs of outer vertices. It is difficult to imagine

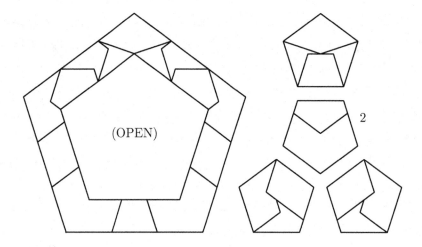

I62d: Improved pentagonal "ring" to five pentagons [GNF]

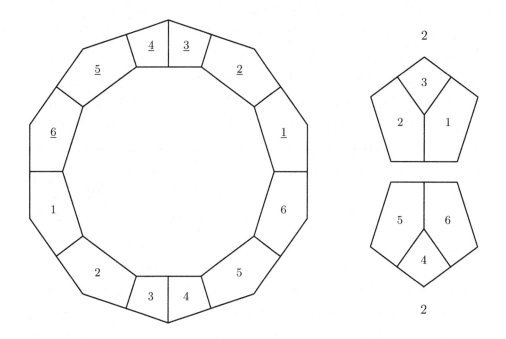

I63a: Hingeable decagonal "ring" to four pentagons

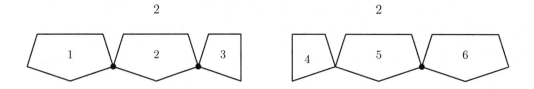

I63b: Hinged pieces for a decagonal "ring" to four pentagons [GNF]

a way to cut a pentagon into fewer than three pieces to fill in the ring. There is a pleasant hinging

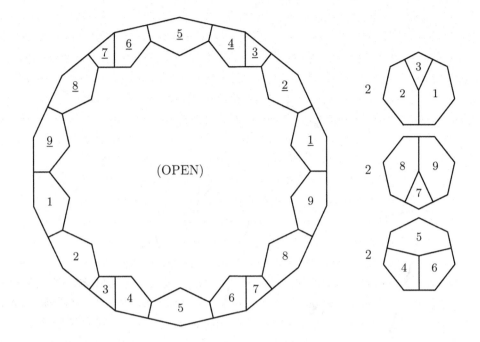

I63c: Hingeable tetradecagonal "ring" to six heptagons [GNF]

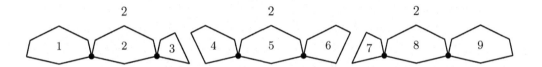

I63d: Hinged pieces for tetradecagonal "ring" to six heptagons [GNF]

of this first dissection, as we see in Figures I63a and I63b.

In the second dissection of this plate, Freese dissected ten pentagons to form a relatively thick ring. It is difficult to imagine a way to pack an uncut pentagon with other pieces to fill up the ring. Thus it is probably not possible to use fewer than twenty pieces.

Are there any other dissections similar to Freese's one in the top of this plate? I have discovered a similar hingeable dissection (Figure I63c) that is based on heptagons. I use six heptagons, cut into three pieces each. Then however the "ring" is a bit different: While a tetradecagon forms the outside boundary of the ring, the star {14/3} forms the inside boundary. This is not so extraordinary, considering that the decagons in Freese's ring have their vertices out of phase with each other. We hinge the pieces in Figure I63d.

This brings us to the end of Freese's chapter on pentagons and pentagrams. We will proceed next to the chapter on hexagons and hexagrams. The attentive reader may have noticed that we have already snuck in a few heptagons when we took our measure of pentagons and pentagon rings. There was nowhere else to put them, because Freese never touched on heptagons! An exploration of those figures would come later with Lindgren, Hanegraaf, Varsady, Theobald, and me.

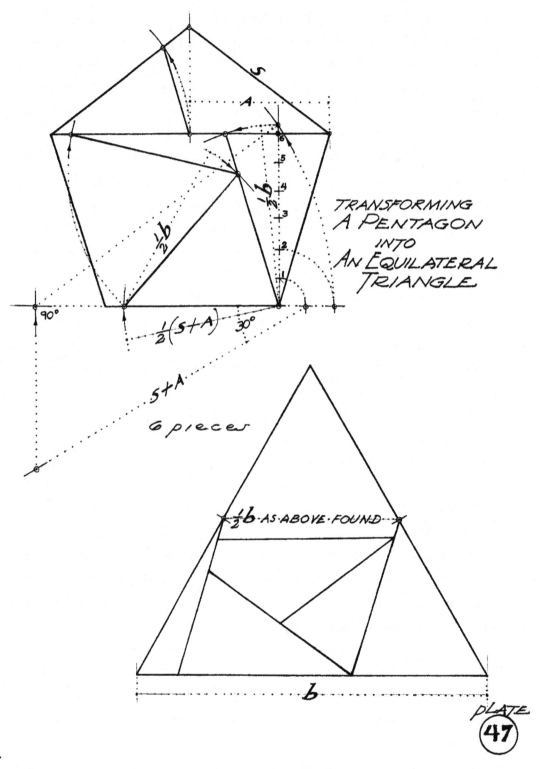

S

A

TRANSFORMING
A PENTAGON
INTO
AN EQUILATERAL
TRIANGLE

$\frac{1}{2}b$

$\frac{1}{2}b$

90°

$\frac{1}{2}(S+A)$ 30°

S+A

6 pieces

$\frac{1}{2}b$ AS ABOVE FOUND

b

PLATE
47

BI P33C
..FREESE..

SQUARING THE PENTAGON
6 PIECES

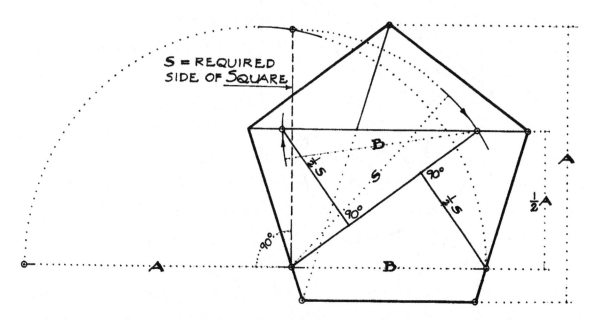

S = REQUIRED
SIDE OF SQUARE

B

S

90°

90°

½A

A

A

B

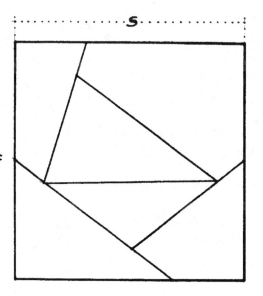

NOTE:
The number of
6-piece SQUARINGS,
by one method or
another, is infinite,
but the method here
shown is simpler and
more direct than any
heretofore known. =

A 5-piece SQUARING
is impossible.

S

The 6 pieces of the above PENTAGON
re-assembled into a
SQUARE

PLATE
48

B3PS14.3
...FREESE...

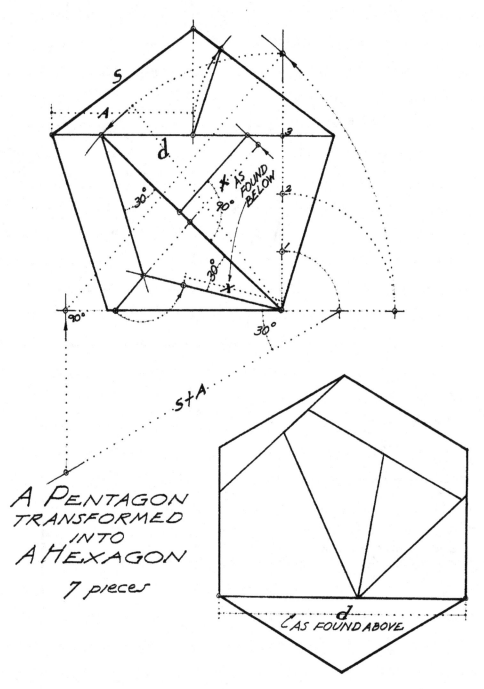

S

A

d

X AS
FOUND
BELOW

30°
90°

30°
X

90°
30°

S+A

A PENTAGON
TRANSFORMED
INTO
A HEXAGON

7 pieces

d

∠ AS FOUND ABOVE

PLATE
49

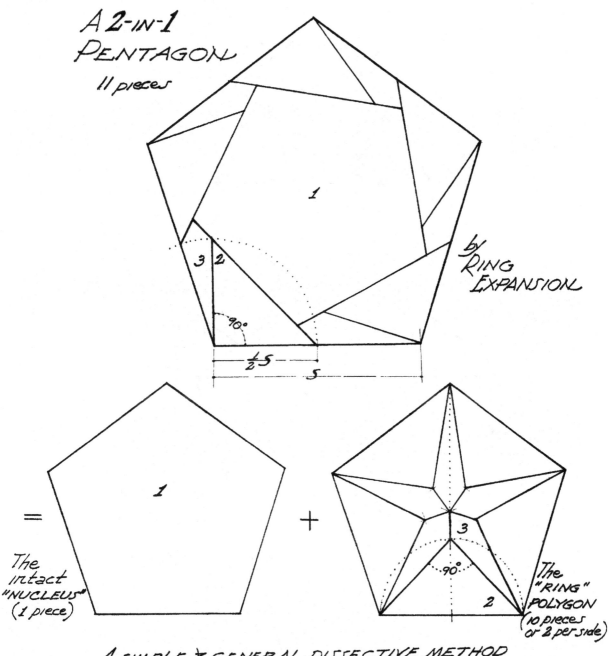

A 2-IN-1 PENTAGON
11 pieces

1

by RING EXPANSION

3 2

90°

½S

S

=

1

The intact "NUCLEUS" (1 piece)

+

3

90°

2

The "RING" POLYGON (10 pieces or 2 per side)

A SIMPLE & GENERAL DISSECTIVE METHOD OF DOUBLING ANY REGULAR POLYGON EXCEPT THE EQUILATERAL TRIANGLE.

B3 PX-2
..FREESE....

IN GENERAL, THIS "RING EXPANSION" METHOD OF DISSECTION REQUIRES THAT ONE OF THE POLYGONS SHALL REMAIN INTACT.

PLATE 50

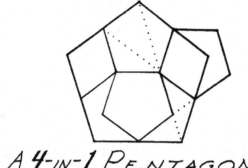

A 4-ɪɴ-1 PENTAGON
6 pieces

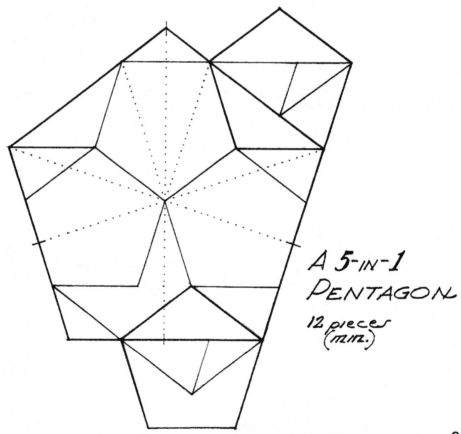

A 5-ɪɴ-1 PENTAGON
12 pieces (min.)

PLATE
51

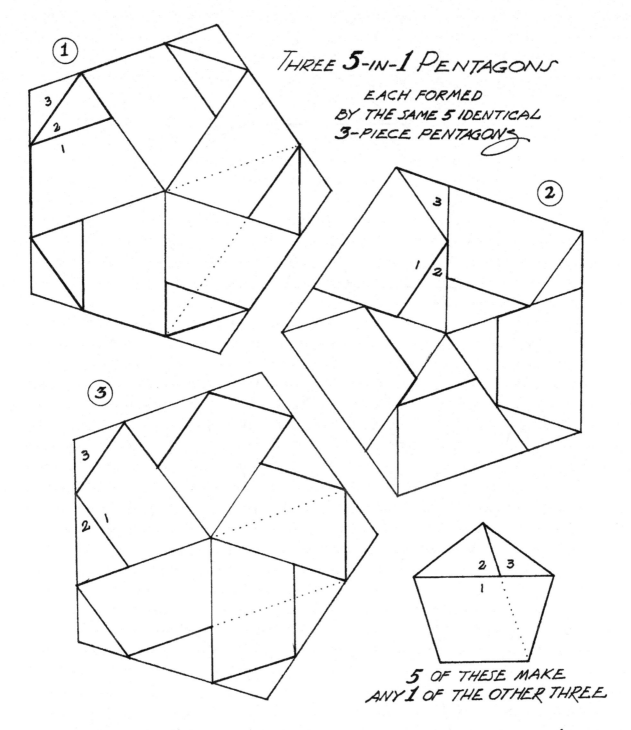

THREE **5-IN-1** PENTAGONS

EACH FORMED
BY THE SAME 5 IDENTICAL
3-PIECE PENTAGONS

5 OF THESE MAKE
ANY **1** OF THE OTHER THREE

PLATE
52

B1 P33C1 & 33C4
..FREESE..

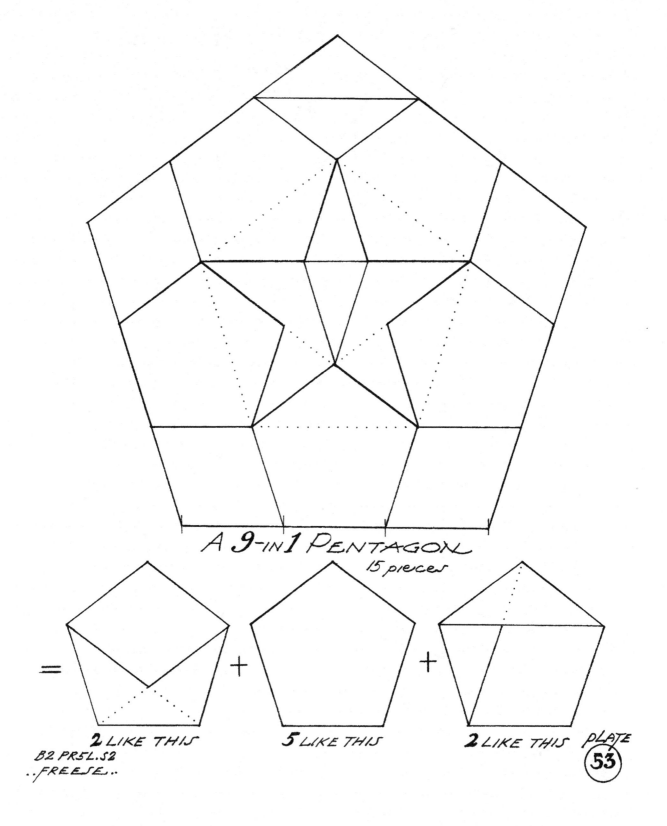

A 9-IN 1 PENTAGON

15 pieces

= 2 LIKE THIS + 5 LIKE THIS + 2 LIKE THIS

B2 PR5L.S2
..FREESE..

PLATE
53

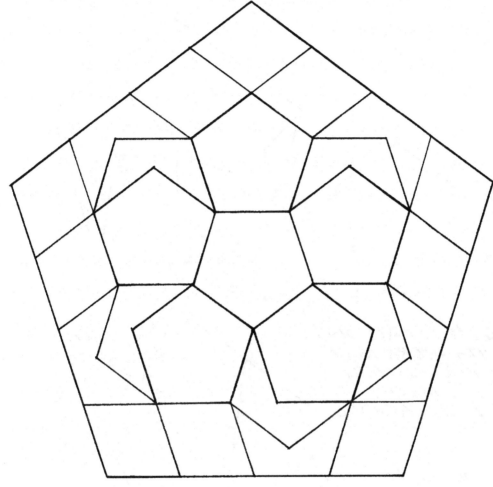

A 16-IN-1 PENTAGON

26 pieces

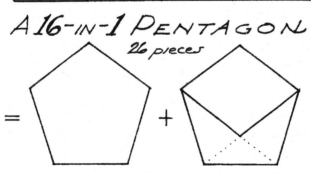

= [pentagon] + [house shape]

6 LIKE THIS   **10 LIKE THIS**

PLATE
(54)

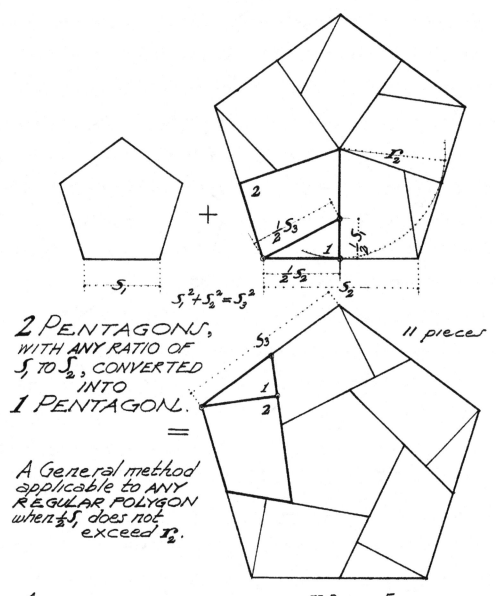

$$S_1^2 + S_2^2 = S_3^2$$

2 PENTAGONS,
WITH ANY RATIO OF
S_1 TO S_2, CONVERTED
INTO
1 PENTAGON.

=

A General method
applicable to ANY
REGULAR POLYGON
when $\frac{1}{2}S_1$ does not
exceed r_2.

11 pieces

A TYPICAL EXAMPLE OF "RING EXPANSION"

PLATE
55

D4 P2
..FREESE..

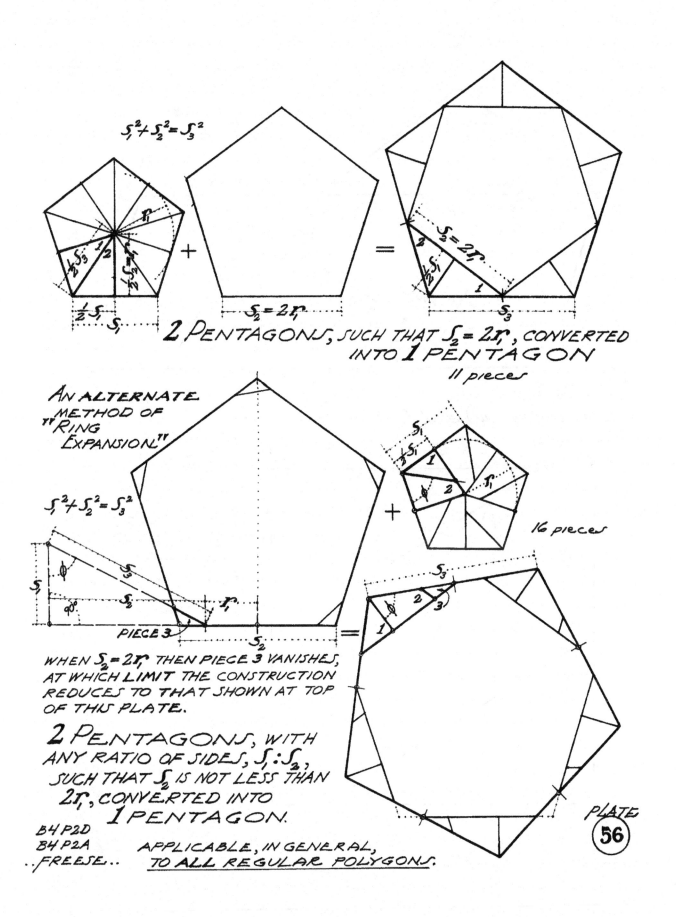

$$S_1^2 + S_2^2 = S_3^2$$

$S_2 = 2r_1$

2 PENTAGONS, SUCH THAT $S_2 = 2r_1$, CONVERTED
INTO 1 PENTAGON

11 pieces

AN ALTERNATE
METHOD OF
"RING
EXPANSION"

$$S_1^2 + S_2^2 = S_3^2$$

16 pieces

PIECE 3

WHEN $S_2 = 2r_1$ THEN PIECE 3 VANISHES,
AT WHICH LIMIT THE CONSTRUCTION
REDUCES TO THAT SHOWN AT TOP
OF THIS PLATE.

2 PENTAGONS, WITH
ANY RATIO OF SIDES, $S_1 : S_2$,
SUCH THAT S_2 IS NOT LESS THAN
$2r_1$, CONVERTED INTO
1 PENTAGON.

B4 P2D
B4 P2A
..FREESE..

APPLICABLE, IN GENERAL,
TO ALL REGULAR POLYGONS.

PLATE
56

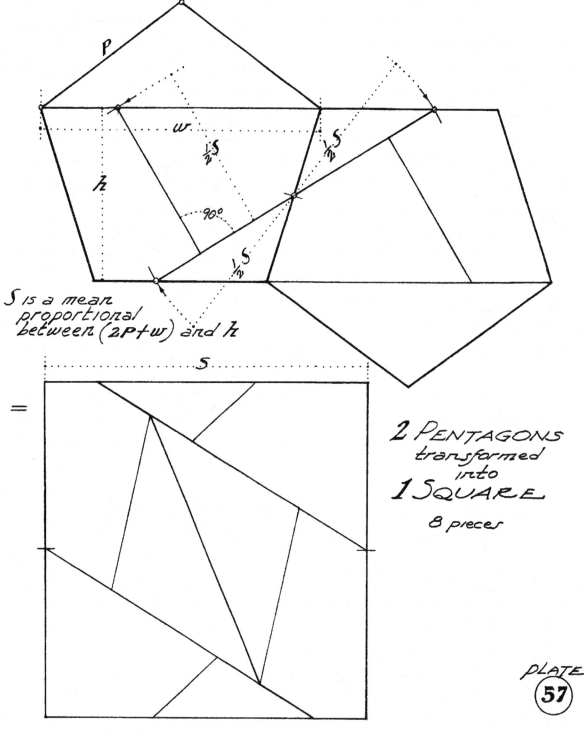

P

w

½S

½S

h

90°

½S

S is a mean
proportional
between (2P+w) and h

S

=

2 PENTAGONS
transformed
into
1 SQUARE
8 pieces

PLATE
57

B2.PR5A
..FREESE..

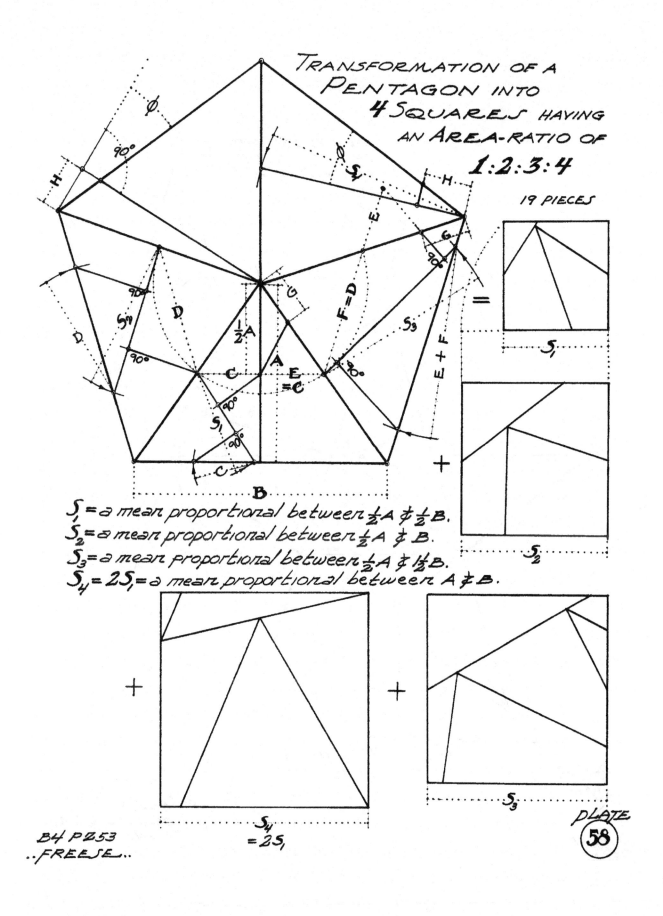

TRANSFORMATION OF A PENTAGON INTO 4 SQUARES HAVING AN AREA-RATIO OF 1:2:3:4

19 PIECES

S_1 = a mean proportional between $\frac{1}{2}A$ & $\frac{1}{2}B$.
S_2 = a mean proportional between $\frac{1}{2}A$ & B.
S_3 = a mean proportional between $\frac{1}{2}A$ & $1\frac{1}{2}B$.
S_4 = $2S_1$ = a mean proportional between A & B.

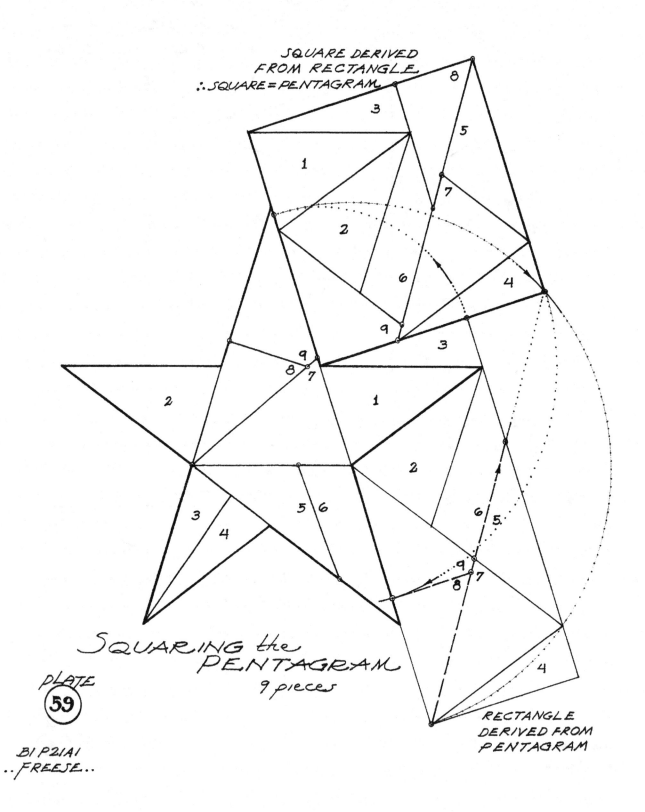

SQUARE DERIVED
FROM RECTANGLE.
∴ SQUARE = PENTAGRAM.

SQUARING the
PENTAGRAM
9 pieces

PLATE
59

BI P21A1
..FREESE..

RECTANGLE
DERIVED FROM
PENTAGRAM

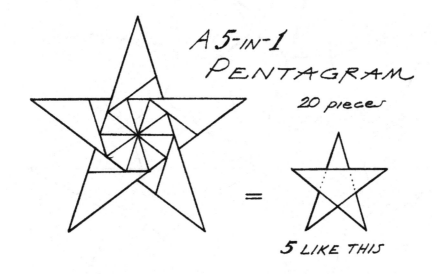

A 5-IN-1 PENTAGRAM

20 pieces

=

5 LIKE THIS

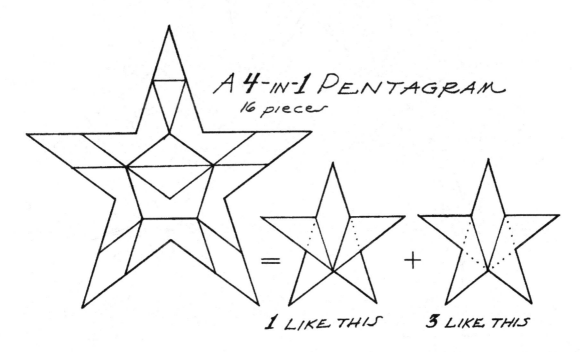

A 4-IN-1 PENTAGRAM

16 pieces

= 1 LIKE THIS + 3 LIKE THIS

B1P28D4
B4PZ29.S1
. .FREESE..

PLATE
60

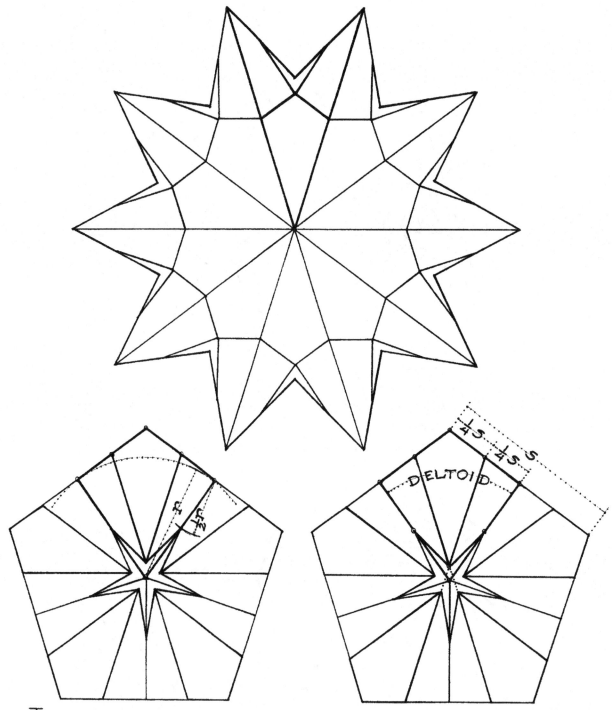

THE TEN 4-PIECE DELTOIDS OF THE 2 PENTAGONS FORM THE
STELLATED DECAGON
by turning the deltoids "inside out"

B4 PZ54
..FREESE..

PLATE
61

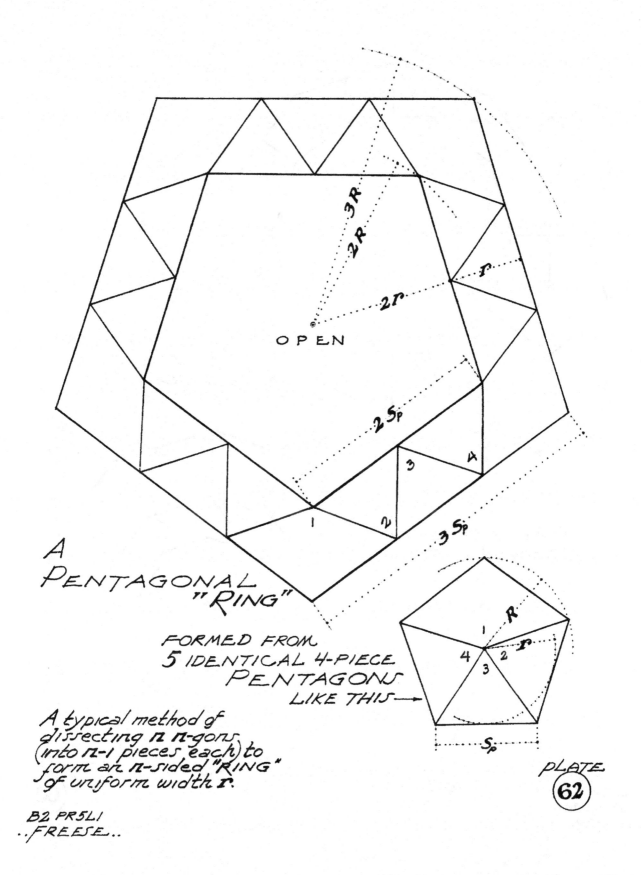

OPEN

A
PENTAGONAL
"RING"

FORMED FROM
5 IDENTICAL 4-PIECE
PENTAGONS
LIKE THIS →

A typical method of
dissecting n n-gons
(into $n-1$ pieces each) to
form an n-sided "RING"
of uniform width r.

B2 PR5LI
..FREESE..

PLATE
(62)

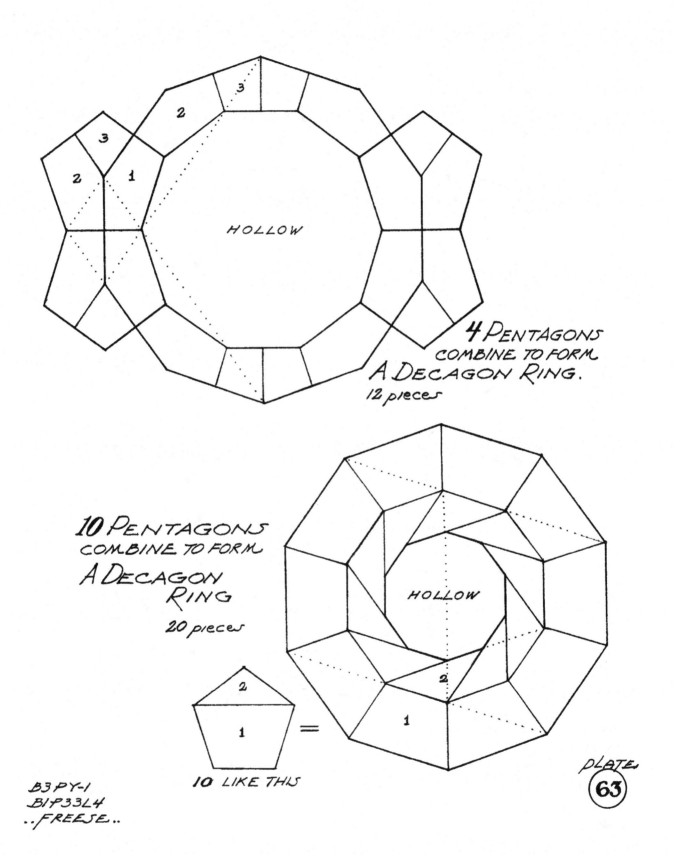

HOLLOW

4 PENTAGONS
COMBINE TO FORM
A DECAGON RING.
12 pieces

10 PENTAGONS
COMBINE TO FORM
A DECAGON
RING
20 pieces

HOLLOW

10 LIKE THIS =

B3PY-1
B1P33L4
..FREESE..

PLATE
63

Chapter 9

Hexagons and Hexagrams (Plates 64–88)

After handling the pentagons and pentagrams in the last chapter, Freese tackled hexagons and hexagrams, those figures with 6-fold rotational symmetry. In Plates 64–67, he explored dissections of a regular hexagon to one, two, or three congruent squares, and also two or three congruent regular hexagons to a square. In Plates 68–73, he focused on dissecting a number of congruent regular hexagons to a single large hexagon, handling in turn the cases of 2, 3, 4, 6, 7, 9, 12, and 13 congruent hexagons. He also handled cases such as 3 hexagons to 4, and 4 hexagons to 9. In Plate 74, Freese gave a marvelous dissection of a pair of an equilateral triangle and a square to a regular hexagon. Plates 72, 75, 76, 81–84 show dissections of various numbers of congruent equilateral triangles to various numbers of congruent regular hexagons. In Plates 77–82, Freese dissected primarily regular hexagons of different sizes to either an equilateral triangle or a regular hexagon. In Plates 85–87, Freese extended his efforts to hexagrams: a hexagram to an equilateral triangle, or to congruent hexagons, or to a square. He also dissected a hexagram to 3 hexagrams or to 4 hexagrams or to a dodecagram. Finally, he dissected 3 congruent hexagons to a hexagonal ring, and a hexagon to a dodecagonal ring.

Many of these dissections are translational, including most of those in the first two thirds of this chapter. Two of the dissections are pure hexstacy! One that Freese created by overlapping tessellations is the dissection of two congruent hexagons to one in Plate 68. And yes, it is translational! A second inspiration, the dissection of three congruent hexagrams to one in Plate 86, is both translational and hingeable.

Plate 64: In this plate Freese reproduced the 5-piece dissection of a regular hexagon to square by Paul-Jean Busschop (1876). Freese's layout of the plate makes it clear that the dissection is translational, though Freese did not point that out. At the bottom of the plate, Freese claimed that there are an unlimited number of such 5-piece dissections, but that no 4-piece dissection is possible. Yes, there are an unlimited number of such dissections, but the technique in the plate gives only one. Lindgren (1964) formed a P-strip differently, slicing the hexagon along a diagonal of length $\sqrt{3}$ times the length of a side, and with that approach one can get an infinite number of 5-piece dissections. The claim, that no 4-piece dissection is possible, is unsubstantiated, though widely believed.

Plate 65: Freese appears to have been the first person to have dissected a regular hexagon to two congruent squares. He employed a P-strip approach, using the same hexagon strip as in the previous plate. It would seem tough to improve on his 7-piece dissection. From the layout it is clear that the dissection is translational, though Freese did not claim it.

Plate 66: Freese appears to have been the first to dissect specifically a square to two congruent regular hexagons. He employed a T-strip approach, using a simple hexagon strip in which he split one of the hexagons in half and did not cut the other one. Note that he misstated the relationship amongst S, w, and d: S should be a mean proportional between $2w$ and d.

It would seem to be difficult to improve on Freese's 7-piece dissection. Its layout clearly shows that the dissection is translational, though Freese did not claim it. Without any discussion or caption, Harry Lindgren (on page 21 of (1964)), gave precisely the same 7-piece dissection.

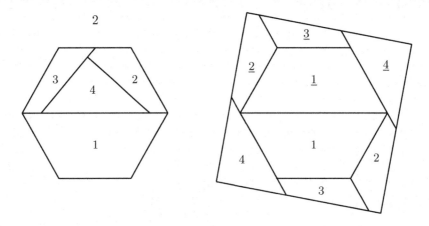

I66a: Hingeable dissection of two hexagons to a square [GNF]

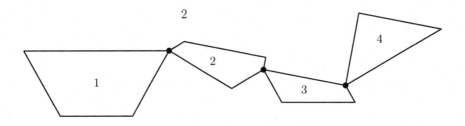

I66b: Hinged pieces for dissection of two hexagons to a square [GNF]

If we use just one more piece, we can modify Freese's dissection to be hingeable. Take the hexagon that Freese split into six pieces, and replace the bottom three pieces by one piece. Then duplicate the cuts in the second hexagon, giving a total of eight pieces, as in Figure I66a. We hinge the pieces in each hexagon as in Figure I66b.

Plate 67: In this plate, Freese dissected three squares to a regular hexagon and three regular hexagons to a square. Again, both of these dissections are translational, although the layout of the first makes it less obvious. Freese appears to have been the first person to attempt either dissection. For the first dissection, he split the hexagon into three rhombuses and used a P-strip to dissect each rhombus to a square. For the second dissection, he created a P-strip for the hexagons by splitting one on its diameter. It appears difficult to improve on the second of these dissections.

However, there is a way to reduce by one the number of pieces for the first. The trick is to leave two of the rhombuses attached together, and then perform a P-strip on the attached rhombuses. This then gives the 8-piece dissection in Figure I67, which is still translational.

Plate 68: Freese was apparently the first to pose the puzzle of dissecting two congruent regular

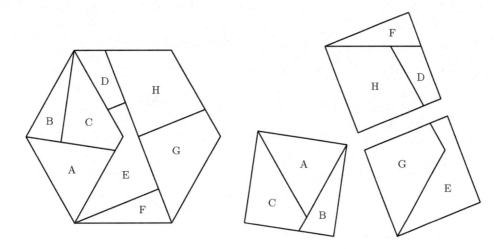

I67: Translational improved hexagon to three squares [GNF]

hexagons to one. We can easily extend his technique for the 9-piece dissection to the case in which the two hexagons are not congruent. Lindgren (1964) independently discovered this dissection and discussed how to derive it by superposing tessellations. You can get a tessellation from the two hexagons by cutting one hexagon into four pieces that form two equilateral triangles. Once again, this dissection is translational, as we see in Figure I68a, though Freese did not show the hexagons suitably oriented. For the case in which one hexagon is sufficiently smaller than the other, Lindgren gave an 8-piece dissection, again based on tessellations.

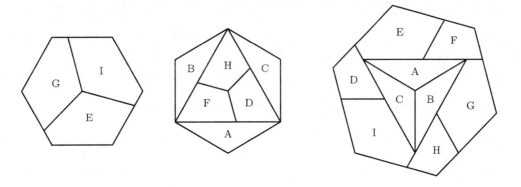

I68a: Showing that Freese's two hexagons to one is translational

Gavin Theobald found a fabulous way of dissecting two congruent regular hexagons to one hexagon, if we turn over two pieces. (Note however that Freese did not approve of turning over pieces.) To accomplish this, Gavin used a modified strip dissection. As shown in Figure I68b, he created a strip for two congruent hexagons (in solid lines) and crossed it with a strip for a single hexagon (in dashed and dotted lines). Basing the latter strip on lining up two halves of a hexagon, Gavin found a 9-piece dissection. Then, by extending the rightmost half down to incorporate the dashed trapezoid beneath it, and thus removing a corresponding trapezoid from the leftmost half, he avoided cutting one piece in half. He was careful to accomplish this in a way that allowed the larger half to mate with the smaller half, after turning over the larger half. Gavin turned over

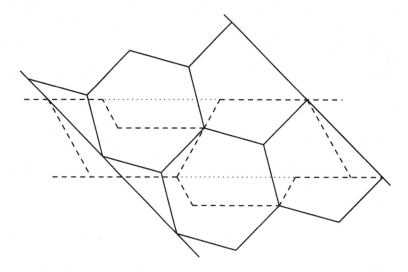

I68b: Crossposition for improved two hexagons to one [Gavin Theobald]

pieces 1, 2, and 3 in the large hexagon. Later he noted that only pieces 3 and 4 need to be turned over. (Identify the smaller hexagon that has pieces 1 and 2 turned over, note that pieces 1, 2, and 4 form a symmetric figure, and then replace pieces 1, 2, and 4 in the symmetric figure by their mirror image.) We see the final result in Figure I68c.

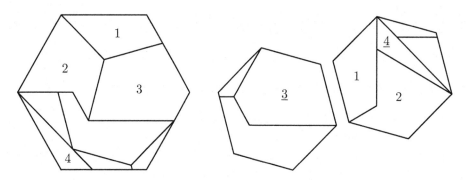

I68c: Improved two hexagons to one with turning over [Gavin Theobald]

Plate 69: Freese was apparently the first to pose the puzzle of dissecting three congruent regular hexagons to one. In (1964) Lindgren also gave the same dissection, which we can derive by superposing tessellations. The dissection is translational, which is fairly easy to see from its presentation. My 2002 book contains a hinging of this dissection.

Freese was also the first to give a 6-piece dissection of four congruent hexagons to one. It is easy to see that this dissection is translational too. Freese's "General Theorem" is not very clear: Does he mean something special by being "geometrically dissected"? The earliest such interesting example of a four-into-one dissection seems to have been the 8-piece dissection of four regular heptagons to one by Lindgren (1953).

Plate 70: Freese found a clever way to produce a dissection of six congruent regular hexagons to one, by using ring expansion. First, he formed the center of a large hexagon by dissecting three

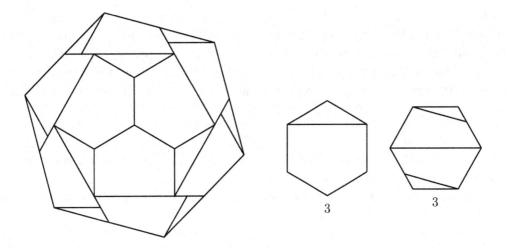

I70a: Improved six hexagons to one [GNF]

hexagons to one. (See Plate 69.) Then he extended each of the six sides, using half of a hexagon for each.

However, we can do better by going back to first principles. Slice each of the last three hexagons in half along a longest diagonal. Then position each half hexagon along a side of the three-in-one central hexagon, with the long side of the half hexagon against the side of the three-in-one central hexagon. Draw lines from one far vertex to the next, and cut along these lines. This gives the 18-piece dissection in Figure I70a. The improved dissection is also translational, though to see that from Figure I70a we must start with each hexagon in a triple of hexagons oriented in a different way.

We can do even better if we turn over pieces (which of course Freese did not approve). Resist cutting the three hexagons that fill in the center, and change how we cut the three remaining hexagons into halves to accommodate the now bumpier center. We thus achieve a 15-piece dissection (Figure I70b), though turning over pieces then seems unavoidable.

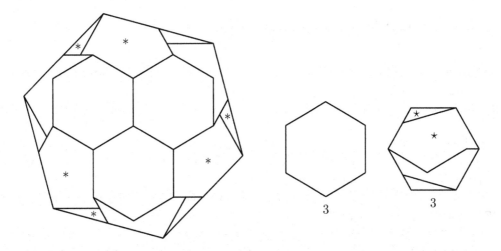

I70b: Improved six hexagons to one (with turning over) [GNF]

Plate 71: Freese was the first to dissect four congruent hexagons to nine congruent hexagons. If the four hexagons need not be identically cut, then the 24-piece dissection that he gave in this plate is overshadowed by the 21-piece dissection that he gave in Plate 73. In this plate he also gave a 12-piece dissection of seven regular hexagons to one. Lindgren (1964) also gave this same dissection. With the seven dissected hexagons suitably oriented, the dissection is translational.

Plate 72: Freese's first dissection on this plate, an 18-piece dissection of seven hexagons to three, is the same as the one that Lindgren (1964) gave. With the dissected hexagons suitably oriented, the dissection is translational. Apparently both men discovered the dissection independently. Freese's 18-piece dissection of twelve hexagons to one appears to have been original, as does his 22-piece dissection of twelve equilateral triangles to a hexagon. The former dissection is easily hingeable. When suitably oriented, both dissections are translational.

Plate 73: The upper dissection on this plate is of four hexagons to nine hexagons. With just 21 pieces, this is clearly better than the 24 pieces in Plate 71. Even so, Freese missed the obvious, that his leftmost large hexagon needs only 7 pieces, with the three in the center merged to give one of the smaller hexagons. We see the resulting 19-piece dissection in Figure I73. You can derive this dissection by superposing a tessellation of large hexagons with a tessellation of small hexagons. Suitably oriented, the dissection is translational.

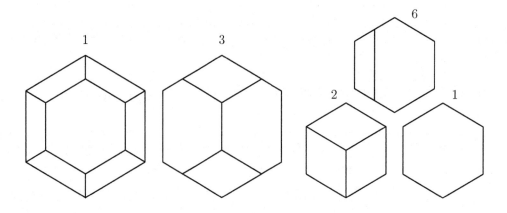

I73: Improved nine hexagons to four hexagons [GNF]

The lower dissection, of thirteen hexagons to one, is also based on tessellations. With just 19 pieces, it appears to be minimal. Freese did not note that the dissection is hingeable. Suitably oriented, it is also translational.

Plate 74: Freese seems to have been the only one to attempt a dissection of a regular hexagon to an equilateral triangle and a square of equal area. His assertion that the dissection is uniquely minimal may well be true, but comes without a proof.

Freese did not identify the dissection as hingeable. We see the dissection with its pieces labeled in Figure I74a and the hinged pieces in Figure I74b.

Plate 75: Freese appears to have been the first to specifically dissect three congruent regular hexagons to four congruent regular hexagons. His 12-piece dissection seems difficult to beat. Interestingly, it is fully hingeable, as we see in Figure I75a. Furthermore it is translational, as we can

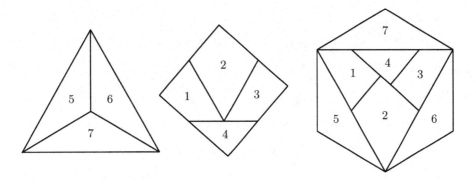

I74a: Hingeable equilateral triangle and a square to a hexagon

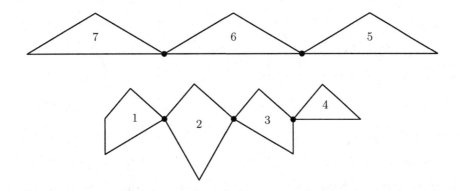

I74b: Hinged pieces for an equilateral triangle and a square to a hexagon

see in Figure I75b, in which the pieces in each of the slightly larger three hexagons have the same labels. Note that each small pointy piece pairs up with a congruent piece of the same label such that one can move to the other without any rotation.

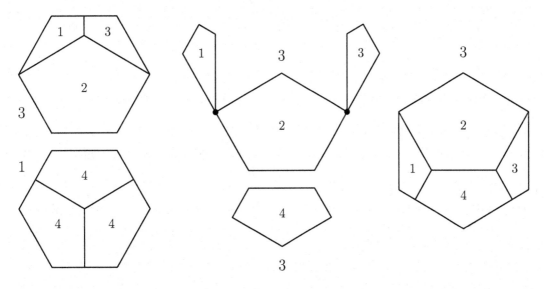

I75a: Freese's dissection of three hexagons to four hexagons is hingeable

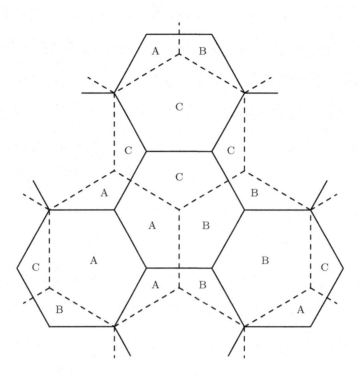

I75b: Freese's three hexagons to four hexagons is translational

Similarly, Freese appears to have been the first to specifically dissect three regular hexagons to eight equilateral triangles. His 14-piece dissection seems hard to beat. It too is translational.

Plate 76: Freese identified several easy dissections in this plate. They are his 12-piece dissection of nine congruent regular hexagons to one, his 9-piece dissection of six congruent regular hexagons to an equilateral triangle, his 12-piece dissection of six congruent regular hexagons to a hexagram, his 12-piece dissection of a regular hexagon to eight equilateral triangles, and his 6-piece dissection of two hexagons to three equilateral triangles. All of these seem difficult to improve on. Yet the first, third, and fourth in the above listing are all hingeable, and are also translational.

Plate 77: Freese seems to have been the only one to have proposed a three-way dissection of a hexagon to 3 hexagons and to 4 hexagons. He didn't push hard enough on this easy dissection, using two more pieces than he needed, as we see in Figure I77. This improved dissection is translational.

Plate 78: In Plate 21, Freese noted that there is a natural dissection of triangles whose side lengths are in the ratio $1:2:3:4:5:6$ to a larger triangle. The desired side length of the larger triangle is $\sqrt{91}$, which is the third side of a triangle that has an angle of $120°$ sandwiched between sides of lengths 5 and 6. In this plate Freese presented the corresponding dissection for hexagons. He split the 5-hexagon and the 6-hexagon into three pieces each, and used those six pieces to form the boundary of the large hexagon. He then cut the remaining hexagons to fill in the interior of the large hexagon. There are several alternate 15-piece dissections, but I have yet to find a 14-piece dissection.

Plate 79: Freese appears to have been the first to dissect a regular hexagon to two squares having an area ratio of $1:2$. He cut the hexagon into four pieces, one of which is a rectangle,

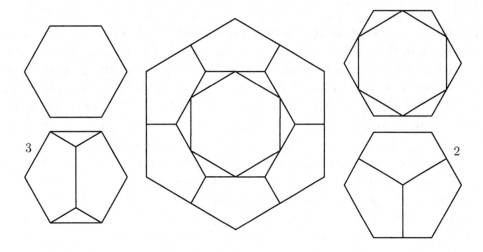

I77: Improved hexagon to 3 hexagons or 4 hexagons [GNF]

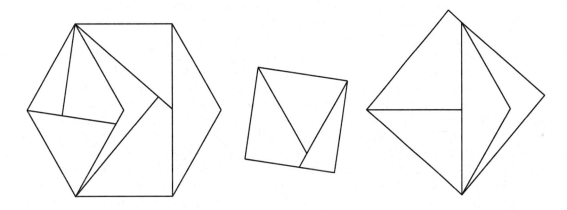

I79: Improved hexagon to squares of area ratio 1:2 [Gavin Theobald]

and the other three form a rectangle. He then used a P-strip to convert the first rectangle to a square, and a P-slide to convert the second rectangle. He could also have used a P-slide on the first rectangle too, and still have produced an 8-piece dissection. Suitably oriented, the dissection is translational.

Freese noted that the pieces of the larger square also form an equilateral triangle. Indeed, without the P-slide the three pieces that form the second rectangle also form that triangle. This dissection problem is the first in an infinite sequence of such problems, as I discuss in the comments for Plate 91.

This dissection problem is similar in spirit to that in Plate 67, so we should not be surprised that Gavin Theobald has found a way to improve Freese's dissection by one piece, as we see in Figure I79.

Plate 80: Freese discovered that regular hexagons of sides 1, 2, and 3 contain the area of an equilateral triangle of side $\sqrt{84}$. Furthermore, $\sqrt{84}$ is a distance that we can realize on a hexagonal grid, by moving distance 2, then making a 120° turn, and then moving distance 8. Freese produced a fairly symmetrical 11-piece dissection by placing the 1-hexagon in the center of the triangle, and then cutting the other two hexagons to produce pieces that fill in the rest of the triangle in a symmetrical fashion. A most straightforward application of his approach would cut each of the other two hexagons into three 60°-rhombuses, and each rhombus into two pieces. This would have cost thirteen pieces, but by swapping and rearranging pieces, Freese got the number down to eleven.

By breaking the symmetry altogether, however, we can save yet another piece, or we can actually save two if we allow turning over. My 9-piece dissection in Figure I80 turns over just one piece. Two ideas turn out to be key to reducing the number of pieces. First, if we place the 2-hexagon and 3-hexagon side by side so that they share one vertex, then the diagonal from one corner of the 3-hexagon to an opposing corner of the 2-hexagon is of length $\sqrt{84}$. Cutting along this diagonal goes a long way towards a 10-piece solution with no turning over. Second, if we cut a cavity in one remnant of the 3-hexagon into which we can insert the 1-hexagon, and expand the cavity to allow various pieces to fit together, we can save one more piece. If you, like Freese, don't want to turn over pieces, just cut a 1-triangle off of the turned-over piece, leaving an isosceles trapezoid. Since both of these pieces have reflection symmetry, you need not turn over either of them in the resulting 10-piece dissection.

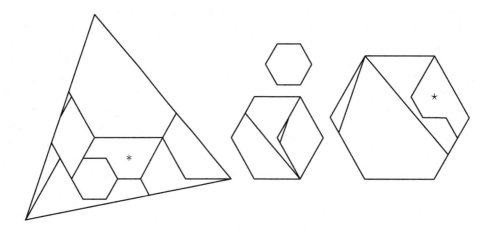

I80: Improved triangle to hexagons with side ratios 1:2:3 [GNF]

In the other dissection on this plate, Freese produced an equilateral triangle from fourteen congruent regular hexagons. He relied on the same basic number identity to give a side of the triangle that we can realize on the hexagonal grid. It would seem to be difficult to do better than his 24-piece dissection.

Plate 81: Freese appears to have been the first to dissect specifically a regular hexagon to three congruent equilateral triangles. He split the hexagon into three congruent rhombuses, and cut each into four pieces that form an equilateral triangle. However, I found it more efficient to use a strip dissection, saving one piece. Then Gavin Theobald saved yet another piece with his strips in Figure 81a. We see the resulting 9-piece dissection in Figure I81b.

Freese's second dissection in Plate 81 is the 4-piece dissection of a regular hexagon to two equilateral triangles. Lindgren (1964) also gave this same dissection, which is surely minimal. The dissection is hingeable, and in a suitable orientation, it is also translational.

Freese gave a third dissection, of congruent hexagons for $1^2 + (\sqrt{3})^2 = 2^2$ for which he used seven pieces. The six pieces from the small hexagon are cyclicly hingeable, and suitably oriented,

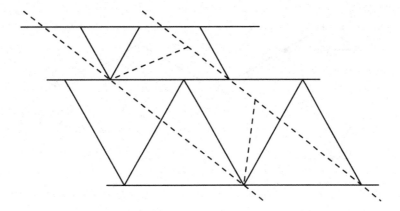

I81a: Crossposition for improved hexagon to three triangles [Gavin Theobald]

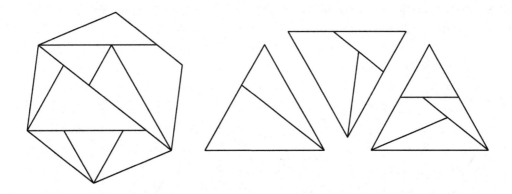

I81b: Improved hexagon to three triangles [Gavin Theobald]

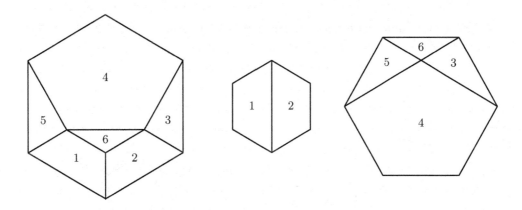

I81c: Improved hexagon to hexagons of area ratio 1:3 [Harry Lindgren]

the dissection is translational. However, Lindgren (1964) saved a piece with his 6-piece dissection (Figure I81c), which is hingeable, as we see in Figure I81d.

Plate 82: Freese seems to have been the first to have dissected equilateral triangles with area ratio 1:2:3 to a regular hexagon. He took advantage of the fact that the hexagon consists of six

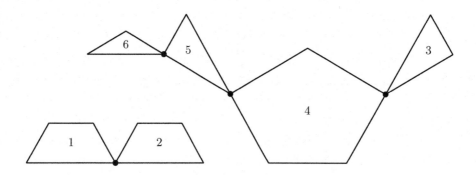

I81d: Hinged pieces for a hexagon to hexagons of area ratio 1:3

small triangles, and that $1 + 2 + 3 = 6$. Unfortunately, his 3-piece dissection of the largest triangle led him to cut the middle triangle into one piece too many. If we cut the large triangle into three pieces that fill half of the hexagon, as bounded by a longest diagonal, then we can realize an 8-piece dissection as in Figure I82a.

I82a: Improved hexagon to triangles of area ratio 1:2:3 [GNF]

I82b: Further improved hexagon to triangles of area ratio 1:2:3 [GNF]

Yet it is a pleasant surprise that we can improve the dissection further. We can glue the tiny (irregular) triangle onto an adjacent piece in the hexagon, and then swap a comparable portion from one piece to another, creating a cavity to accommodate the glued-on piece. We thus arrive at the 7-piece dissection in Figure I82b!

Plate 83: Freese was the first to dissect a regular hexagon to four congruent equilateral triangles. This dissection is similar in spirit to the one in Plate 68, because it uses tessellations similarly. In this plate Freese cut two equilateral triangles into pieces that form a regular hexagon, and then tessellated using the hexagon and the two remaining triangles. It would seem difficult to improve on his 10-piece dissection. This exquisite dissection is hingeable, with each of the three triangles that we cut into three pieces having its three pieces hinged together, so that these hinged assemblages wrap around the uncut triangle, as we see in Figures I83a and I83b. Suitably oriented, the dissection is also translational.

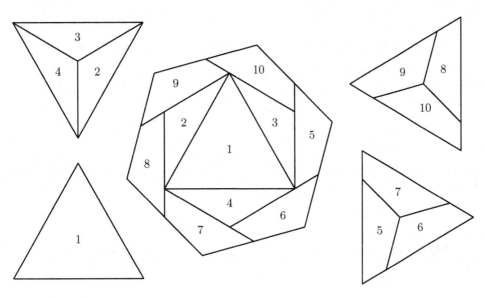

I83a: Hingeable dissection of 4 triangles to a hexagon

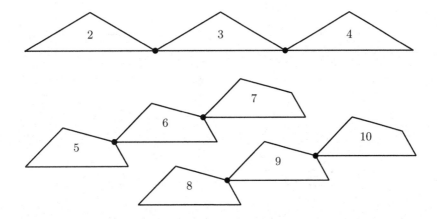

I83b: Hinged pieces for 4 triangles to a hexagon [GNF]

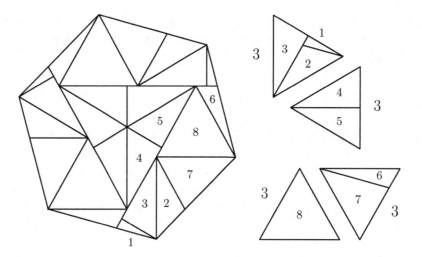

I84a: Hingeable dissection of 12 triangles to a hexagon [GNF]

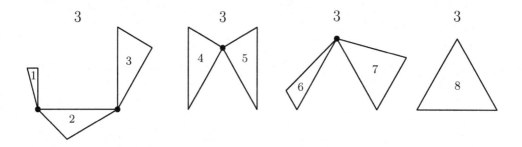

I84b: Hinged pieces for dissection of 12 triangles to a hexagon [GNF]

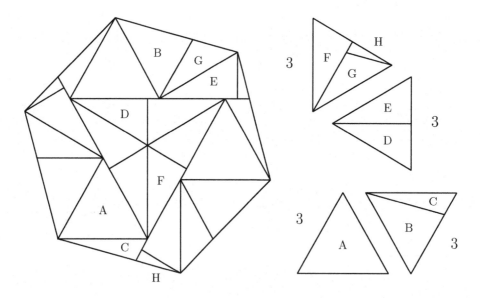

I84c: Translational dissection of 12 triangles to a hexagon [GNF]

Plate 84: Freese dissected twelve equilateral triangles to a regular hexagon by forming two hexagons consisting of six triangles each. He then dissected the two hexagons to one using the analog of the dissection for pentagons that he gave in Plate 50. Curiously, he could have achieved the same number of pieces (24) by using his dissection of two hexagons to one in Plate 68, while at the same time achieving both a hingeable dissection (Figures I84a and I84b) and a translational dissection (Figure I84c). To make Figures I84a and I84c easier to understand, I have labeled pieces from just one copy of each of the four triangles. The remaining pieces have labels consistent with an obvious symmetric pattern.

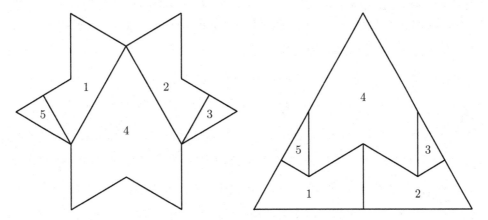

I85a: Freese's dissection of a hexagram to an equilateral triangle

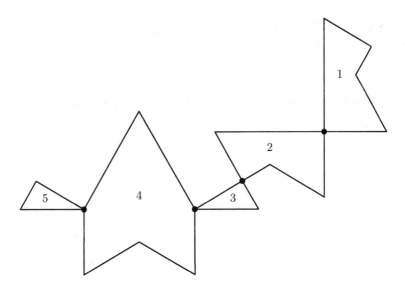

I85b: Hinging a dissection of hexagram to an equilateral triangle [GNF]

Plate 85: In 1961 Harry Lindgren published the top dissection in this plate, a 5-piece dissection of a hexagram to an equilateral triangle. Although Freese did not note that this dissection is hingeable, it clearly is, as we see in Figures I85a and I85b. To transform the triangle to the star, first swing pieces 1 and 2 as a unit around the hinge between pieces 2 and 3. Then swing pieces

1, 2, and 3 as a unit around the hinge between pieces 3 and 4. Next swing piece 1 around the hinge between pieces 1 and 2. Finally swing piece 5 around the hinge between pieces 4 and 5. (By the way, the purported hinging in Figure 15.14 of my second book is not a hinging as accepted by Freese, because it is only "wobbly-hinged." Sigh!)

As noted in our Chapter 1, Geoffrey Mott-Smith (1946) had already found several different 5-piece dissections, and a 6-piece dissection had appeared in the 700-year-old anonymous Persian manuscript, *Interlocks of Similar or Complementary Figures*.

Lindgren (1964) also published the middle dissection of this plate, of a hexagram to two regular hexagons in four pieces. It is trivially translational. Harry Bradley (1921) first published the dissection at the bottom of the plate, a 5-piece dissection of a hexagram to a square.

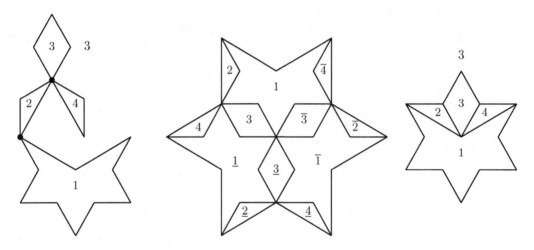

I86a: Hinged pieces for Freese's three hexagrams to one [GNF]

Plate 86: Freese was the first to find a 12-piece dissection of three hexagrams to one. Lindgren (1964) gave a 13-piece dissection, which he soon improved to the same 12-piece dissection as Freese's. Interestingly, Freese's dissection is hingeable, as we see in Figure I86a. It is also translational, as in Figure I86b.

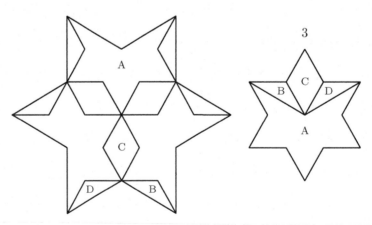

I86b: Freese's three hexagrams to one is translational

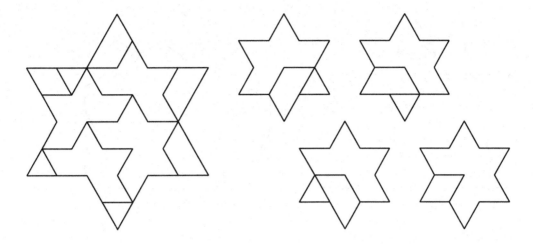

I86c: Improved four hexagrams to one [Robert Reid]

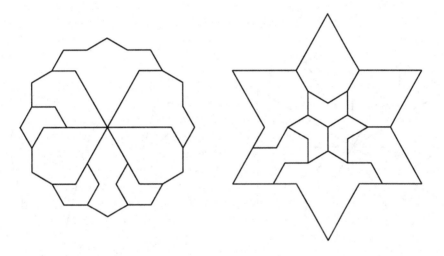

I86d: Improved dodecagram to hexagram

Freese was the first to tackle four hexagrams to one, which when suitably arranged is translational. However, this was improved on by Robert Reid, who designed the 11-piece dissection in Figure I86c.

For the final dissection on this plate, Freese found a 13-piece dissection of a hexagram to a dodecagram. Yet Lindgren (1964) found a scrumptious 9-piece dissection, a variation of which is in Figure I86d. Even so, we see in Figure I86e that Freese's 13-piece dissection is fully hingeable, only one piece more than the hingeable dissection in my 2002 book.

Plate 87: Freese gave two ways to form a hexagonal ring from six identical regular hexagons. Both are translational, and the second is hingeable. Freese also remarked that these lead to a 7-piece dissection of four hexagons. However, they are not as good as the 6-piece dissection in Plate 69.

Plate 88: Freese noted that you can slice a regular hexagon into six equilateral triangles, and each of those into two right triangles, and half of the right triangles into a pair consisting of an isosceles triangle and an equilateral triangle, in such a way that they assemble into a dodecagon

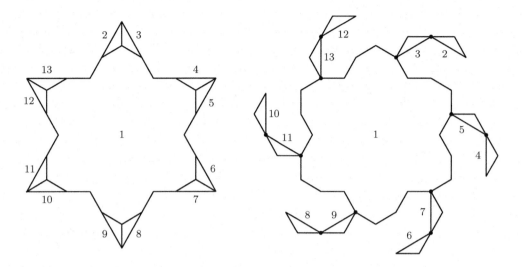

I86e: Hinging of Freese's dodecagram to hexagram [GNF]

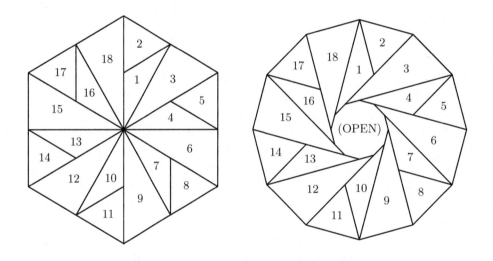

I88a: Labeling Freese's dissection of a hexagon to dodecagon ring

ring. It is splendid that the dissection splits up into the three most fundamental triangles: isosceles, right, and equilateral. Although Freese did not mention it, his dissection is hingeable. Labeling the pieces as in Figure I88a, we get the hinged pieces as in Figure I88b. Enjoy the rendition of this dissection as the cover art of this book!

Freese could have assembled the twelve right triangles into a dodecagon ring, if he were willing to flip half of the pieces, as we see in Figure I88c. Yet we can do better than Freese's 18-piece dissection without turning over pieces. Note how, in Figure I88d, we position eight of the twelve right triangles in the dodecagon without turning over any of them. We then split the remaining four into two pieces each to fill in the remaining area. Thus we easily produce a 16-piece dissection.

By merging pieces together and cutting cavities into which we fit these merged pieces, we further reduce the number of pieces without resort to turning over. The resulting dissection (Figure I88e) has just thirteen pieces, with none turned over. Is there some way to reduce this number to twelve?

The general construction that Freese exploited in Plate 88 works for any other $(2n)$-sided regular

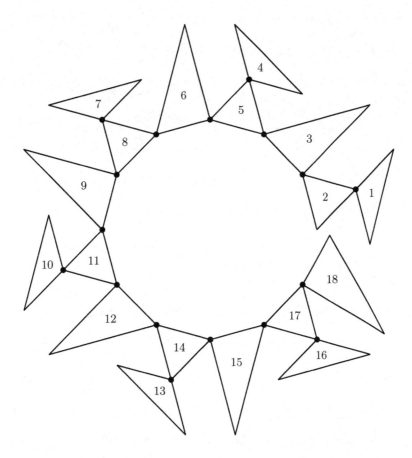

I88b: Hinging Freese's dissection of a hexagon to dodecagon ring [GNF]

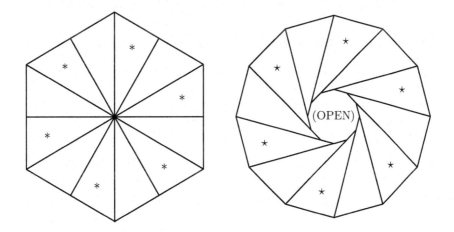

I88c: Improved hexagon to dodecagon ring, with pieces turned over [GNF]

polygon as well, producing a $3n$-piece dissection, which is hingeable(!). The number of pieces reduces to $2n$ if we allow turning over, though we lose hingeability. Furthermore, we can apply the same tricks as in Figures I88d and I88e to have fewer than $3n$ pieces with no turning over. The case of $n = 2$ is special, in that we can cut a square into eight isosceles right triangles, which need no further cutting because turning over an isosceles triangle gives no benefit. Freese presented this

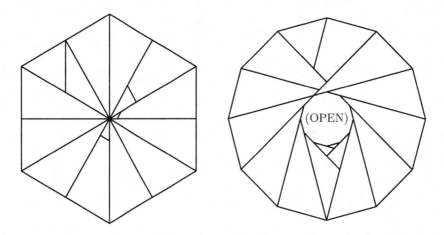

I88d: Improved hexagon to dodecagon ring, with no pieces turned over [GNF]

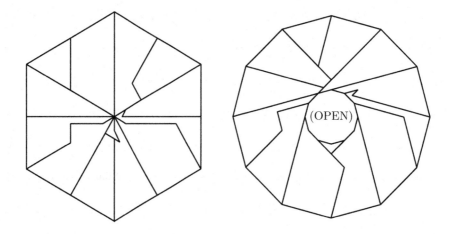

I88e: Further improved hexagon to dodecagon ring, no pieces turned over [GNF]

special case in a disguised form in Plate 165.

We have now come to the end of a long, sweet chapter on hexagons and hexagrams. It is long because of how well hexagons fit together, not only with themselves, but also with equilateral triangles, dodecagons, and dodecagrams. As already mentioned, Freese did not proffer any dissections of heptagons or heptagrams. So we will proceed directly to octagons, of which he found some truly pleasing dissections. Yet as we move on to polygons with more than six sides, we note that Freese found progressively fewer interesting dissections.

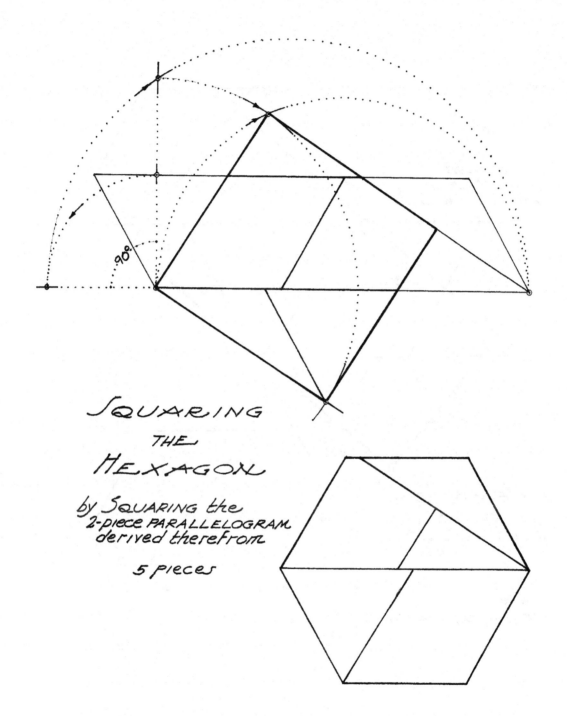

SQUARING

THE

HEXAGON

by Squaring the
2-piece PARALLELOGRAM
derived therefrom

5 pieces

90°

NOTE:
There are an unlimited
number of 5-piece SQUARINGS.
A 4-piece SQUARING is impossible.

PLATE
64

B1P29
..FREESE

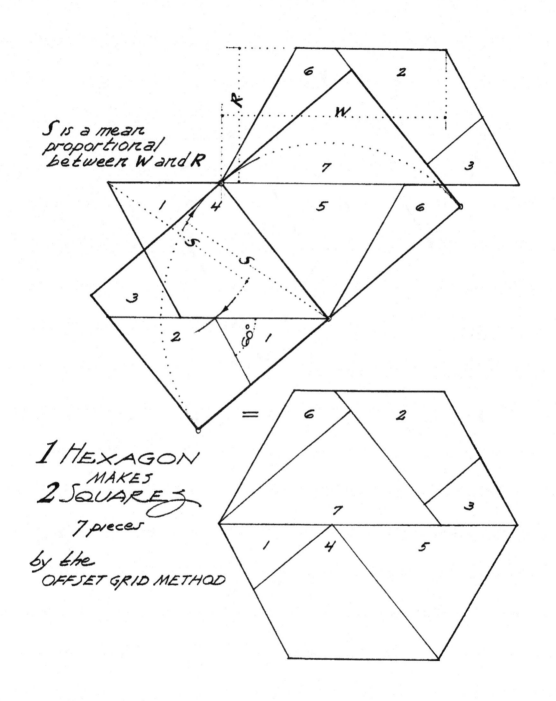

S is a mean
proportional
between W and R

1 HEXAGON
MAKES
2 SQUARES
7 pieces

by the
OFFSET GRID METHOD

PLATE
65

B2 PRGD.51
..FREESE..

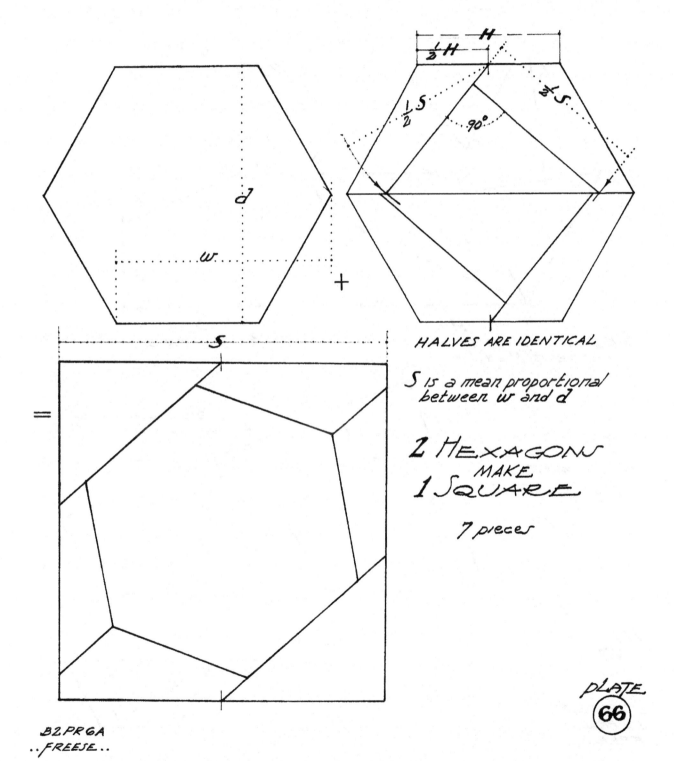

H

$\frac{1}{2}H$

$\frac{1}{2}S$ 90° $\frac{1}{2}S$

d

w

$+$

S

$=$

HALVES ARE IDENTICAL

S is a mean proportional
between w and d

2 HEXAGONS
MAKE
1 SQUARE

7 pieces

PLATE
66

B2.PR6A
..FREESE..

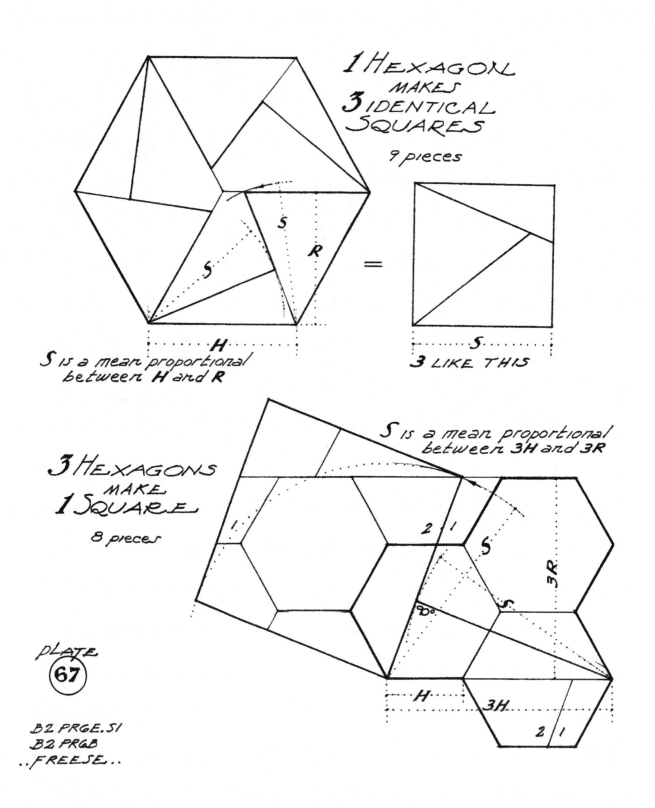

1 HEXAGON
MAKES
3 IDENTICAL SQUARES

9 pieces

=

3 LIKE THIS

S is a mean proportional between H and R

S is a mean proportional between 3H and 3R

3 HEXAGONS
MAKE
1 SQUARE

8 pieces

PLATE
(67)

B2 PRGE.SI
B2 PRGB
..FREESE..

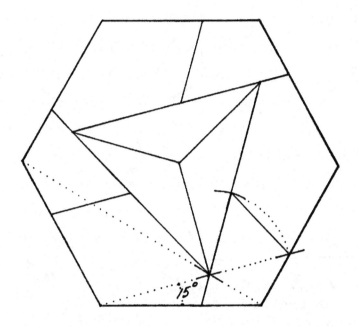

A 2-IN-1 HEXAGON

9 pieces,
3 of each shape.

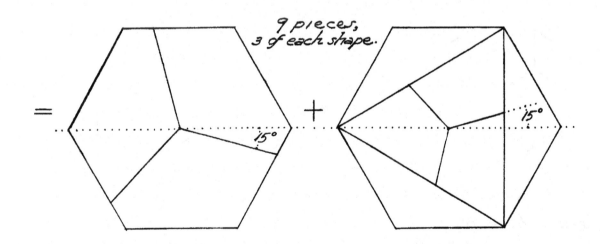

PLATE
68

B3PTG.21
..FREESE..

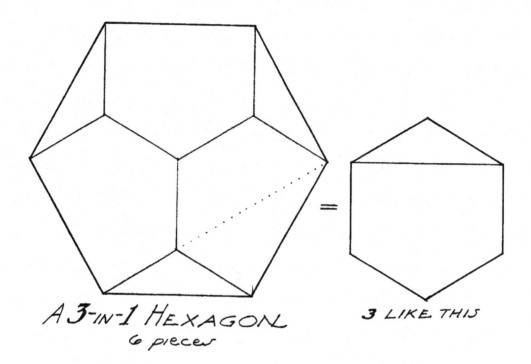

A 3-IN-1 HEXAGON
6 pieces

3 LIKE THIS

A 4-IN-1 HEXAGON, 6 pieces.
(min.)

General Theorem:

Any polygon whatsoever, whether regular, irregular, convex or concave, can be geometrically dissected to make m^2 similar equiareal polygons, m being any integer greater than unity.

PLATE
69

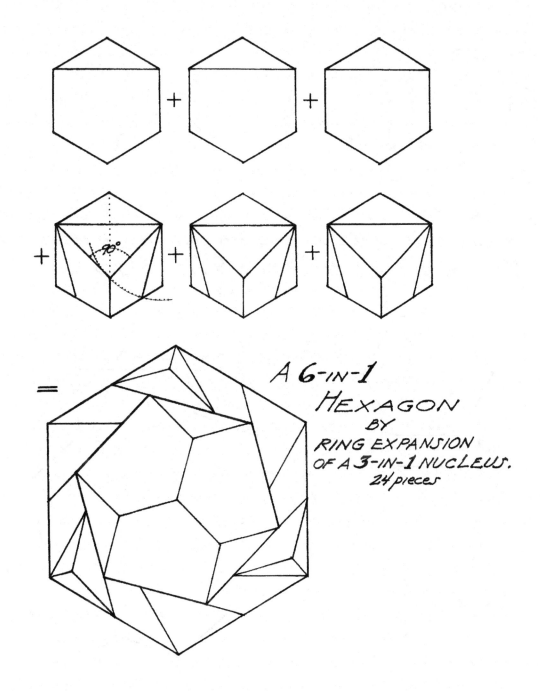

A 6-IN-1
HEXAGON
BY
RING EXPANSION
OF A 3-IN-1 NUCLEUS.
24 pieces

90°

PLATE
70

B4 P6J
..FREESE..

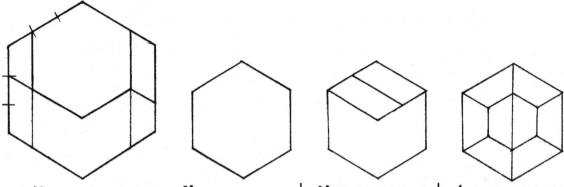

4 LIKE THIS **= 4** LIKE THIS **+ 4** LIKE THIS **+ 1** LIKE THIS

4 IDENTICAL **H**EXAGONS MAKE **9 H**EXAGONS

24 pieces

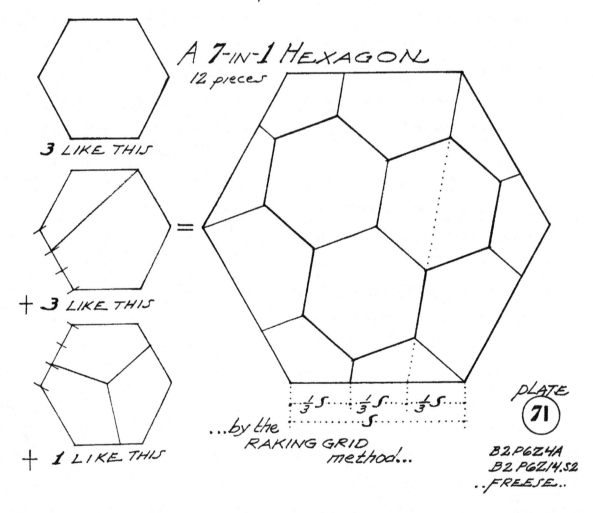

A **7**-IN-**1** HEXAGON

12 pieces

3 LIKE THIS

+ 3 LIKE THIS

+ 1 LIKE THIS

=

...by the
RAKING GRID
method...

$\frac{1}{3}s$ $\frac{1}{3}s$ $\frac{1}{3}s$

s

PLATE
71

B2.P6Z4A
B2.P6Z14.S2
..FREESE..

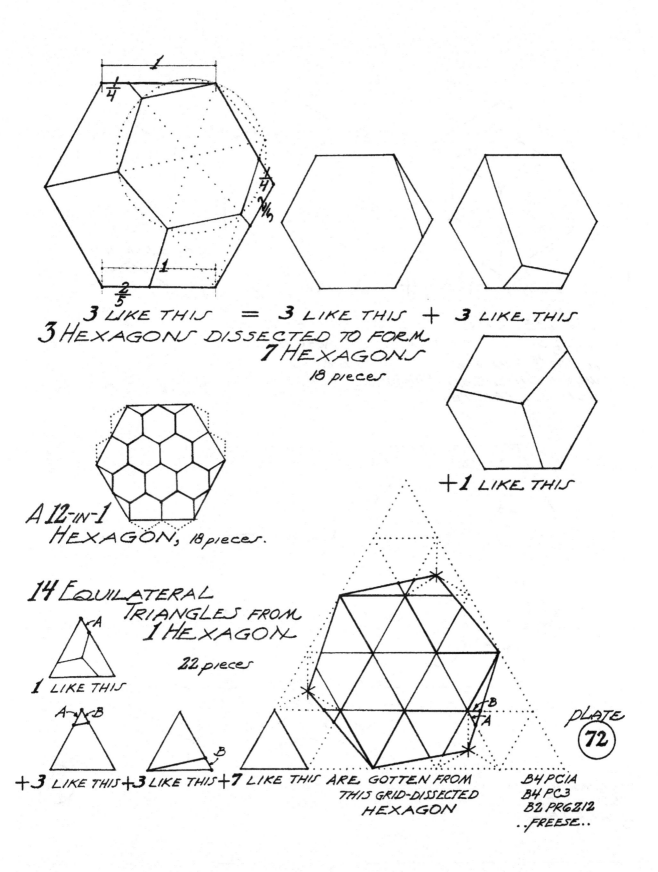

3 LIKE THIS = 3 LIKE THIS + 3 LIKE THIS

3 HEXAGONS DISSECTED TO FORM
7 HEXAGONS
18 pieces

+1 LIKE THIS

A 12-IN-1
HEXAGON, 18 pieces.

14 EQUILATERAL
TRIANGLES FROM
1 HEXAGON
22 pieces

1 LIKE THIS

+3 LIKE THIS +3 LIKE THIS +7 LIKE THIS ARE GOTTEN FROM
THIS GRID-DISSECTED
HEXAGON

PLATE
72

B4 PCIA
B4 PC3
B2 PR6Z12
..FREESE..

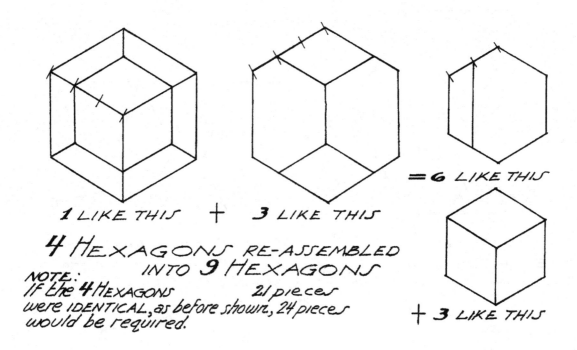

1 LIKE THIS $+$ **3** LIKE THIS $=$ **6** LIKE THIS

$+$ **3** LIKE THIS

4 HEXAGONS RE-ASSEMBLED INTO **9** HEXAGONS

NOTE:
If the **4** HEXAGONS 21 pieces
were IDENTICAL, as before shown, 24 pieces
would be required.

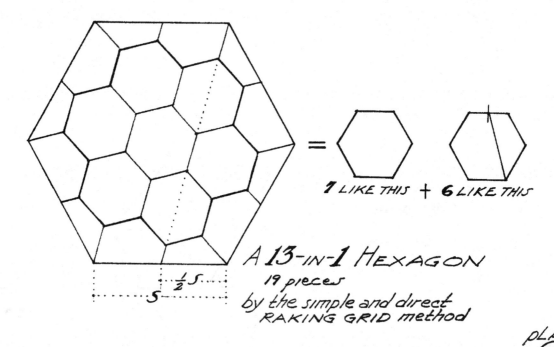

$=$ **1** LIKE THIS $+$ **6** LIKE THIS

A **13**-IN-**1** HEXAGON
19 pieces
by the simple and direct
RAKING GRID method

PLATE
73

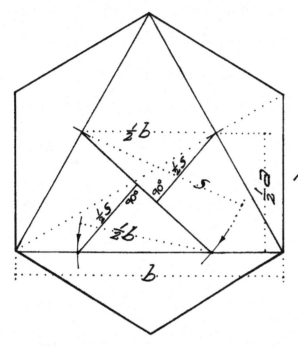

AN EXAMPLE OF
COMPOSITE
DISSECTION
that transforms
A HEXAGON
into
AN EQUILATERAL
TRIANGLE
plus a
SQUARE
of equal area

with **7** *pieces.* †

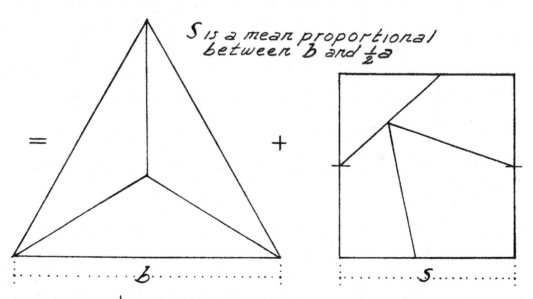

S is a mean proportional
between b and $\frac{1}{2}a$

† This transformation of 2 non-similar
equiareal polygons into 1 non-similar polygon
IS UNIQUE in that it is the only such composite
combination that can be dissected with so few pieces.

B2 PRGN.S2
..FREESE..

PLATE
74

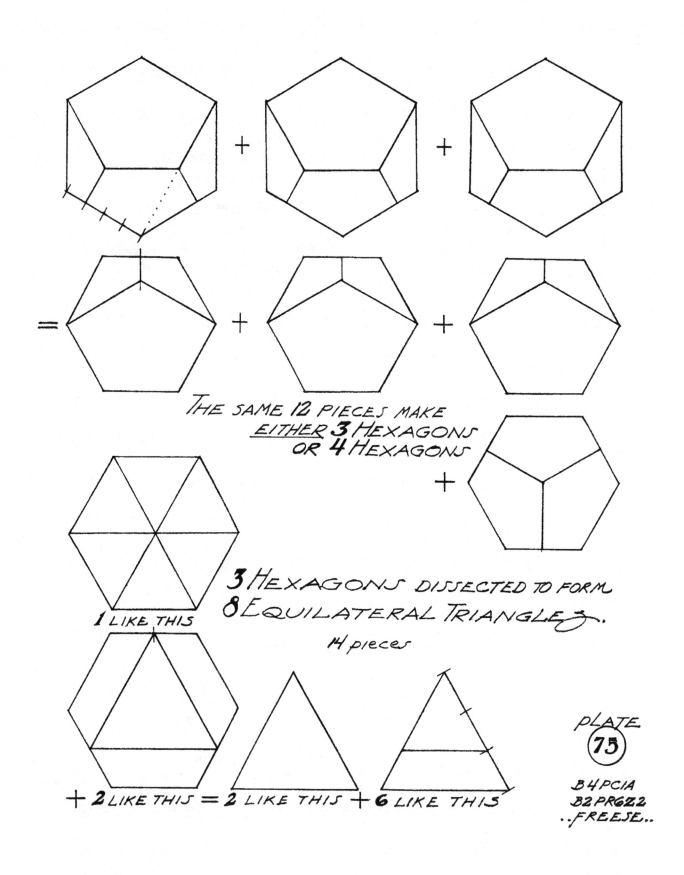

THE SAME 12 PIECES MAKE
EITHER 3 HEXAGONS
OR 4 HEXAGONS

1 LIKE THIS

3 HEXAGONS DISSECTED TO FORM
8 EQUILATERAL TRIANGLES.
14 pieces

+ 2 LIKE THIS = 2 LIKE THIS + 6 LIKE THIS

PLATE
75

B4PCIA
B2PRGZ2
..FREESE..

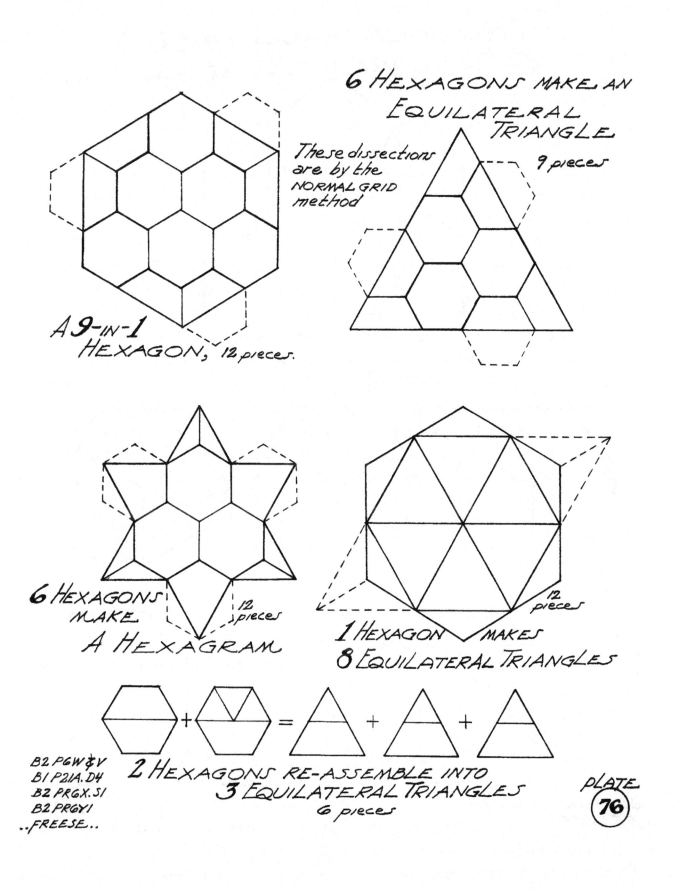

6 HEXAGONS MAKE AN EQUILATERAL TRIANGLE

9 pieces

These dissections are by the NORMAL GRID method

A 9-IN-1 HEXAGON, 12 pieces.

6 HEXAGONS MAKE A HEXAGRAM

12 pieces

1 HEXAGON MAKES 8 EQUILATERAL TRIANGLES

12 pieces

2 HEXAGONS RE-ASSEMBLE INTO 3 EQUILATERAL TRIANGLES

6 pieces

B2 PGW $V
B1 P21A. D4
B2 PRGX. S1
B2 PRGY1
..FREESE..

PLATE 76

1 HEXAGON MAKES EITHER
3 HEXAGONS OR 4 HEXAGONS

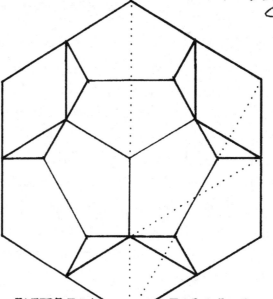

FIFTEEN PIECES

= ⬡ + ⬡ + ⬡

OR

= ⬡ + ⬡ + ⬡ + ⬡

B2 PRG21,51
..FREESE..

PLATE
77

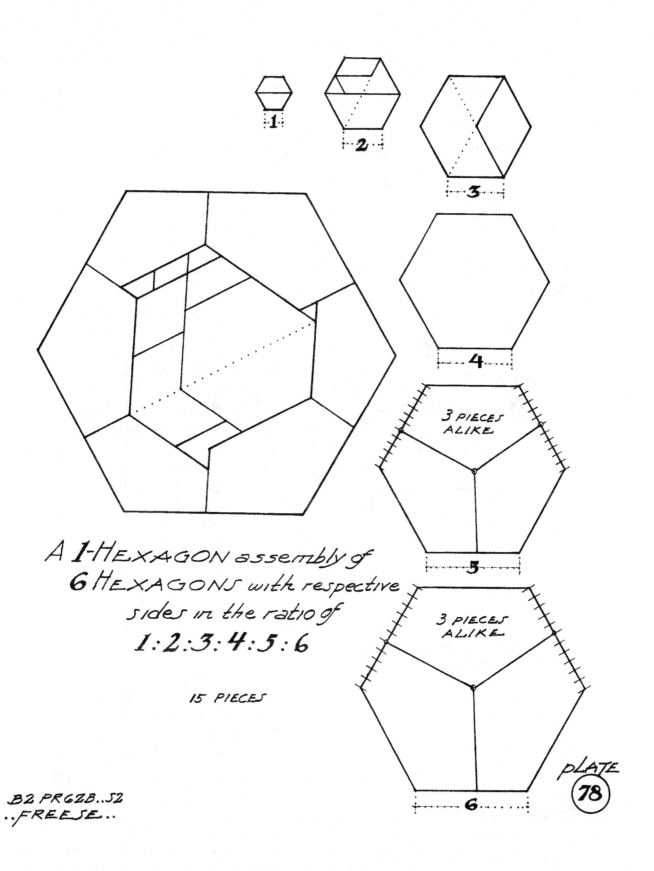

1

2

3

4

5

6

3 PIECES ALIKE

3 PIECES ALIKE

A *1*-Hexagon assembly of
6 Hexagons with respective
sides in the ratio of
1:2:3:4:5:6

15 PIECES

B2 PR62B..S2
..FREESE..

pLATE
78

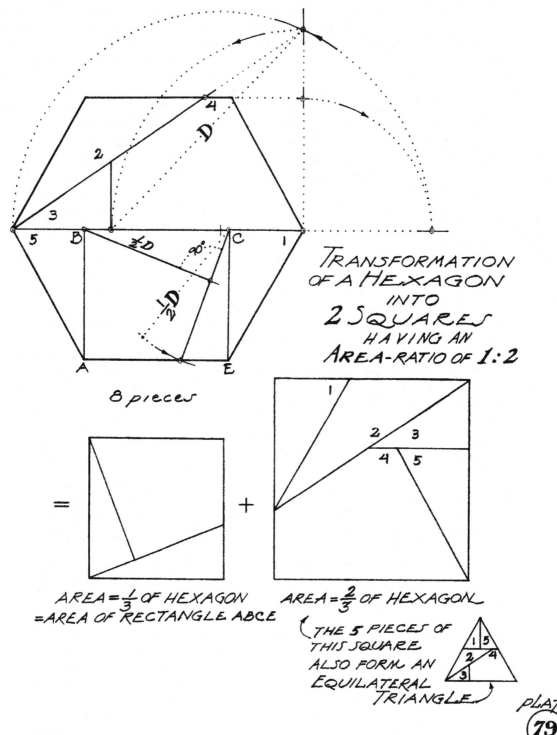

TRANSFORMATION
OF A HEXAGON
INTO
2 SQUARES
HAVING AN
AREA-RATIO OF 1:2

8 pieces

AREA = $\frac{1}{3}$ OF HEXAGON
= AREA OF RECTANGLE ABCE

AREA = $\frac{2}{3}$ OF HEXAGON

THE 5 PIECES OF
THIS SQUARE
ALSO FORM AN
EQUILATERAL
TRIANGLE

PLATE
79

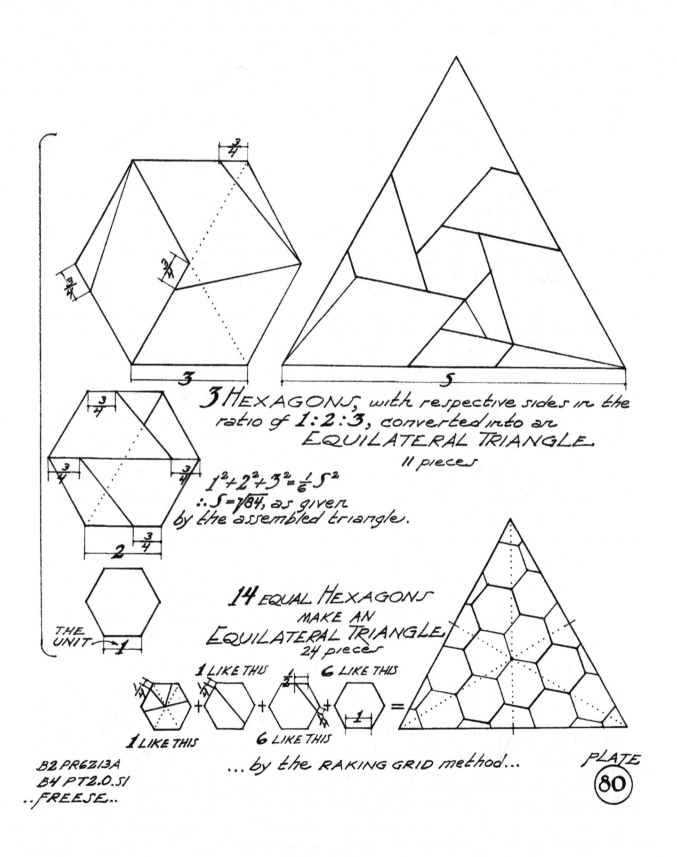

$\frac{3}{4}$

$\frac{3}{4}$

$\frac{3}{4}$

$\frac{3}{4}$

3

5

$\frac{3}{4}$

$\frac{3}{4}$

$\frac{3}{4}$

2   $\frac{3}{4}$

3 HEXAGONS, with respective sides in the ratio of 1:2:3, converted into an EQUILATERAL TRIANGLE.

11 pieces

$1^2 + 2^2 + 3^2 = \frac{1}{6} S^2$
∴ $S = \sqrt{84}$, as given
by the assembled triangle.

THE UNIT — 1

14 EQUAL HEXAGONS MAKE AN EQUILATERAL TRIANGLE.
24 pieces

1 LIKE THIS   1 LIKE THIS $\frac{1}{2}$ 6 LIKE THIS

1 LIKE THIS   6 LIKE THIS   1   =

... by the RAKING GRID method ...

B2 PR6Z13A
B4 PT2.0.51
..FREESE..

PLATE
80

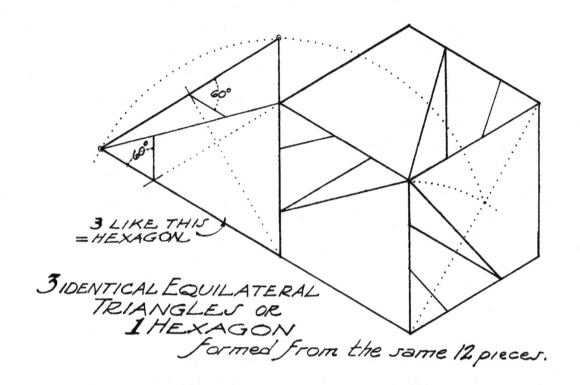

60°

60°

3 LIKE THIS
= HEXAGON

3 IDENTICAL EQUILATERAL
TRIANGLES OR
1 HEXAGON
formed from the same 12 pieces.

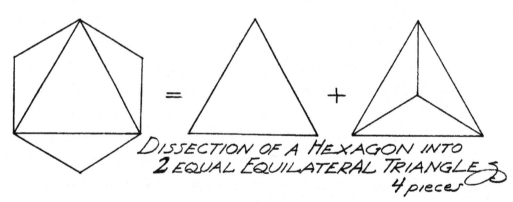

= +

DISSECTION OF A HEXAGON INTO
2 EQUAL EQUILATERAL TRIANGLES
4 pieces

PLATE
81

B2 PRGR.S2
B2 PRGK.S1
B4 P2C
..FREESE..

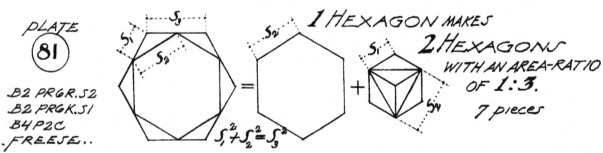

S_3
S_1
S_2
S_2
S_1
S_4

1 HEXAGON MAKES
2 HEXAGONS
WITH AN AREA-RATIO
OF 1:3.
7 pieces

$$S_1^2 + S_2^2 = S_3^2$$

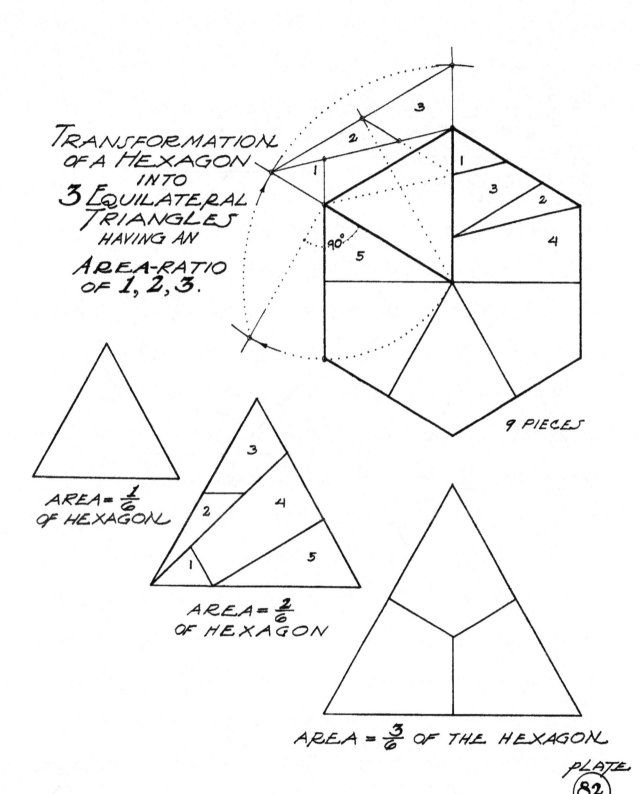

TRANSFORMATION OF A HEXAGON INTO 3 EQUILATERAL TRIANGLES HAVING AN AREA-RATIO OF 1, 2, 3.

90°

9 PIECES

AREA = $\frac{1}{6}$ OF HEXAGON

AREA = $\frac{2}{6}$ OF HEXAGON

AREA = $\frac{3}{6}$ OF THE HEXAGON

PLATE
82

B4 PZ 51
..FREESE..

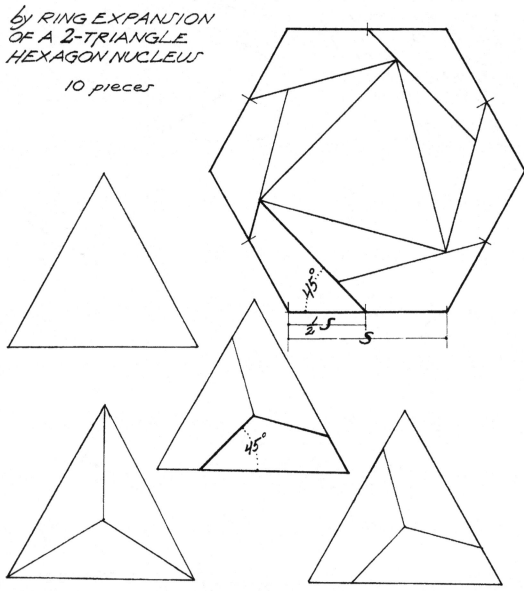

A HEXAGON TRANSFORMED INTO
4 EQUILATERAL TRIANGLES
by RING EXPANSION
OF A 2-TRIANGLE
HEXAGON NUCLEUS

10 pieces

45°

½S

S

THE 4 COMPONENT TRIANGLES

45°

PLATE
83

B3 PT6.7
..FREESE..

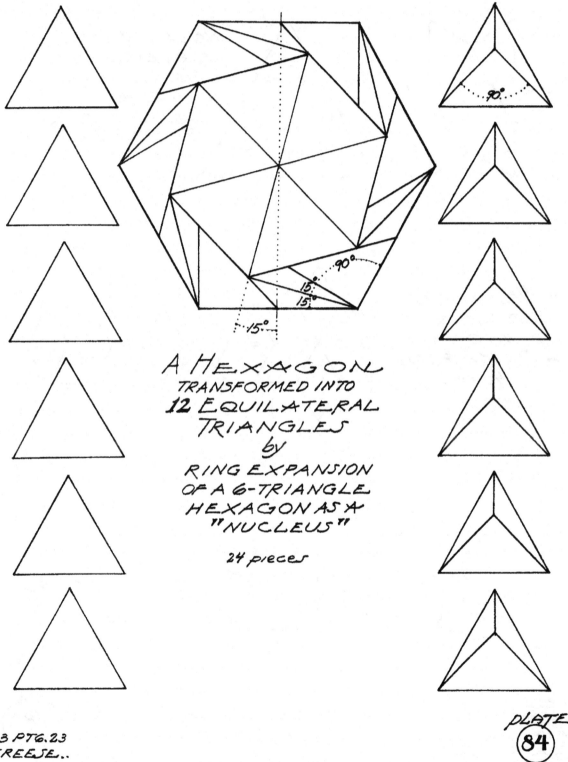

A HEXAGON
TRANSFORMED INTO
12 EQUILATERAL
TRIANGLES
by
RING EXPANSION
OF A 6-TRIANGLE
HEXAGON AS A
"NUCLEUS"

24 pieces

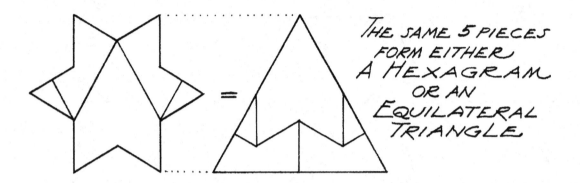

THE SAME 5 PIECES FORM EITHER A HEXAGRAM OR AN EQUILATERAL TRIANGLE

A HEXAGRAM CONVERTED INTO 2 HEXAGONS 4 pieces

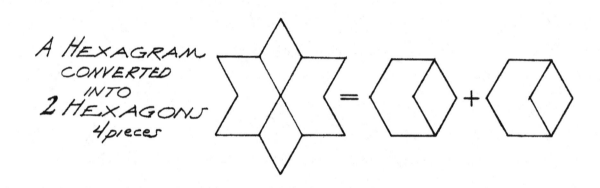

SQUARING the HEXAGRAM 5 pieces

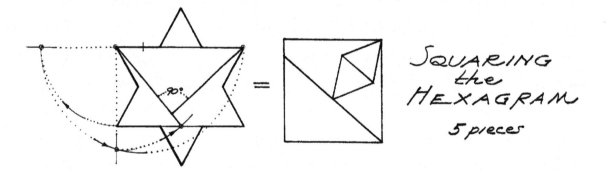

90°

B4 PZ30
B4 PZ30
B1 P21A
..FREESE

PLATE
85

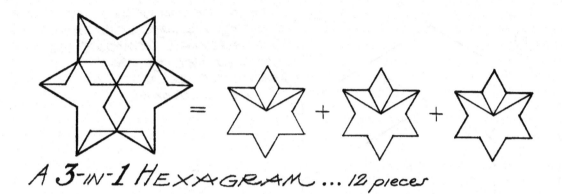

A 3-IN-1 HEXAGRAM ... 12 pieces

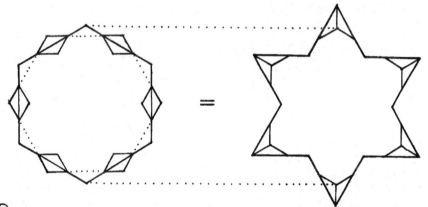

A 4-IN-1
HEXAGRAM
12 pieces

A DODECAGRAM transformed into
A HEXAGRAM
13 pieces

PLATE
86

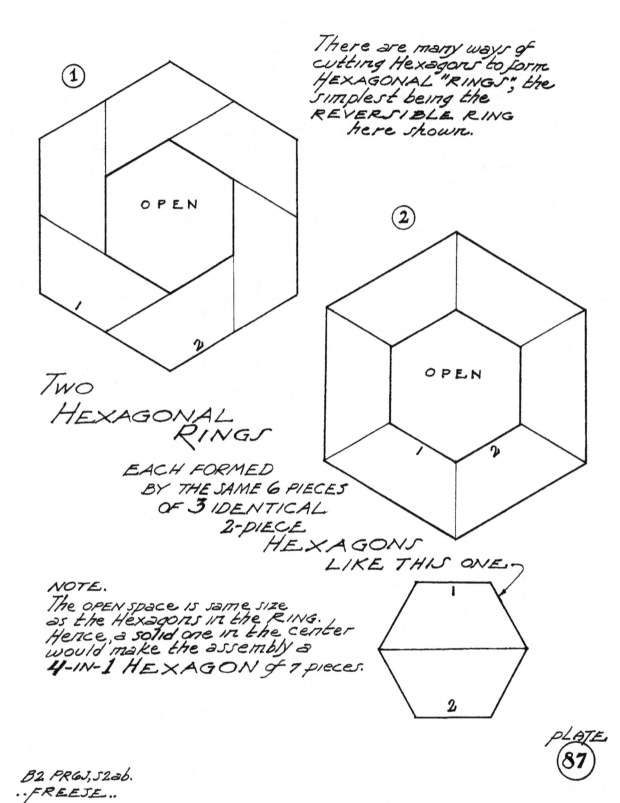

① OPEN

There are many ways of cutting Hexagons to form HEXAGONAL "RINGS", the simplest being the REVERSIBLE RING here shown.

② OPEN

TWO
HEXAGONAL
RINGS

EACH FORMED
BY THE SAME 6 PIECES
OF 3 IDENTICAL
2-PIECE
HEXAGONS
LIKE THIS ONE.

NOTE.
The OPEN space is same size as the Hexagons in the RING. Hence, a solid one in the center would make the assembly a 4-IN-1 HEXAGON of 7 pieces.

PLATE
87

B2 PROJ, S2ab.
..FREESE..

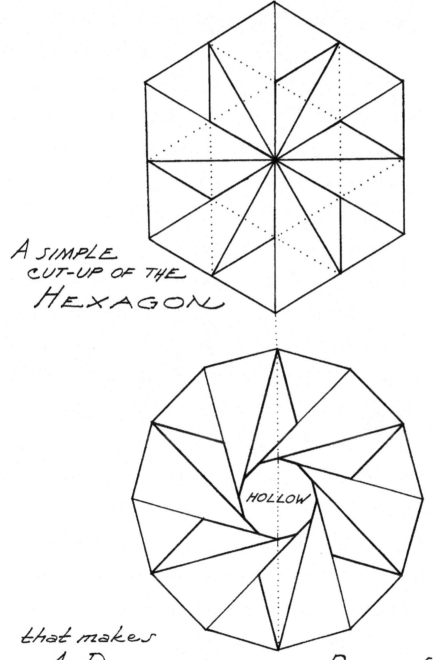

A SIMPLE
CUT-UP OF THE
HEXAGON

HOLLOW

that makes
A DODECAGON RING, from the
same 18 pieces.

PLATE
88

B3 PY-8
..FREESE..

Chapter 10

Octagons and Octagrams (Plates 89–98)

Freese's first four plates in this chapter involve both regular octagons and squares. Plate 89 displays the derivation of the legendary 5-piece dissection of a regular octagon to a square and three variations of it. Plate 90 is an interesting dissection of an octagon and small square to a large square, Plates 91 and 92 tackle dissections of an octagon to two congruent squares, and an octagon to four congruent squares. The next five plates explore n octagons to one, where n is, respectively, 2, 3, 4, 8, 5, and 9. Finally Freese served up his pièce de résistance on Plate 98, a 3-way dissection of an octagon to 2 octagons to 4 octagons. Although he started the chapter with relatively few dissections that have interesting properties, Freese concluded the chapter with seven dissections that are translational, even if he failed to mention that fact.

Of especial note is Plate 89, which includes that remarkable 5-piece symmetric dissection of a regular octagon to a square. Also worthy of particular mention is the bilateral and rotationally symmetric dissection of eight congruent octagons to a large octagon in Plate 95.

Plate 89: In this plate Freese presented several related 5-piece dissections of an octagon to square. One, which we have already encountered in Figures 1.2 and 1.3, is the highly symmetrical dissection that appeared in the 700-year-old anonymous Persian manuscript, *Interlocks of Similar or Complementary Figures*. The others result from shifting one tessellation relative to the other. Freese made unsupported claims that there are no such 4-piece dissections and that the only possible 5-piece dissections result from that tessellation method. He presented these dissections in terms of the tessellations with which we can generate them. Freese later used this insight to discover the 6-piece dissection of a regular dodecagon to a regular hexagon, which you will see in Plate 116. I discussed these as well as other examples of this technique in Chapter 13 of my 1997 book. Lindgren (1951) first described the tessellation method for producing the highly symmetrical version.

In Plate 160 Freese gave a related dissection to a more challenging problem involving regular octagons, in which he indicated a partial hinging. I have given a completely hingeable dissection for the example in Plate 160, one which relies on a hingeable dissection of a regular octagon to a square. Let's now examine my 7-piece hingeable dissection of a regular octagon to a square, which I first described in my 2002 book.

Freese must surely have recognized that the symmetric dissection of an octagon to a square has four points at which we can position hinges. The challenge is to also hinge the small square with the four other pieces. If we split one of those four pieces as on the left of Figure I89a, and then glue the small square to one of the resulting pieces, we get the 7-piece hingeable dissection on the right in Figure I89a. We see how to hinge the pieces in Figure I89b. If you swing the pieces around piece 1 in a counter-clockwise manner, you will obtain the octagon, while if you swing the pieces around piece 1 in a clockwise manner, you will obtain the square! I applied a similar technique to produce a number of related dissections in Chapter 12 of my 2002 book.

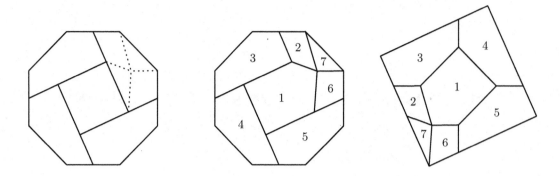

I89a: Splitting a piece to get a hingeable dissection of an octagon to a square [GNF]

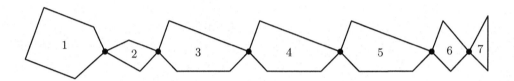

I89b: Hinged pieces for an octagon to a square [GNF]

Plate 90: Freese was the first to dissect a regular 1-octagon and $\sqrt{2}$-square to a square. By choosing a $\sqrt{2}$-square, he was able to cut away four isosceles right triangles that he could add to the octagon to give a $(1 + \sqrt{2})$-square. He then applied Perigal's method (on Plate 35) to get his 9-piece dissection. Freese chose a 9-piece dissection that also allowed him to illustrate a related 9-piece dissection of a regular octagon to a square. Neither of these dissections is minimal; however they do follow the same general form as those in Plate 115 that involve the dodecagon.

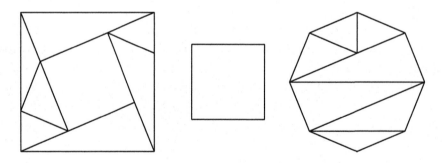

I90: Improved 1-octagon and $\sqrt{2}$-square to a square [GNF]

Freese recognized that the long diagonal of the octagon equals the side length of the large square. Yet we can save a piece, as we see in the 8-piece dissection in Figure I90. The key approach seems to be to cut the octagon into pieces with which we can surround an uncut small square.

Freese of course knew that the 9-piece dissection of a regular octagon to square at the bottom of the plate is not minimal, since he had described a 5-piece dissection on the previous plate. Interestingly, the 9-piece dissection appeared in the same circa-1300 CE anonymous manuscript that gave the 5-piece dissection. Freese was probably unfamiliar with that source.

Plate 91: Freese seems to have been the first to dissect a regular octagon to two congruent squares. Noting that rectangle ABCD in the plate is half the total area, he produced an 8-piece piecemeal dissection. One square comes from applying a P-slide to the rectangle, and the other comes from applying a P-strip to the remaining pieces. The dissection is translational, and it would seem difficult to reduce the number of pieces in this dissection.

Why did Freese decide to attempt this dissection? We can view the octagon as comprised of eight congruent pie-shaped pieces, which Freese termed sectors. Rectangle ABCD contains two whole sectors, and has its remaining area filled in by two more sectors that are each bisected. Since an octagon contains eight sectors, the rectangle contains half the area of the octagon. It then seems natural to dissect the octagon into two squares of equal area.

With good success in this particular case, Freese may have been motivated to try other related dissections. For example, a regular hexagon would have an analogous rectangle with area equivalent to four of its six sectors. In Plate 79, Freese found a dissection of the hexagon into two squares whose areas are in the ratio of 4 to 2, i.e., 2 to 1. Carrying this idea further, Freese may have realized that a "central rectangle" of a regular decagon would have an area equal to four out of its ten sectors. Then a natural dissection would be to dissect the decagon into two squares of area ratio 2 to 3. And indeed, Plate 109 displays his corresponding dissection.

Freese's next step might have been with a dodecagon, which would have a central rectangle with area equal to four out of its twelve sectors, suggesting a dissection into squares of area ratio 1 to 2. And once more, Freese did not disappoint, with a corresponding dissection in Plate 124. By now you may guess that these dissection problems are the first few in an infinite sequence, which probably gets more challenging as the number of sides in the regular polygon increases. Indeed, Freese's dissections in Plates 109 and 124 are not minimal, as you will see when you read about those plates in later chapters. And what about the succeeding problems, namely a regular 14-sided polygon to squares of area ratio 2 to 5, and a regular 16-sided polygon to squares of area ratio 1 to 3? Perhaps enterprising readers will take up that challenge.

Plate 92: Freese appears to have been the first to dissect a regular octagon to four identical squares. Yet Freese's dissection is less creative than it first appears. He based it on the symmetric 5-piece dissection of an octagon to a square (Plate 89). With the five pieces forming the square, he made two cuts that produced the four small squares. If he had shown the four small squares arranged to highlight that fact, his dissection would be obvious. Note the implicit use of the completing the tessellation technique.

Plate 93: Freese gave two different 8-piece dissections of two congruent regular octagons to a larger octagon. The first is an example of what he called the "rhombic method." As he pointed out on Plate 118, this method works whenever the number of sides of a regular polygon is a multiple of 4, and it produces a number of pieces equal to the number of sides. The method imposes a structure of rhombuses on the polygon, and cuts both along their boundaries and along the diagonals of some of the squares that are present.

The second dissection in this plate uses what Freese called the "analytical method." By this, he probably meant that analysis of the geometry produced the cuts of the appropriate lengths. C. Dudley Langford (1960) gave the same 8-piece dissection. This dissection is translational.

Plate 94: To obtain a dissection of three regular octagons to one, Freese started with a dissection of two octagons to one (from Plate 93). He then applied a standard ring expansion to add the third octagon, giving a total of 24 pieces. His dissection is translational, since he used the second

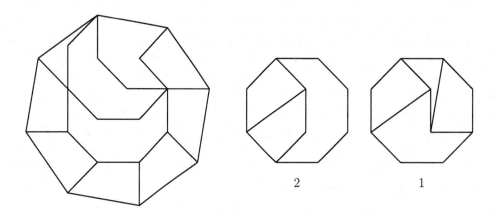

I94: Improved three octagons to one [GNF 1974]

dissection from the previous plate, which is translational, and used a ring in which pieces 1 and 2 share shortest sides in the large octagon, and piece 1 has a short side parallel to the short side of a different right triangle.

Freese could have done much better if he had realized that a length of $\sqrt{3}$ appears rather naturally in an octagon: It is the hypotenuse of a right triangle whose legs are 1 and $\sqrt{2}$. I exploited this fact in (1974) to design the 10-piece dissection in Figure I94. I doubt that Freese had this dissection in mind when he stated that there was a transformation that used fewer pieces by less elegant methods, because the dissection in Figure I94 is more elegant than Freese's.

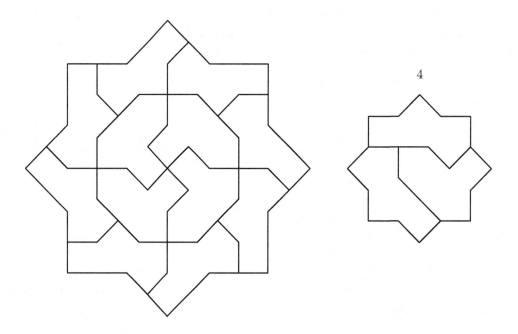

I95a: Translational dissection of four octagrams to one [GNF]

Plate 95: The dissection of four regular octagons to one is translational. It conforms to the pattern that if a regular polygon has a number n of sides that is even, then there is an n-piece dissection of four of the polygons to one. This dissection is Freese's fourth example of his 4-IN-1 dissections of regular polygons and stars. For polygons, there appears to be a simple rule. If the number n of sides is even then the number of pieces is n, whereas if n is odd then the number of pieces is $n + 1$, as we have already seen in Plate 51 for regular pentagons.

Such dissections are more challenging for stars. We have already seen the cases for $\{5/2\}$ and $\{6/2\}$ in Plates 60 and 86, respectively. Freese did not include a 4-IN-1 dissection for regular octagrams, but let's see what we can do with them. Using just twelve pieces, my translational dissection in Figure I95a is an unanticipated gem.

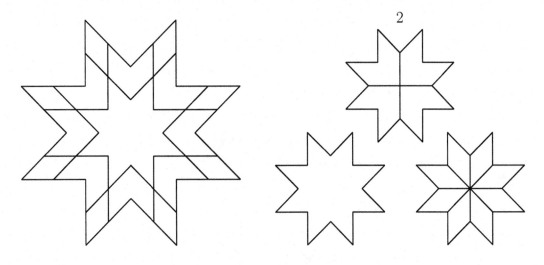

I95b: Translational dissection of four $\{8/3\}$s to one [GNF]

I95c: Translational dissection of four $\{7/2\}$s to one [GNF]

Of course, there is another 8-pointed star, called an $\{8/3\}$, for which every point is connected to the third point around from it. Freese called this star a "co-octagram" in Plate 171. We see a 17-piece 4-IN-1 dissection of this star in Figure I95b, which is a blueprint for a whole class of related figures: For any n-sided polygon, its co-polygram $\{n/((n-2)/2)\}$ will have a 4-IN-1 dissection of

$2n+1$ pieces if n is even. Similarly for any regular polygon of n sides when n is odd, its co-polygram will have a 4-IN-1 dissection of $2n+2$ pieces. Isn't it great to have a simple rule for 4-IN-1 dissections of co-polygrams analogous to the rule for polygons? This rule seems to give a minimal number of pieces for a 4-IN-1 dissection for co-polygrams whenever $n > 6$.

Of course, such rules do not apply to all cases that we might hope to handle. As a final example that will leave us a bit stranded, we consider a 4-IN-1 dissection of a heptagram. Figure I95c shows a 16-piece dissection for which we have not found a pattern.

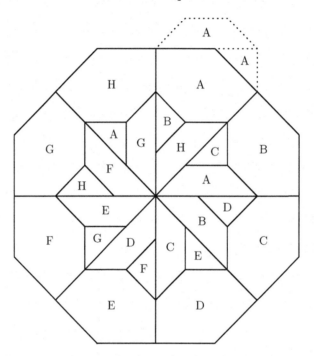

I95d: Translational dissection of eight octagons to one

Freese was the first to propose a dissection of eight congruent regular octagons to one, for which he gave a precious 24-piece dissection. His dissection is not translational, and yet a similar, rotationally symmetric dissection (Figure I95d) is translational. We see one small octagon with smaller pieces indicated by dotted lines at the top of the large octagon. We verify that we can slide those two small pieces into the corresponding positions in the large octagon with no rotation. It is easy to see where the small pieces for the remaining small octagons would lie on the outside of the figure, and again how to slide them without rotation into the designated positions in the interior of the large octagon.

Plate 96: To dissect five octagons to one, Freese used an approach similar to that in Plate 94 for three to one. He started with a dissection of four octagons to one that is similar to that in Plate 95. He then applied a standard ring expansion to add the fifth octagon, giving 24 pieces. Indeed, his dissection is translational for the same reason that the dissection in Plate 94 is translational.

Yet he would have done much better if he had realized that the distance from a corner of a unit square to the midpoint of an opposite side is $\sqrt{5}/2$. I exploited this fact in (1997) to design the 17-piece dissection in Figure I96a.

On the other hand, Freese employed a "ring expansion" that derives from Harry Hart's (1877) dissection of similar polygons. As I discussed with respect to pentagons in Plate 55, the ring

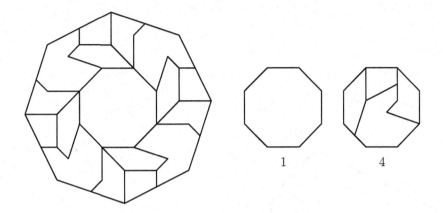

I96a: Improved five octagons to one [GNF]

expansion portion is hingeable. Yet the rest of Freese's dissection in Plate 96 is not hingeable. To get a fully hingeable dissection of five octagons to one, we need only one more piece, as in Figure I96b, borrowed from my 2002 book.

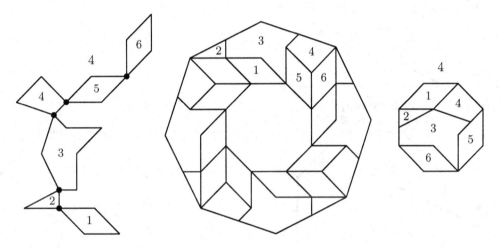

I96b: Hinged dissection of five octagons to one [GNF]

Plate 97: Freese was the first to dissect nine octagons to one. His 17-piece dissection would appear to be hard to beat. It is also translational, when suitably oriented.

Plate 98: Freese noted that he could dissect four octagons to two octagons to one octagon by combining two copies of his rhombic dissection in Plate 93 with his dissection of four octagons to one in Figure I95. His method uses no more than the total number of pieces for both of the two octagon-to-one dissections. This dissection is translational, although once again Freese did not point out this fact.

This chapter has ended in a flurry of dissections of the form: some number of congruent regular octagons to a larger regular octagon. This sets the stage for similar approaches in dissections of many congruent regular polygons to a larger similar polygon that we will see in the next two chapters, involving enneagons and decagons.

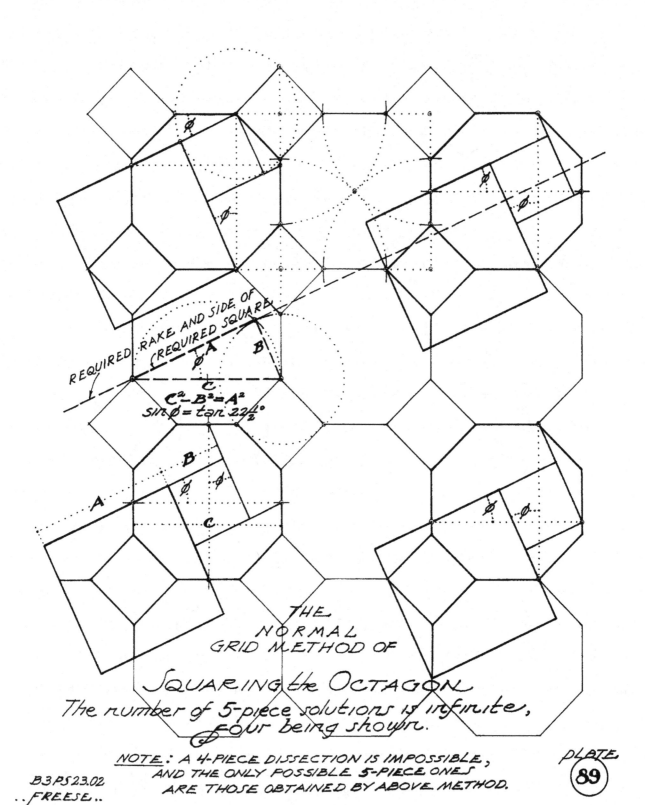

REQUIRED RAKE AND SIDE OF
(REQUIRED SQUARE)

ϕ A   B

C

$C^2 - B^2 = A^2$
$\sin \phi = \tan 22\frac{1}{2}°$

A   B
ϕ ϕ
C

THE
NORMAL
GRID METHOD OF

SQUARING the OCTAGON

The number of 5-piece solutions is infinite,
four being shown.

NOTE: A 4-PIECE DISSECTION IS IMPOSSIBLE,
AND THE ONLY POSSIBLE 5-PIECE ONES
ARE THOSE OBTAINED BY ABOVE METHOD.

B.3.P.S.23.02
..FREESE..

PLATE
89

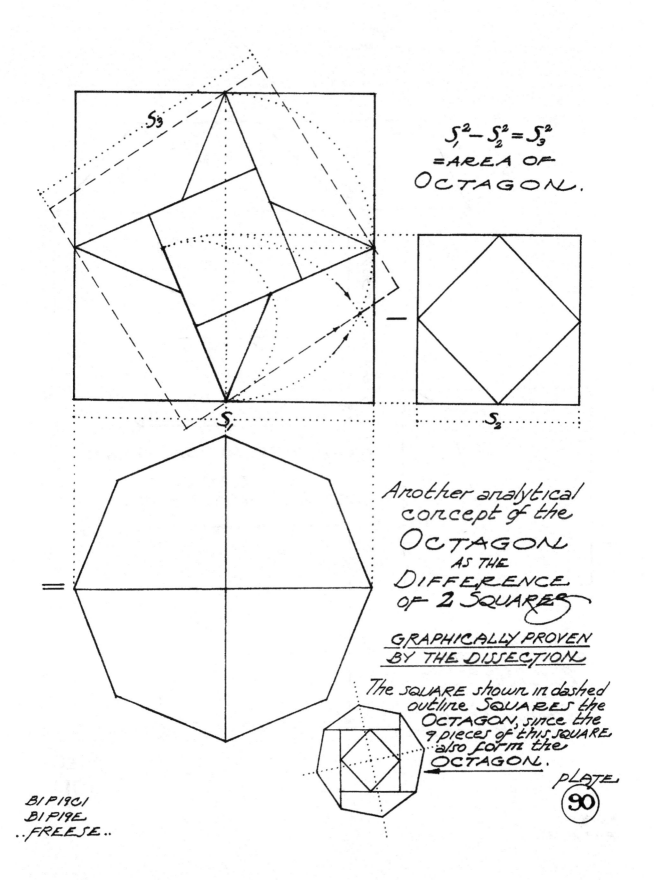

$$S_1^2 - S_2^2 = S_3^2$$
=AREA OF
OCTAGON.

S_3

S_1

S_2

Another analytical
concept of the
OCTAGON
AS THE
DIFFERENCE
OF 2 SQUARES

GRAPHICALLY PROVEN
BY THE DISSECTION

The SQUARE shown in dashed
outline SQUARES the
OCTAGON, since the
9 pieces of this SQUARE
also form the
OCTAGON.

PLATE
90

B1 P19C1
B1 P19E
..FREESE..

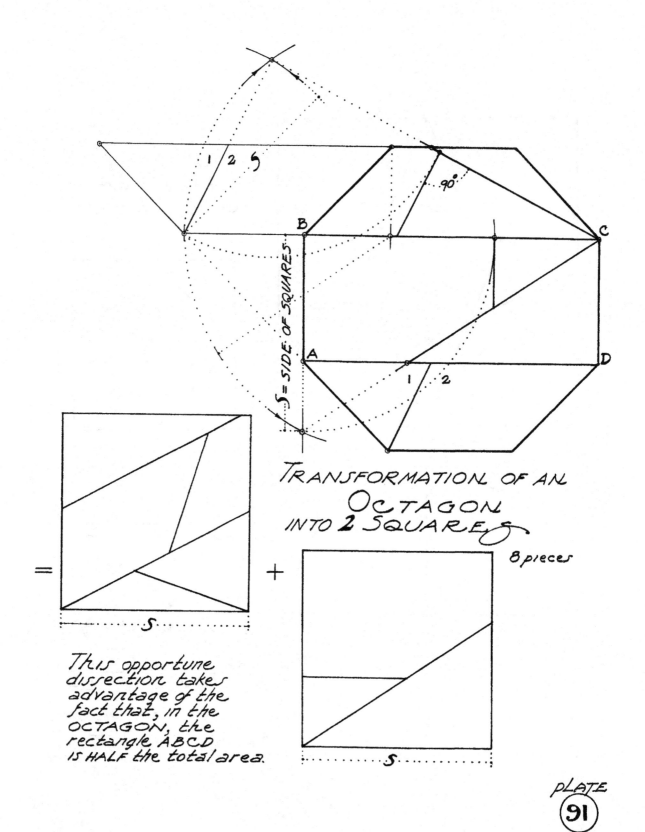

1 2 S

90°

B

C

S = SIDE OF SQUARES

A

D

1 2

TRANSFORMATION OF AN OCTAGON INTO 2 SQUARES

8 pieces

=

S

+

S

This opportune dissection takes advantage of the fact that, in the OCTAGON, the rectangle ABCD is HALF the total area.

PLATE
91

B2 PR7K
..FREESE..

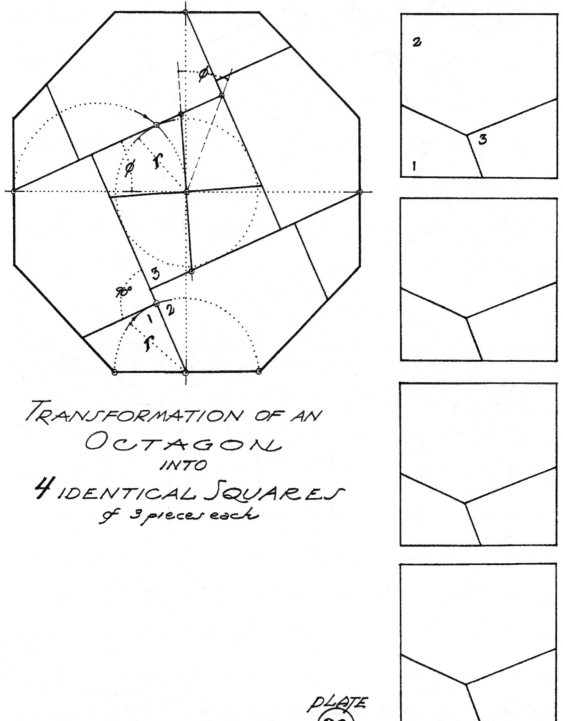

TRANSFORMATION OF AN
OCTAGON
INTO
4 IDENTICAL SQUARES
of 3 pieces each

PLATE
92

B3 PS 23.0
.,FREESE..

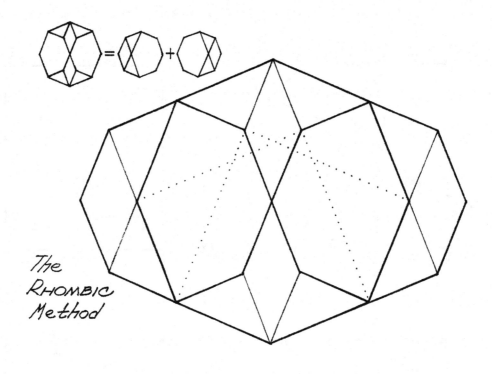

The
RHOMBIC
Method

The
ANALYTICAL
Method

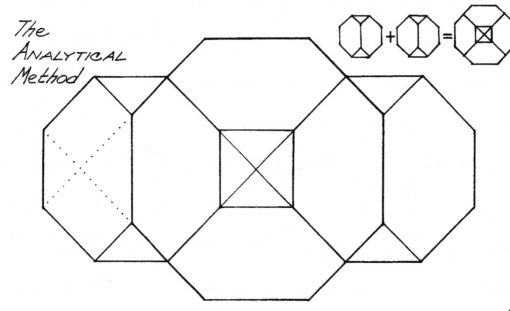

PLATE
93

TWO METHODS OF DISSECTION, EACH
YIELDING AN 8-PIECE 2-IN-1 OCTAGON

B2PR7C.S2
B2PR7C.S1
..FREESE..

A 3-in-1 Octagon
by Ring Expansion of a 2-Octagon Nucleus
24 pieces

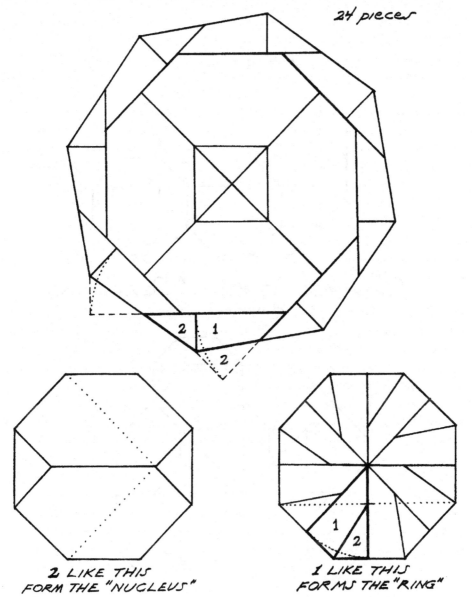

2 LIKE THIS FORM THE "NUCLEUS"

1 LIKE THIS FORMS THE "RING"

NOTE:
THIS TRANSFORMATION CAN
BE MADE WITH FEWER PIECES
BY MORE INVOLVED AND
LESS ELEGANT METHODS.

BH P8
..FREESE..

PLATE 94

A 4-in-1 Octagon and an 8-in-1 Octagon

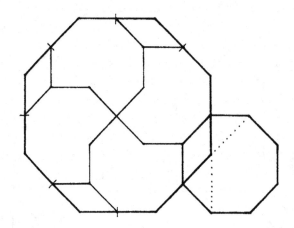

4 Octagons make 1 Octagon
8 pieces

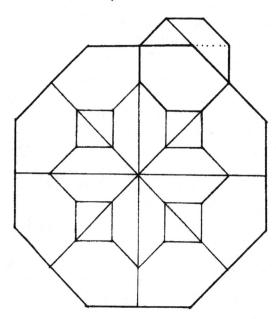

8 Octagons make 1 Octagon
24 pieces

B2.PR7D.S1
B2.PR7F.S1
..FREESE..

PLATE

95

A 5-in-1 Octagon
by Ring expansion of a 4-octagon nucleus

24 pieces

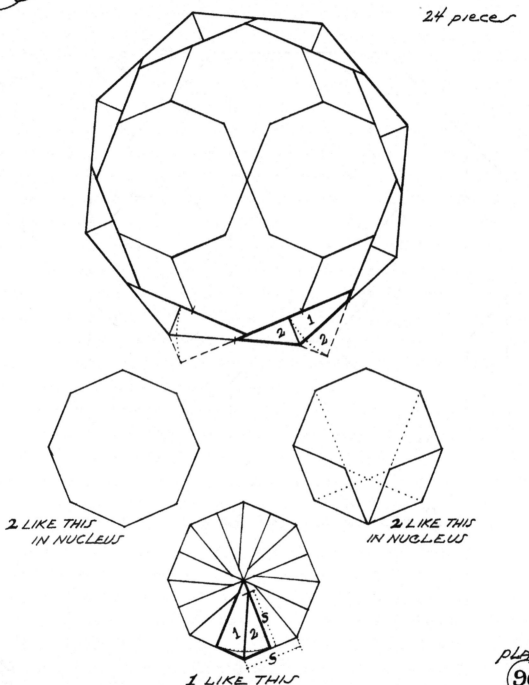

2 LIKE THIS
IN NUCLEUS

2 LIKE THIS
IN NUCLEUS

1 LIKE THIS
IN THE ADDED RING

B4P8A
..FREESE..

PLATE
96

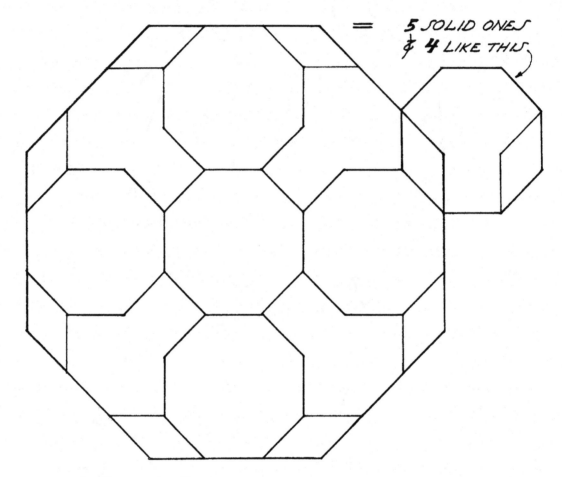

= 5 SOLID ONES
& 4 LIKE THIS

A 9-IN-1 OCTAGON... 17 pieces

PLATE
97

B2PR7G.S2
..FREESE..

1 OCTAGON MAKES _EITHER_
2 IDENTICAL OCTAGONS OR
4 IDENTICAL OCTAGONS

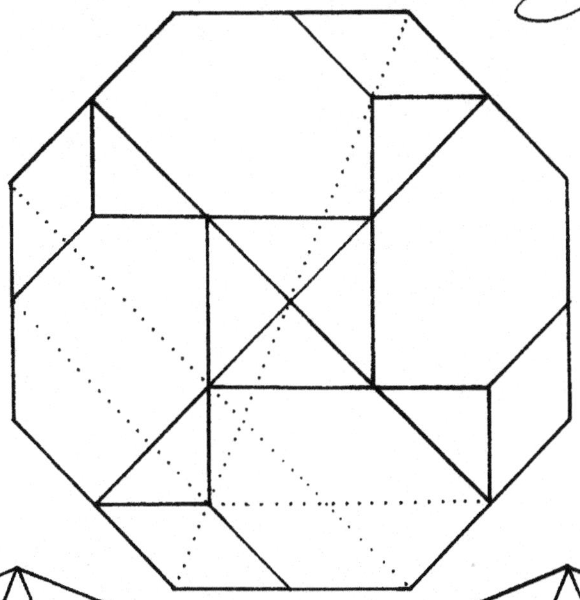

16 pieces

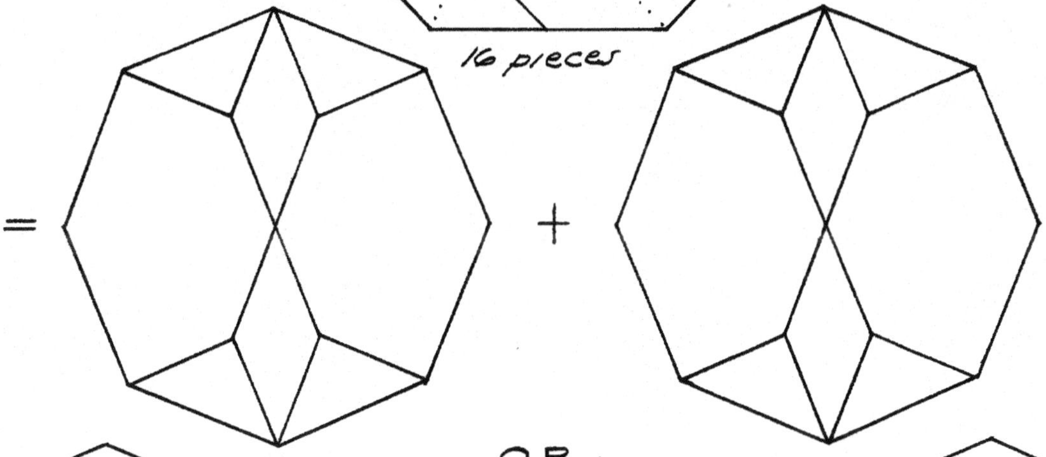

=   +

OR

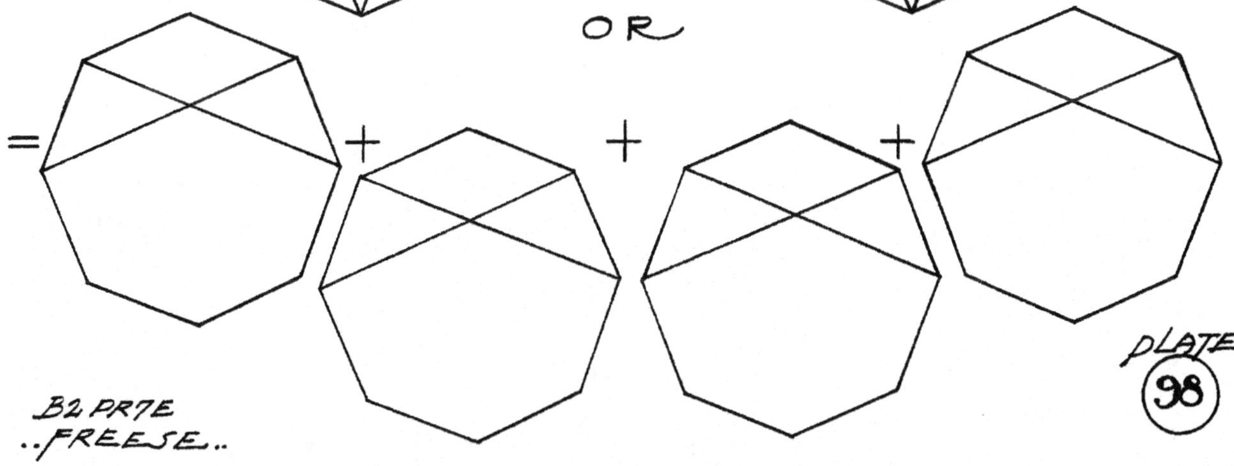

=   +   +   +

B2 PR7E
..FREESE..

PLATE
98

Chapter 11

Enneagons (Nonagons)　(Plates 99–103)

Judging by the paucity of enneagons in his manuscript, Freese apparently found these 9-sided polygons more challenging to dissect. In Plate 99 he started with a strip dissection of a regular enneagon to a square. It is not clear why he didn't include his clever dissection of an enneagon to an equilateral triangle in this chapter, instead delaying it until Plate 166. The remainder of this chapter consists of dissections (Plates 100-103) of n congruent regular enneagons to a larger enneagon, for n equal to 2, 3, 4, and 5. For $n = 2$, or 3, or 5, he relied on Hart's 1877 dissection or his own variation of it to go from $n - 1$ enneagons to n enneagons.

The case for $n = 4$ is easier, but Freese still needed to introduce a clever new trick. He chose crescent-moon-shaped pieces to fill in where rhombuses would not help.

Plate 99: Freese was the first to dissect specifically a regular enneagon to a square. He cut the enneagon into eight pieces which he rearranged to form a strip element. He then crossposed the strip with a strip of squares to get a 12-piece dissection. While Freese did find a correct dissection of the enneagon, he made a mistake in drawing how the pieces fit together in the square. The four triangles that Freese drew as meeting together at a point in the square do not actually meet at a point. This is because the altitude of the triangle with angles of 80°, 80°, and 20° is not equal to the altitude of the triangle with angles of 100°, 60°, and 20°. Several years later, Lindgren (1964) also found a 12-piece dissection via a strip dissection.

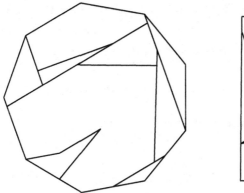 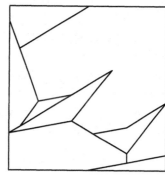

I99a: Improved enneagon to a square [Gavin Theobald]

Gavin Theobald found a 9-piece dissection which I reproduced in my 1997 book, but here I present Gavin's more recent and quite ingenious 9-piece dissection in Figure I99a. Gavin cut the enneagon into five pieces, upon which he based a T-strip element. The first two pieces are similar

113

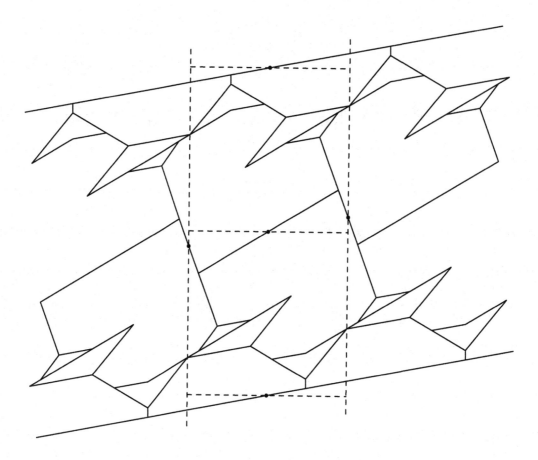

I99b: Crossposition for an improved enneagon to a square [Gavin Theobald]

isosceles triangles, one four times the area of the other. Gavin then cut the remaining piece into three pieces. He could have sliced each triangle into a pair of triangles, with each triangle in one pair similar to a triangle in the other pair. He could have assembled those six pieces into a T-strip element and crossposed the resulting strip with a strip of squares to give a 10-piece dissection. Instead, Gavin cut neither isosceles triangle, but instead cut a V-shaped piece out of the largest piece, and used the V-shaped piece to fill one hole of the strip element while he filled the remaining hole with the two isosceles triangles. A second clever trick of Gavin's is how he chose to cut the two largest pieces in the enneagon strip. His choice minimizes the number of pieces when he overlays the two strips. Crossposing with a strip of squares then gives the 9-piece dissection in Figure I99b. Note that the V-shaped piece is exactly what one would get if one had split the two isosceles triangles and merged the smaller piece from the smaller triangle with the larger piece from the larger triangle.

Plate 100: In this plate Freese applied his variation of the ring expansion method (Plate 50) to dissect two congruent enneagons to an enneagon, producing a 19-piece dissection.

In (2006b) I did better by turning over pieces. I based my 16-piece dissection on my approach in Figure I135 in which I cut isosceles right triangles off some pieces, merged them with other pieces, and matched together the mirror image of one piece with itself. Cutting an isosceles right triangle out of piece 1 in Plate 100 will not work, because the piece is too thin. However, if we merge two copies of piece 1 together, we can then cut an isosceles right triangle out of the result. I

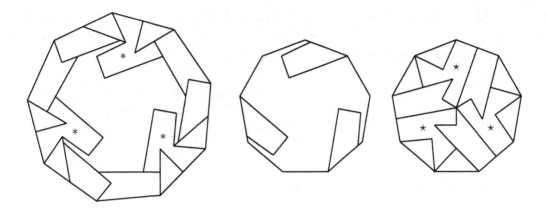

I100: Improved two enneagons to one, with turned-over pieces [GNF]

then merged an isosceles right triangle onto the merged piece, but at a different position, and cut a cavity out of the inner enneagon into which to fit the new result.

Plate 101: In this plate Freese applied his variation of the ring expansion method (Plate 50) to dissecting two enneagons to one, and then used Hart's dissection to make a second ring around the first one. This gives a 37-piece dissection. Note that if Freese had used Hart's dissection to also produce the first ring, then he would have had a fully hingeable dissection.

However, Freese missed something much better, apparently not realizing that he could break up a regular enneagon into three equilateral triangles, three 80°-rhombuses, and three 40°-rhombuses. C. Dudley Langford was the first to make use of this fact, producing the 21-piece dissection in ([Cundy and Langford (1960)]). In (1972a), I swapped around portions of triangles to give a 15-piece refinement of that dissection. Finally, Robert Reid and Anton Hanegraaf independently found 14-piece dissections (Figure I101).

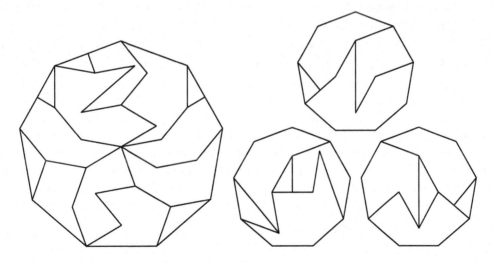

I101: Improved three enneagons to one [Anton Hanegraaf, Robert Reid]

Plate 102: I know of no previous dissection of four regular enneagons to one. With the small enneagons suitably oriented, Freese's dissection is translational. However, Lindgren (1964) reproduced a different 10-piece dissection (unpublished) that he attributed to C. Dudley Langford. Langford's dissection had a small enneagon in the center of the large one, and cut each of the remaining small enneagons identically.

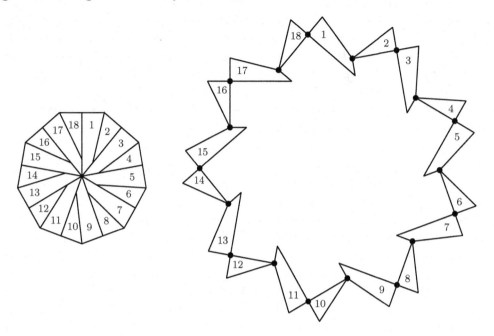

I103: Hinging one enneagon for five enneagons to one [GNF]

Plate 103: Freese was the first person to attempt a dissection of five identical enneagons to one. He converted four to one, using the dissection in Plate 102, and then used Hart's dissection to cut the remaining enneagon into pieces that form a ring that encloses the one large enneagon.

It is possible to cyclicly hinge the enneagon that Freese cut up into 18 pieces, in a fashion similar to what I describe for pentagons in my comments for Plate 55. I show this hinging in Figure I103. Furthermore, I have described a hinged dissection of four enneagons to one on page 158 of my 2002 book. That takes 16 pieces, which means that there is a 34-piece fully hinged dissection of five enneagons to one.

This concludes the thinnest chapter of Freese's manuscript, reflecting perhaps the situation that he was challenged to find interesting dissections of 9-sided objects. In part that was due to the enneagon having a geometry less cooperative than that of squares, pentagons, hexagons, octagons, decagons, and dodecagons. And in part that was due to Freese's greater reluctance to approach stars other than the pentagram, hexagram, octagram, decagram, and dodecagram. The former part of the challenge was admirably addressed by Gavin Theobald, and I addressed the 9-pointed stars in an article in 1972, as well as in my (1997) book.

At least Freese showed us a variety of ways to apply Hart's dissection and his own variation of it to dissect n congruent enneagons to one for integral values of n up to 5. In the next chapter Freese explores a wider range of dissections once again, made possible by the more cooperative shape of a regular decagon.

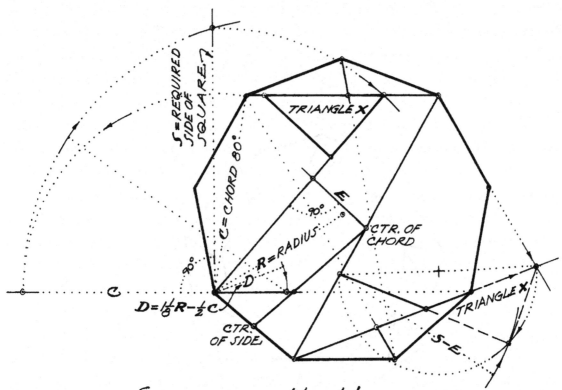

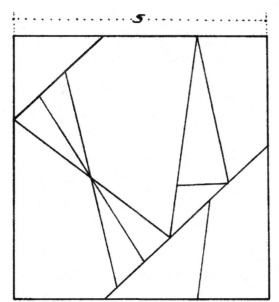

SQUARING the NONAGON

12 pieces

NOTE:
The central angle subtended by a side of the NONAGON is 40°, which can not be constructed by Euclidean geometry. However, the 9-point division of the circumcircle is a simple operation with a protractor.

THE 12 PIECES OF THE NONAGON, AS FOUND ABOVE, MAKE THIS SQUARE

BY PZ7
..FREESE..

PLATE
99

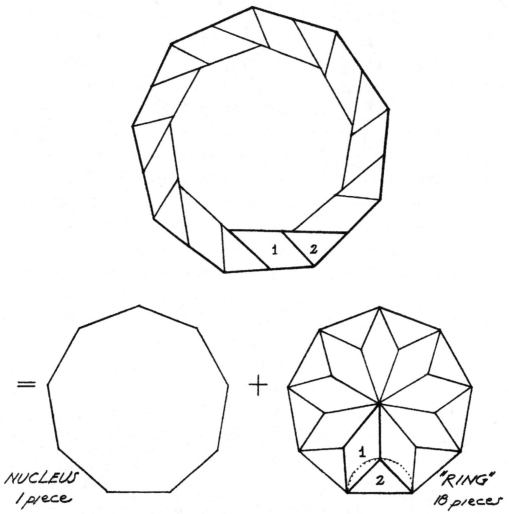

NUCLEUS
1 piece

"RING"
18 pieces

A 2-IN-1 NONAGON
by RING EXPANSION of a 1-NONAGON NUCLEUS
19 pieces

B4 PZ.57
..FREESE..

PLATE
100

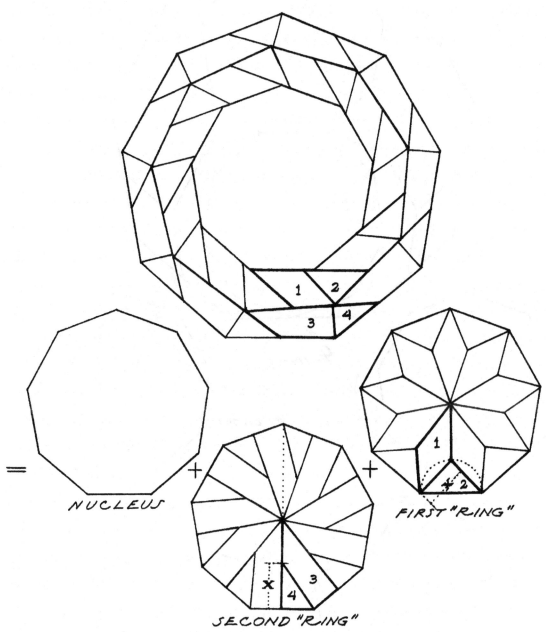

NUCLEUS

FIRST "RING"

SECOND "RING"

A 3-IN-1 NONAGON by
double RING EXPANSION of a 1-NONAGON NUCLEUS
37 pieces

B4 P57
..FREESE..

PLATE
101

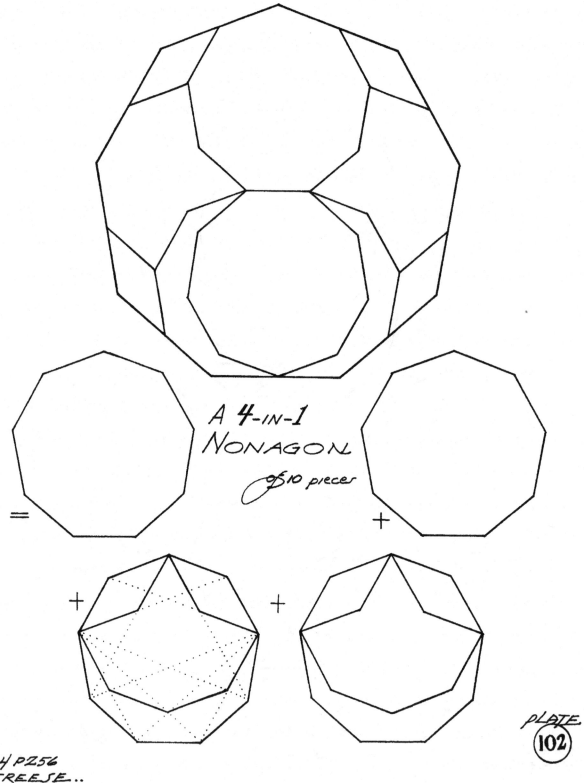

A *4*-ᴵᴺ-*1*
NONAGON
of 10 pieces

=

+

+

+

BY PZ56
..FREESE..

PLATE
102

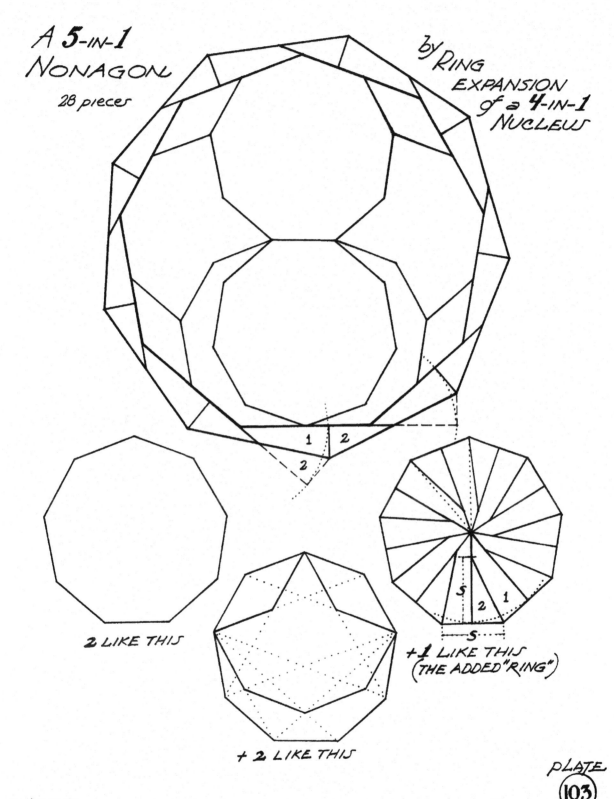

A 5-IN-1 NONAGON
28 pieces

by RING EXPANSION of a 4-IN-1 NUCLEUS

1 2

2

2 LIKE THIS

+ 2 LIKE THIS

S

2 1

S

+1 LIKE THIS
(THE ADDED "RING")

PLATE
103

B4 PZ56
..FREESE..

Chapter 12

Decagons and Decagrams (Plates 104–111)

After displaying relatively few dissections in the previous chapter, Freese picked up his pace again in this chapter. He found a useful partition of a regular decagon into a figure that would fill a strip, and thus found reasonably efficient dissections of a decagon to an equilateral triangle and also to a square. In addition to using ring expansion to produce dissections of a regular decagon to either two or three congruent decagons, he used structural properties of the regular decagon to create dissections to either four or five congruent decagons. He also dissected a regular decagon to two squares whose areas are in the ratio of 2:3 and to four squares whose areas are in the ratios of 1:2:3:4. The chapter closes with a dissection of a decagram to four congruent decagrams.

Be sure to enjoy two outstanding dissections in this group. First we have Freese's inventive dissection of a decagon to a square in Plate 105. Second, Freese gave us the first dissection of four congruent decagrams to a larger one, in Plate 111.

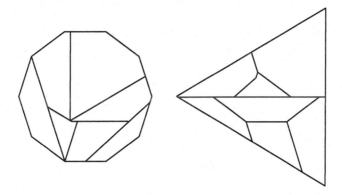

I104a: Improved decagon to a triangle [Gavin Theobald]

Plate 104: Freese was the first to dissect a regular decagon to an equilateral triangle. He sliced the top quarter off of the triangle to form one strip, then cut the decagon into four pieces that form a second strip element. He crossposed these two strips to produce his 10-piece dissection.

Freese's dissection of a decagon to a square inspired Lindgren (1964) to find additional strip elements for the decagon. One of them, in combination with the T-strip for the triangle, led him to an 8-piece dissection. Years later, Gavin Theobald topped them both with the 7-piece dissection in Figure I104a. Gavin started with yet another decagon strip, a T-strip, that gave him an 8-piece dissection. Then he "customized" this strip (See my 1997 book) by looking to see how it would cross the triangle strip, as in Figure I104b.

Plate 105: Freese was the first to dissect a regular decagon to a square. He crossposed the

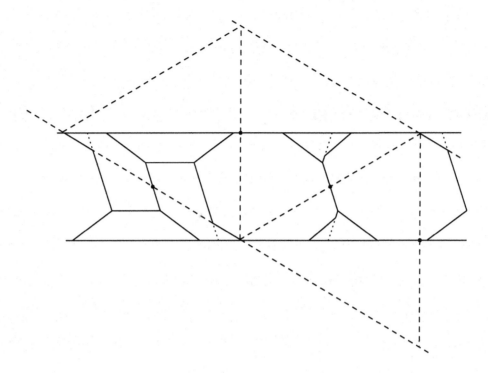

I104b: Crossposition of strips for decagon and triangle [Gavin Theobald]

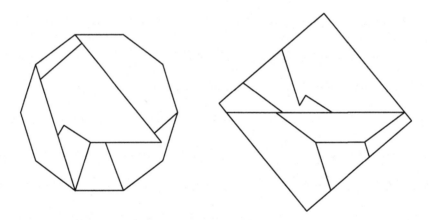

I105a: Improved decagon to square [Gavin Theobald]

decagon strip that he had used in Plate 104 with a strip of squares to get his 8-piece dissection.

Then Theobald found an improvement by creating yet another customized decagon strip. This time Theobald created a crossposition with P-strips and adapted it as in Figure I105b. This gives the 7-piece dissection in Figure I105a.

Plate 106: To produce a 21-piece dissection of two regular decagons to one, and similar to the dissections of pentagons in Plate 55, Freese applied Hart's technique. Freese could have identified this dissection as hingeable. It is also translational.

However, Gavin Theobald designed the inventive 16-piece dissection in Figure I106a, which uses the crossposition of two strips that we see in Figure I106b. To find the strip for the large decagon,

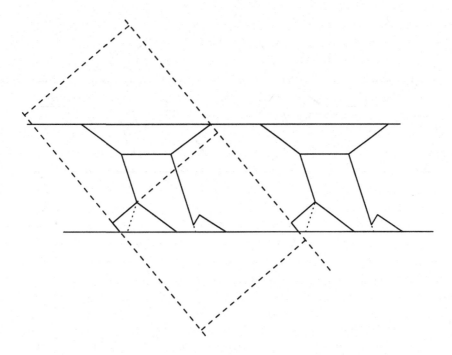

I105b: Crossposition of strips for decagon and square [Gavin Theobald]

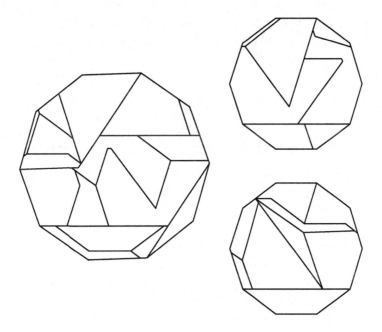

I106a: Improved two decagons to one [Gavin Theobald]

Gavin cut the decagon into just three pieces. To get the strip for the pair of small decagons Gavin incorporated Harry Lindgren's decagon strip from (1964), but then customized it to save an additional piece or two. The customization forces the latter strip to share a line segment with the strip for the large decagon. The shared line segment appears as a "thickened" line segment in the crossposition.

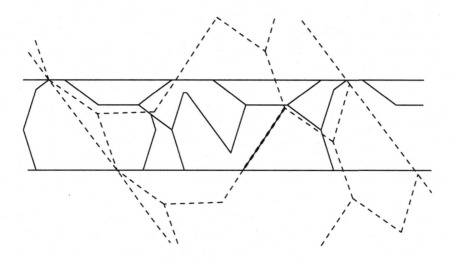

I106b: Crossposition of strips for two decagons to one

Plate 107: To produce this 41-piece dissection of three decagons to one, Freese applied Hart's technique twice, using the dissection of two decagons to one in Plate 106. Similar to my comments on the dissection of two decagons to one in Plate 106, I note that the dissection here is also hingeable and translational. However, I found a 24-piece dissection by crossposing strips, with the strip for the large decagon being related to the strip in Figure I105b and the strip for three decagons being a variation of a strip from Lindgren (1964). Yet Gavin Theobald found an improvement of one piece! He created yet another customized strip and used it in his crossposition (Figure I107a). Gavin's nifty 23-piece dissection is in Figure I107b.

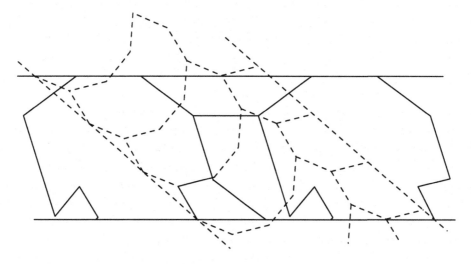

I107a: Crossposition of strips for three decagons to one [Gavin Theobald]

Plate 108: The top dissection, of four decagons to one, conforms to the pattern that if a regular polygon has a number of sides n that is even, then there is an n-piece dissection of four of them to one. It is also translational.

The bottom dissection, of five decagons to one, is a sumptuous beauty that uses 25 pieces. It is

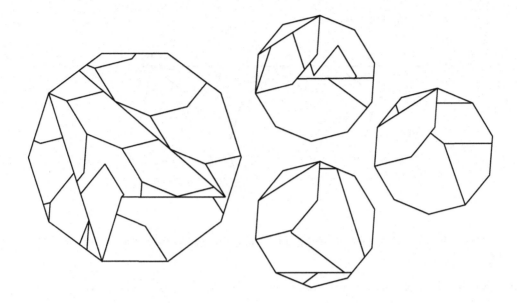

I107b: Improved three decagons to one [Gavin Theobald]

translational, too. Langford (1960) gave precisely the same dissection.

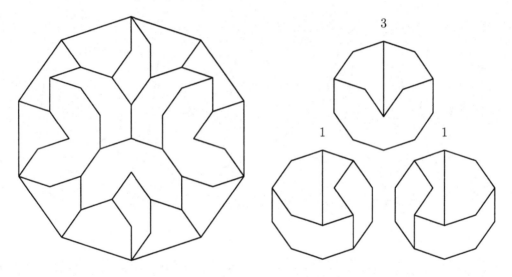

I108: Improved five decagons to one [GNF]

However, we can do considerably better by noting that the diameter of the decagon is $(1 + \sqrt{5})$ times its sidelength. This led in (1974) to my 17-piece dissection in Figure I108.

Plate 109: This dissection problem is the third in an infinite sequence of similar problems, as explained in my discussion of Plate 91. Freese noticed that the rectangle that spans between opposite sides of a decagon contains 2/5 of the decagon's area. This led him to dissect the decagon to the two squares, of areas 2/5 of the decagon and 3/5 of the decagon. Although he could dissect the rectangle in three pieces, using a P-slide, the remaining task is trickier. Freese produced a 6-piece strip element, which he crossposed with a strip of squares to give ten pieces. For the two squares together, he got thirteen pieces.

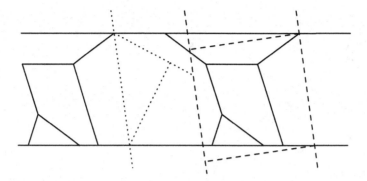

I109a: Crosspositions for improved decagon to two squares [GNF]

We can do considerably better if we create one strip element (solid edges) for the whole decagon. If we use Theobald's strip element for the decagon, we can crosspose a strip of the larger squares (dashed edges), taking six pieces. (See Figure I109a.) This leaves us with a parallelogram which we can convert (dotted edges) to a smaller square, using four pieces. We see the resulting 10-piece dissection in Figure I109b.

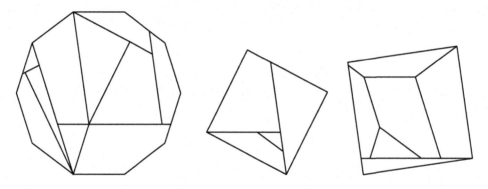

I109b: Improved decagon to two squares [GNF]

Plate 110: Continuing with his observation from Plate 109, Freese noticed that the area beneath the central rectangle of the decagon is 3/10 of the total area, and similarly for the area above the central rectangle. This led him to dissect a decagon to four squares, of area 1/10, 2/10, 3/10, and 4/10 of the area of the decagon. He performed these dissections in a piecemeal fashion, isolating areas of the desired size and then converting each to the corresponding square, and using a total of 24 pieces. By crossposing strips, I reduced the number of pieces by five.

Yet Gavin Theobald found a surprising 17-piece dissection by crossposing two strips. He took a fat strip consisting of decagon elements (solid edges) and crossed it with a thin strip consisting of the four squares (dashed edges), as we see in Figure I110a. Gavin chose the width of the thin strip to be the side length of the largest square and then cut the three smaller squares into just seven pieces that assemble into a rectangle whose width is the width of the thin strip. In packing the thin strip, Gavin cut the second largest square into just three pieces, the third largest into just two pieces, and the smallest square into just two pieces. The last cut accommodated beautifully the irregular profile

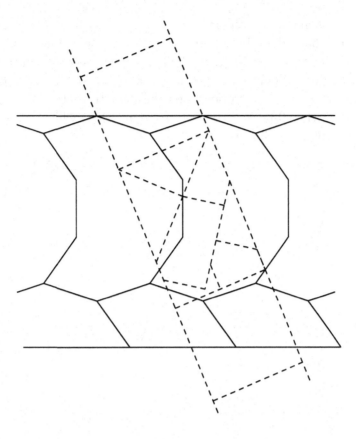

I110a: Crossposition for improved decagon to four squares [Gavin Theobald]

of the cut-up second smallest square. This approach is doubly clever, in that Gavin minimized the number of additional pieces produced when he went ahead and actually crossposed the two strips in Figure I110a. Really lovely! Admire the resulting 17-piece dissection in Figure I110b.

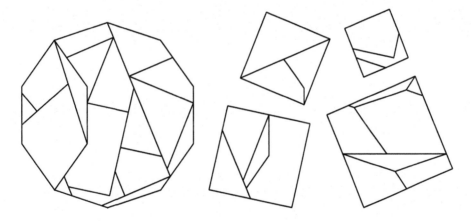

I110b: Improved decagon to four squares [Gavin Theobald]

Plate 111: As Freese noted on Plate 69, we can always find a reasonable dissection of four congruent figures to a larger one of the same shape. In Plate 111 he gave a 24-piece dissection of four decagrams to one, relying on the technique of cutting a portion out of one small decagram so that it would fit up against another in the large decagram. His dissection is translational.

However, Freese could have done better by extending his technique a bit. Instead of fitting together four half-rhombuses to make a larger half-rhombus, he could have combined two half-rhombuses together to make a figure with two concave angles, and he could then have cut out whole rhombuses rather than half-rhombuses to produce cavities into which to fit the pieces with concave angles. This gives the 20-piece dissection in Figure I111, which is also translational, with half of the corresponding pieces so labeled.

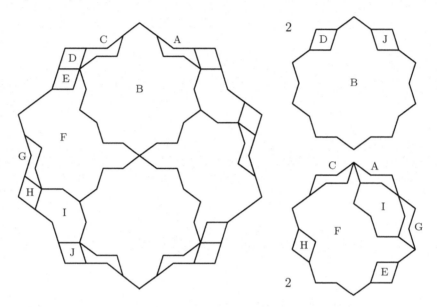

I111: Improved (and still translational) four decagrams to one [GNF]

This chapter for 10-sided figures has concluded on a high note, with some astonishing dissections. Yet don't let down your guard, lest you be blown away by Freese's chapter on dodecagons and dodecagrams.

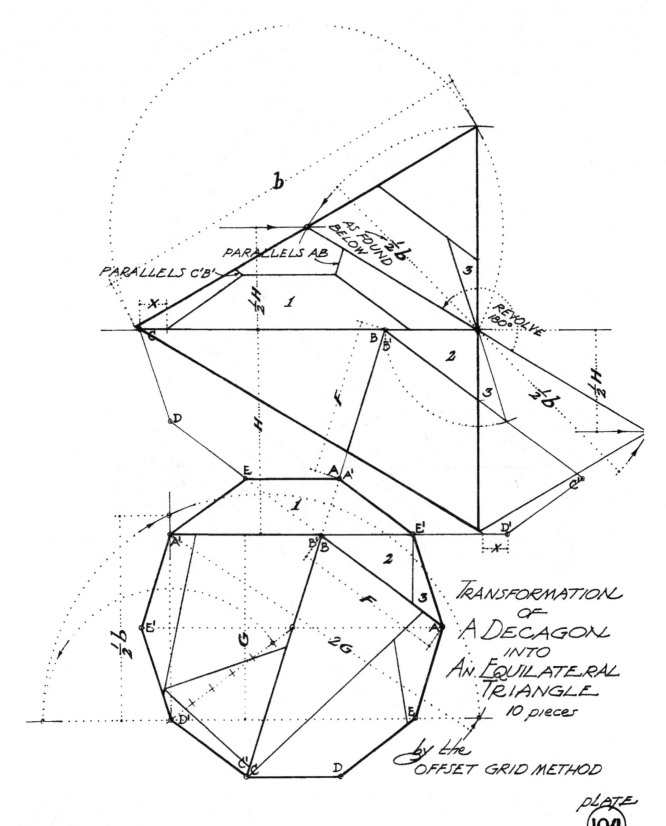

b

PARALLELS C'B'

PARALLELS AB

AS FOUND BELOW

½b

REVOLVE 180°

½H

½H

3

3

3

½b

X

C

1

½H

H

A A'

E

B B'

F

2

D

D

E'

D'

X

TRANSFORMATION
OF
A DECAGON
INTO
AN EQUILATERAL
TRIANGLE
10 pieces

by the
OFFSET GRID METHOD

½b

A'

E'

B'B

2

F

3

G

2G

A

D'

C'C

E

D

PLATE
104

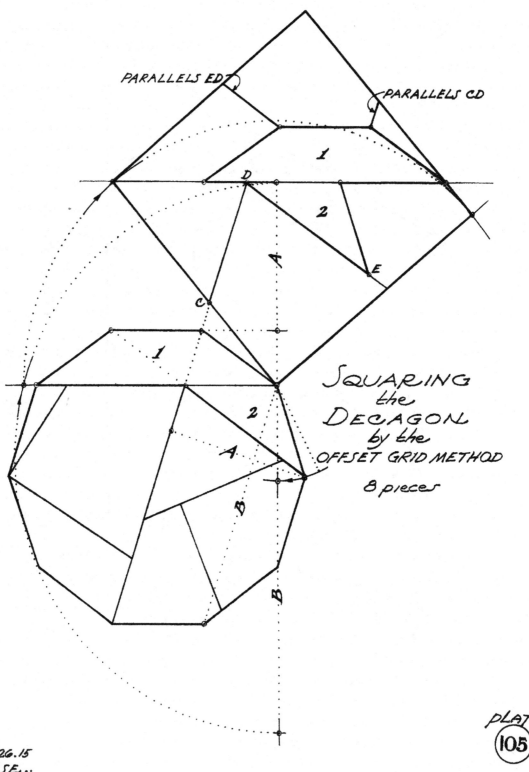

PARALLELS EDT

PARALLELS CD

1

D

2

A

E

C

1

2

A

B

B

SQUARING
the
DECAGON
by the
OFFSET GRID METHOD

8 pieces

B3 PS 26.15
..FREESE..

PLATE
105

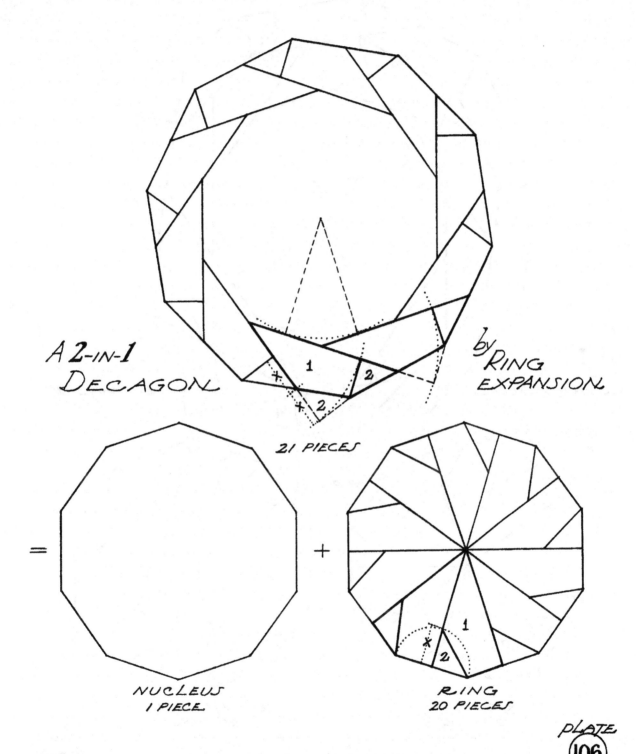

A *2*-IN-*1*
DECAGON

by RING
EXPANSION

21 PIECES

= NUCLEUS
1 PIECE

+ RING
20 PIECES

B3 PX-5
..FREESE..

PLATE
106

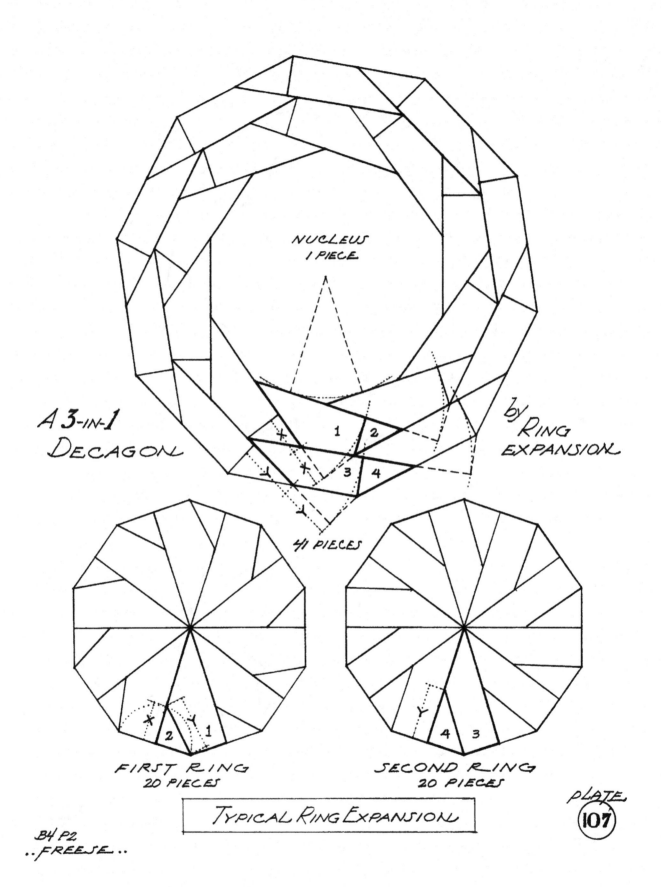

NUCLEUS
1 PIECE

A 3-IN-1
DECAGON

by RING
EXPANSION

1 2

3 4

41 PIECES

FIRST RING
20 PIECES

SECOND RING
20 PIECES

2 1

4 3

TYPICAL RING EXPANSION

B4 P2
..FREESE..

PLATE
107

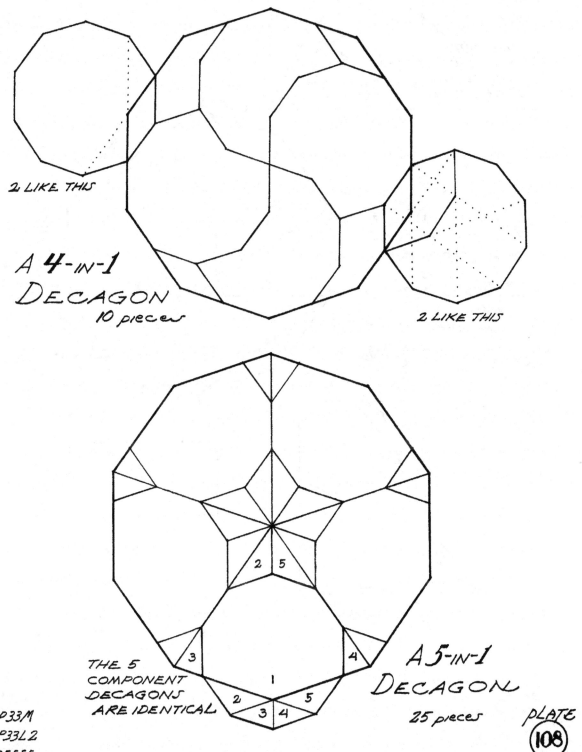

2 LIKE THIS

A **4**-ɪɴ-1
DECAGON
10 pieces

2 LIKE THIS

A **5**-ɪɴ-1
DECAGON
25 pieces

THE 5
COMPONENT
DECAGONS
ARE IDENTICAL

BIP33M
BIP33L2
..FREESE..

PLATE
⬤108

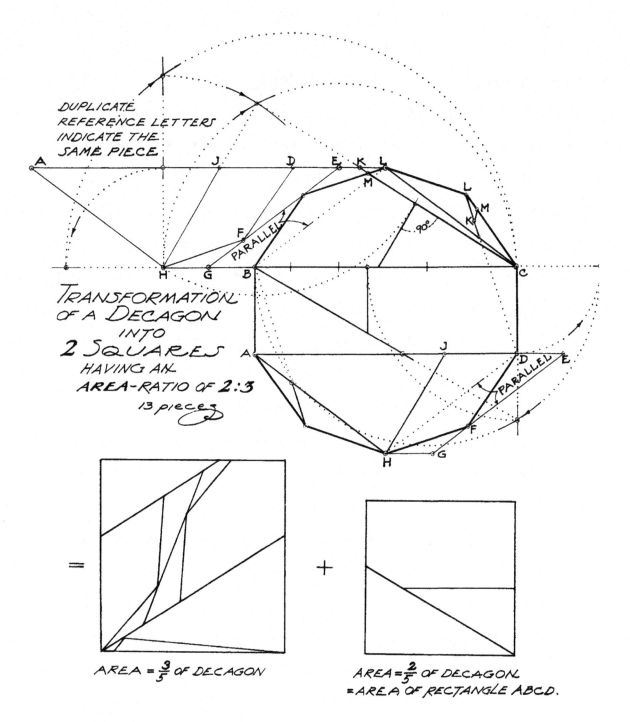

DUPLICATE
REFERENCE LETTERS
INDICATE THE
SAME PIECE

PARALLEL

90°

PARALLEL

TRANSFORMATION
OF A DECAGON
INTO
2 SQUARES
HAVING AN
AREA-RATIO OF 2:3

13 pieces

=

AREA = $\frac{3}{5}$ OF DECAGON

+

AREA = $\frac{2}{5}$ OF DECAGON
= AREA OF RECTANGLE ABCD.

PLATE
109

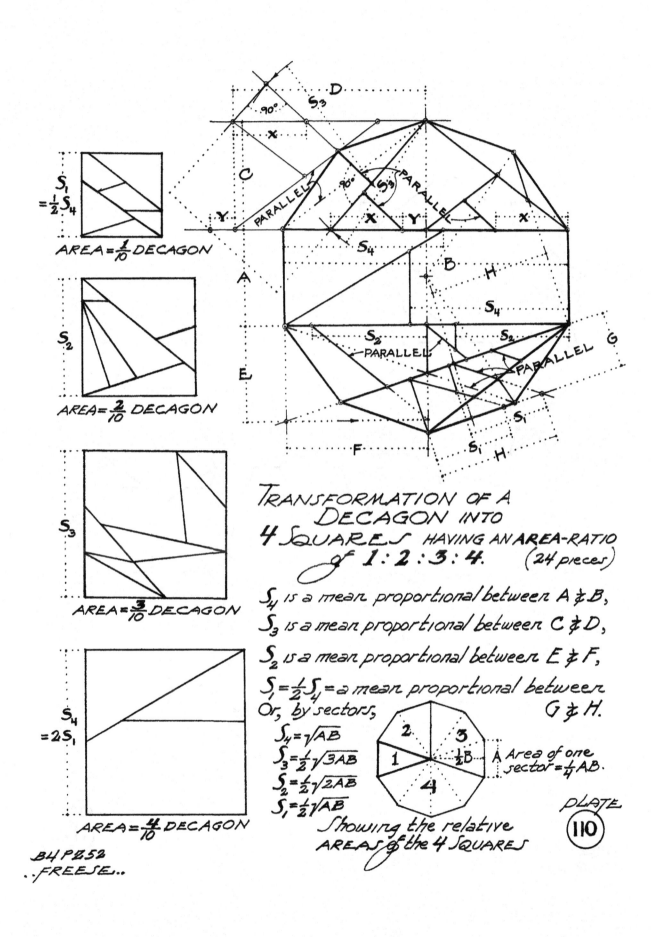

S_1
$= \frac{1}{2}S_4$

AREA $= \frac{1}{10}$ DECAGON

S_2

AREA $= \frac{2}{10}$ DECAGON

S_3

AREA $= \frac{3}{10}$ DECAGON

S_4
$= 2S_1$

AREA $= \frac{4}{10}$ DECAGON

D
S_3
90°
x
C
PARALLEL
90°
S_3
PARALLEL
Y
X
Y
X
S_4
A
B
H
S_4
E
S_2
S_2
G
PARALLEL
PARALLEL
S_1
F
S_1
H

TRANSFORMATION OF A
DECAGON INTO
4 SQUARES HAVING AN AREA-RATIO
of 1 : 2 : 3 : 4.   (24 pieces)

S_4 is a mean proportional between A & B,
S_3 is a mean proportional between C & D,
S_2 is a mean proportional between E & F,
$S_1 = \frac{1}{2}S_4 =$ a mean proportional between
Or, by sectors,   G & H.

$S_4 = \sqrt{AB}$
$S_3 = \frac{1}{2}\sqrt{3AB}$
$S_2 = \frac{1}{2}\sqrt{2AB}$
$S_1 = \frac{1}{2}\sqrt{AB}$

2   3
1    $\frac{1}{2}$B   A Area of one
                     sector $= \frac{1}{4}AB$.
    4

Showing the relative
AREAS of the 4 SQUARES

PLATE
110

B4 PL52
. . FREESE . .

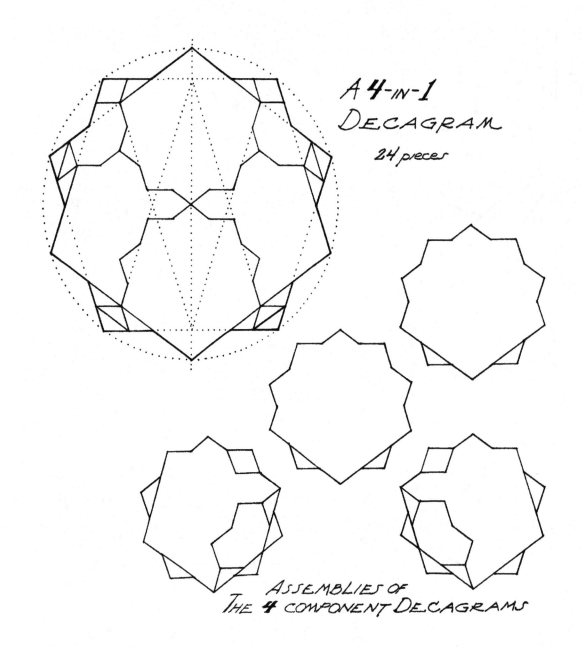

A **4**-ɪɴ-**1**
Dᴇᴄᴀɢʀᴀᴍ
24 pieces

Assemblies of
The **4** component Decagrams

B4 PZ32
..FREESE..

Chapter 13

Dodecagons and Dodecagrams   (Plates 112–133)

Rebounding from the relatively sparse chapters for octagons, enneagons, and decagons, Freese moved energetically on regular dodecagons. In Plates 112-116 he explored dissections of a dodecagon to equilateral triangles, to squares, and to a regular hexagon. In Plates 117-122 he pursued dissections of a dodecagon to different numbers of dodecagons. In Plates 123-131 he explored dissections of some number of congruent dodecagons to some number of congruent squares. In Plate 132 he dissected a dodecagon to a Greek Cross and also to a Latin Cross. Finally, in Plate 133, he converted a dodecagram to a regular hexagon.

While many of these dissections are really good, there are two that are fantastic. One is the dissection of two congruent equilateral triangles to a regular dodecagon, in Plate 113, and the other is the dissection of a regular hexagon to a regular dodecagon, in Plate 116. Both use the technique of completing the tessellation, and it's a tough choice to decide which one is more awesome. Go look at them right now, and see what you think!

Plate 112: There was no earlier published dissection of a regular dodecagon to an equilateral triangle. Lindgren (1964) also found an 8-piece dissection.

Plate 113: Freese was the first to dissect a regular dodecagon to two identical equilateral triangles. His 10-piece beauty uses the technique of completing the tessellation and is closely related to his dissection of a dodecagon to a hexagon in Plate 116.

Plate 114: The dissection at the top of this plate is a visual proof that the area of a dodecagon equals the area of three squares whose sidelengths equal the radius of the dodecagon. Such a dissection appeared as an illustration that the eighteenth-century scholar Tai Chen prepared for the palace edition of the *Chiu Chang Suan Shu* (Nine Chapters on the Mathematical Art). When suitably arranged, it is translational.

The lower dissection results from superposing tessellations, where the tessellation element of the dodecagon was published by Lindgren (1951). Lindgren recognized that there are many 6-piece dissections of the dodecagon to the square. Freese's assertion that 5-piece dissections are impossible has not yet been supported by a rigorous proof.

Plate 115: Freese was the first to dissect a 2-square to a 1-square and a regular dodecagon. He chose a 12-piece dissection that also allowed him to illustrate a related 12-piece dissection of a dodecagon to a square. Neither of these dissections is minimal, but they do follow the same general form as those in Plate 90 that involve the octagon.

Freese could have easily saved three pieces in the dissection of the 2-square to a 1-square and a dodecagon by rearranging the four constituent squares of the 2-square so that the four pieces that

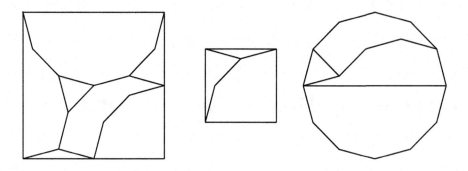

I115a: Improved 2-square to 1-square and a dodecagon [GNF]

form the dodecagon need not be cut apart. He could also have changed the way he cut the interior of the small square, saving one piece, and thus obtaining an 8-piece dissection.

However, further improvement is possible, as we see in the 7-piece dissection in Figure I115a. The key approach seems to be to cut from the dodecagon a small triangle plus a "tail" consisting of a square, a 60°-rhombus, and a 30°-rhombus. Indeed, Freese used this approach on the preceding dissection, in Plate 114.

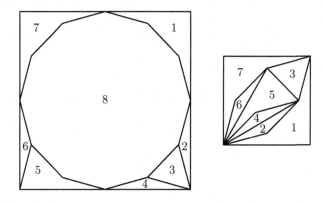

I115b: Hingeable dissection of a 2-square to 1-square and a dodecagon [GNF]

If we want a hingeable dissection of a 2-square to a 1-square and a dodecagon, then we needs at most one more piece, as we see in Figure I115b. We avoid cutting the dodecagon, and just dissect the 1-square into seven pieces that we hinge as in Figure I115c so that they will wrap around the dodecagon to form the 2-square.

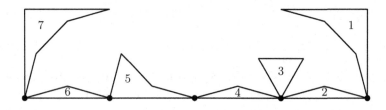

I115c: Hinged pieces for the dissection of a 2-square to 1-square and a dodecagon [GNF]

Plate 116: The 6-piece dissection of a regular dodecagon to a regular hexagon is one of the loveliest dissections. Freese was the first to discover it, extending the completing the tessellation technique that Harry Lindgren (1951) used to derive the 5-piece octagon to square. (See Plate 89).

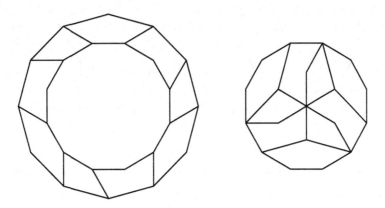

I117: Two dodecagons to one [Harry Lindgren]

Plate 117: Freese's 11-piece dissection of two regular dodecagons to one relies in spirit on the 13-piece dissection by Joseph Rosenbaum (1947). Freese left one dodecagon uncut and arranged the pieces from the other around it. However, Rosenbaum turned over six of the pieces, to which Freese would have objected. Furthermore, Freese identified two pieces from the smaller dodecagon which supply two edges of the larger dodecagon. In labeling the number of pieces as minimal, Freese probably meant that the number of pieces is smaller than for the dissection in Plate 118. Unfortunately for Freese, Lindgren (1964) found a way to cut the smaller dodecagon so that three of its pieces each supply two sides of the larger dodecagon. Thus Lindgren found the 10-piece dissection in Figure I117.

Plate 118: As C. Dudley Langford (1967) observed, Freese was the first to note that any regular polygon with a number of sides that is a multiple of four will have a dissection of two of them to one, such that the number of pieces equals the number of sides of the polygon. Freese relied on the natural division of such regular polygons into rhombuses, some of which are squares, to achieve this result. He had already given an 8-piece dissection of octagons in Plate 93, similar to his 12-piece dissection of dodecagons in this plate. Freese delighted in the further application of this principle in Plates 137, 139, and 141.

Plate 119: Freese appears to have been the first to specifically dissect regular dodecagons of area ratio 1 : 2 to a larger regular dodecagon. He created a symmetric ring from the dodecagon of area 2 to surround the smaller dodecagon, using 25 pieces altogether. When suitably arranged, the dissection is translational.

Evans Valens (1964) found a less symmetrical yet more efficient way to create a ring around the smaller dodecagon, using just 12 pieces altogether (Figure I119a).

A variation of Freese's dissection, as shown in Figure I119b, is hingeable. We see the hinged pieces in Figure I119c, where the dashed line indicates that a corner of piece 12 and a corner of piece 13 should actually be connected by a hinge. A bit surprising is that the idea behind Freese's dissection works for any regular n-sided polygon. We get a dissection of the n-sided polygon to a

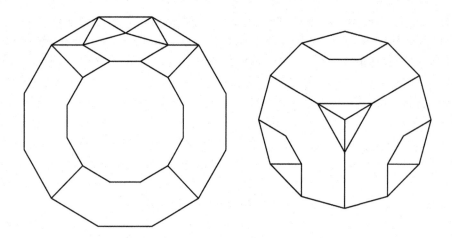

I119a: Improved dodecagons for $1^2 + (\sqrt{2})^2 = (\sqrt{3})^2$ [Evans Valens]

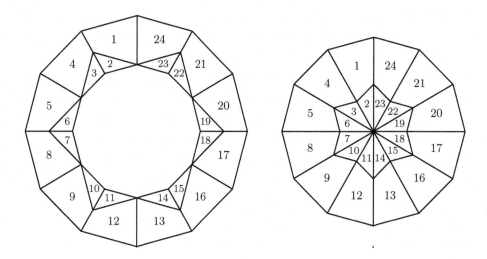

I119b: Hingeable dissection of dodecagons for $1^2 + (\sqrt{2})^2 = (\sqrt{3})^2$ [variation]

pair of n-sided polygons having an area ratio of $1 : (4\sin^2(360°/n)/\sin^2(90-540°/n))$. When $n = 12$ the ratio evaluates to $1 : 2$, and when $n = 8$ the ratio evaluates to $1 : 16\cos^2(22.5°) \approx 1 : 13.6569$. Moreover, we can make a hingeable dissection for any such n-sided regular polygon. The dissection will have $2n + 1$ pieces.

Plate 120: When suitably arranged, Freese's dissection of three congruent dodecagons to one is translational. We can easily improve the dissection from 24 pieces to 21 pieces, or even to 18. To reduce it to 21, just rotate each square in the large dodecagon by 90° and merge two adjacent pieces. To reduce the dissection to 18 pieces, cut an equilateral triangle out next to each square, and fit the triangle plus two of the 30°-rhombuses (uncut) into the resulting cavity. Lindgren (1964) had a further improvement, to 15 pieces, and Anton Hanegraaf and Robert Reid each independently discovered 14-piece dissections, such as the one in Figure I120.

Plate 121: Freese was the first to dissect four dodecagons to one, for which he discovered a 12-piece dissection. When suitably arranged, his dissection is translational. H. Martyn Cundy and

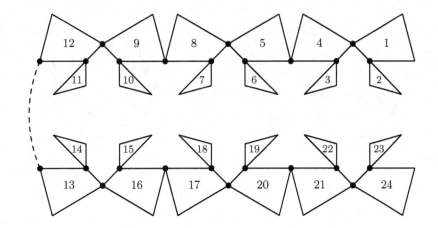

I119c: Hinged pieces for the variation of dodecagons for $1^2 + (\sqrt{2})^2 = (\sqrt{3})^2$

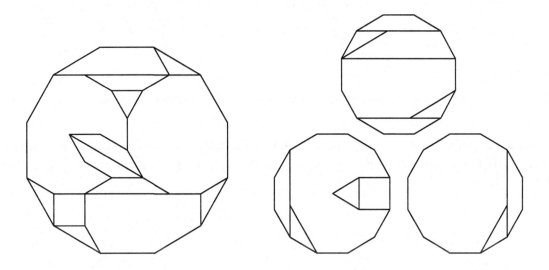

I120: Improved three dodecagons to one [Robert Reid, Anton Hanegraaf (variations)]

C. Dudley Langford (1960) mentioned three different ways to accomplish the same feat.

Plate 122: Freese recognized that since there is a convenient dissection of three dodecagons to one, there should also be a convenient dissection of $4 * 3 = 12$ dodecagons to one. His 78-piece dissection is wonderfully symmetrical, and even translational, but uses many more pieces than necessary. For example, we could modify the wedge-shaped pieces in the six small dodecagons on the lower right so that instead of two adjacent right triangles we would have one equilateral triangle. With additional incremental improvements, we can push the number of pieces down to 48.

Yet, one can do better. In unpublished work from the 1990's, Robert Reid found a remarkable 47-piece dissection, a variation of which I display in Figure I122.

Plate 123: Freese seems to have been the first person to dissect a regular dodecagon to two congruent squares. The 7-piece dissection results from superposing two tessellations.

Freese also seems to have been the first person to dissect two congruent regular dodecagons to a

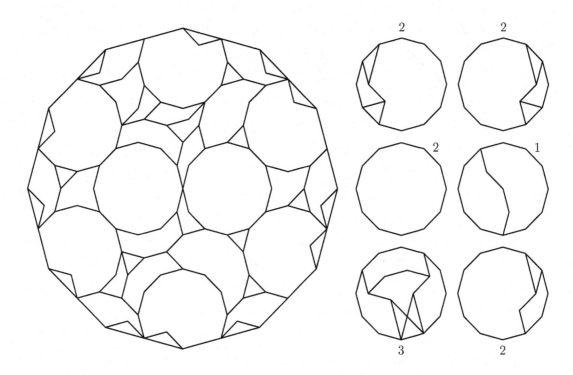

I122: Improved twelve dodecagons to one [Robert Reid (variation)]

square, which is similarly based on a superposition of tessellations. The resulting 8-piece dissection most likely has the minimum number of pieces. Harry Lindgren (1964) (on page 38) gave a slight variation of this same 8-piece dissection.

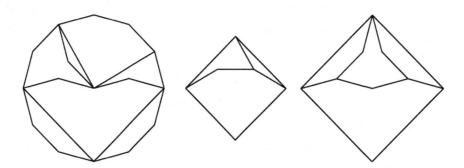

I124: Improved squares with area ratio 1 : 2 to a dodecagon [GNF]

Plate 124: This dissection problem is the fourth in an infinite sequence of similar problems, as discussed in the comments for Plate 91. Freese gave two different ways to dissect a dodecagon to two squares of an area ratio of 1 : 2. Both use nine pieces and are relatively symmetric. In my (2006) book I gave a 7-piece dissection (Figure I124) that is also reasonably symmetric. It is related to the second of Freese's two dissections, and relies on cutting parts off of certain pieces and gluing them onto other pieces. This dissection takes advantage of the internal structure of the dodecagon.

Plate 125: Freese seems to have been the first to dissect a regular dodecagon to an isosceles right triangle, which he did in just five pieces.

He also appears to have been the first to specifically dissect a regular dodecagon to six congruent squares. Using just fourteen pieces, his dissection would seem to be difficult to beat. In labeling the number of pieces as minimal, Freese probably meant that the number of pieces was smaller than for the dissection in Plate 126.

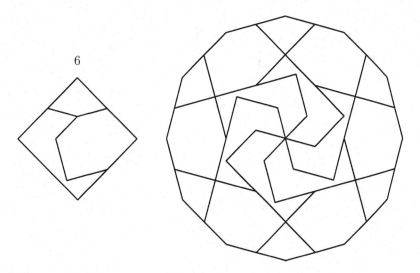

I126: Improved six identically cut squares to a dodecagon [GNF]

Plate 126: After producing a 14-piece dissection of six squares to a regular dodecagon in Plate 125, Freese gave a 24-piece dissection, but with each square being cut identically. While he did not note it, his dissection is hingeable. When suitably arranged, it is also translational. However, if we set out to identically dissect six congruent squares to a regular dodecagon, then we need no more than eighteen pieces for this task, as we see in the modified version of Plate 126 that I give in Figure I126. I have glued together the two mirror-image pieces in each of the six squares from Plate 126. The resulting bulge and cavity in each central piece fit together perfectly. Unfortunately, this dissection is not hingeable, although it misses by just a whisker.

Plate 127: Freese was the first to dissect a regular dodecagon to twelve squares. He missed a simple way to improve his dissection from 28 to 24 pieces: Just pair each half of the squares on the lower left with a half of the squares that are sliced diagonally. Then two of the pieces in each of the four resulting squares will be adjacent in both the squares and the dodecagon. We can then merge these pieces, giving a 24-piece dissection.

With a bit more work I was able to produce the 23-piece dissection in Figure I127. Note that the hypotenuse of the smaller of the isosceles right triangles is the side length of the largest equilateral triangle that we can inscribe inside a square.

Plate 128: Freese was the first to dissect three regular dodecagons to a square. He based his 16-piece dissection on the fact that the side length of such a square is one and a half times the diameter of one of the dodecagons.

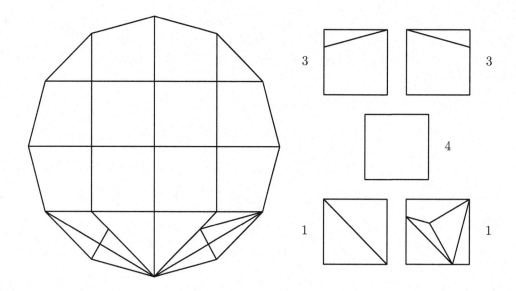

I127: Improved dodecagon to twelve squares [GNF]

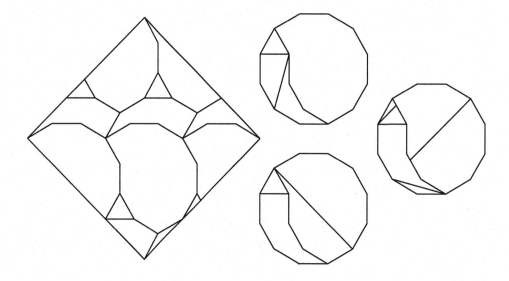

I128: Improved three dodecagons to a square [Lindgren 1964]

In (1964), Harry Lindgren gave a 14-piece dissection (Figure I128). He used his tessellation B13 from that same source, superposing it with a tessellation of squares. There are other 14-piece dissections. Is it possible that there is a 13-piece dissection?

In this plate Freese also gave a 14-piece dissection of two regular dodecagons to three squares, which seems difficult to beat.

Plate 129: Freese seems to have been the first person to have dissected four regular dodecagons to a square. This dissection shares much in common with the second dissection in Plate 123. It is based on the superposition of tessellations, using the same 3-piece tessellation element. It seems likely that this dissection is minimal, since the outline of the square may well require eight different pieces from the dodecagons.

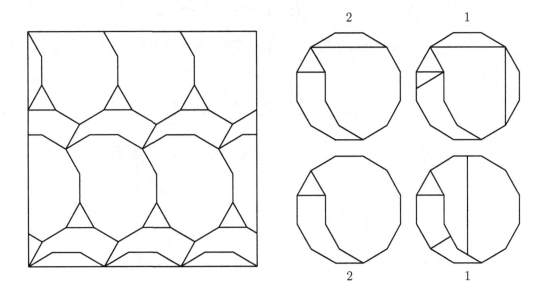

I130: Improved six dodecagons to a square [Harry Lindgren, 1964]

Plate 130: Freese was the first to dissect six regular dodecagons to a square. His 29-piece dissection has a high degree of symmetry.

Again, however, Harry Lindgren found a better dissection based on tessellations. Using the same tessellation for dodecagons as in Figure I128, Lindgren produced a 25-piece dissection.

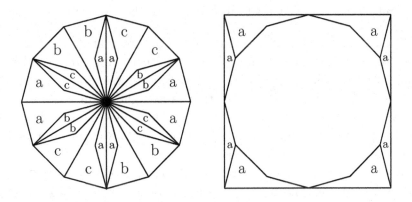

I131a: Key idea for translational four dodecagons to three squares

Plate 131: Freese was the first to dissect four regular dodecagons to three squares. His 27-piece dissection has a high degree of symmetry. When suitably arranged one way, it is hingeable, and when arranged in another way, it is translational. In Figure I131a we see how to shift the pieces from one of the three squares to a third of the dodecagon, moving the pieces marked by the letter a. The pieces marked with the letter b shift from a square that would be oriented 60° clockwise, and the pieces marked by the letter c shift from a square oriented 60° counterclockwise.

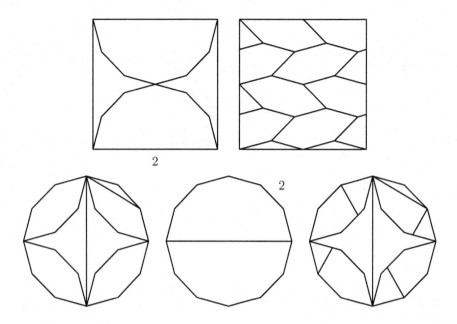

I131b: Improved four dodecagons to three squares [Robert Reid]

Yet again, however, Robert Reid found a better dissection based on tessellations. Reid cut eight "flat hexagons" from two of the dodecagons, producing pieces that along with the uncut dodecagons tile the plane. Overlaying a tessellation of squares, he got two of the squares. The flat hexagons also tile the plane with a repetition pattern that is a square, yielding the third square. Thus Reid produced a 22-piece dissection.

Freese seems to have been the first to dissect one dodecagon, two regular hexagons, and three squares to get a regular hexagon. When suitably arranged, the dissection is translational.

Plate 132: When he included his 7-piece dissection of a regular dodecagon to a Greek Cross, Freese was unaware of the 6-piece dissection by Lindgren (1953). And Freese had missed discovering a 6-piece dissection just by a whisker. All Freese needed to do was glue together the right triangle and isosceles triangle in the dodecagon, and cut that same right triangle out of one piece and glue it onto another piece. Then he could have achieved a 6-piece dissection as shown in Figure I132a. Being symmetric, Lindgren's dissection is much prettier, but the variant of Freese's dissection may make a more challenging "put-together" puzzle.

Freese's alternate 6-piece dissection of a dodecagon to a square indicated by dashed lines seems original, though he should have written that "the non-isosceles right triangles vanish."

Freese was the first to dissect a regular dodecagon to a Latin Cross. However, his 9-piece dissection was eclipsed by the 7-piece dissection of Lindgren (1962) in Figure I132b.

Plate 133: Freese was the first to dissect a {12/2} (a dodecagram) to a hexagon. It is easy to reduce his 15 pieces to 12. Just merge pieces 1 and 5, enlarge piece 2 as we decrease piece 4 to accommodate the enlarged piece 5. Note that Freese's dissection is translational.

However, Lindgren (1964) gave a 10-piece dissection, which I improved to eight pieces in (1972b), as shown in Figure I133.

This last dissection brings us to the end of a transcendent chapter centered on 12-sided figures. We have seen how regular dodecagons are compatible with equilateral triangles, squares, and reg-

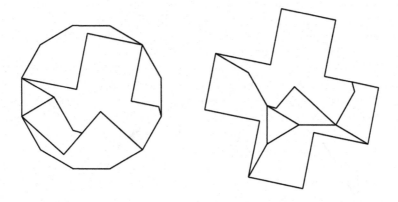

I132a: Six-piece variant of Freese's dodecagon to a Greek Cross

I132b: Improved dodecagon to a Latin Cross [Harry Lindgren 1962]

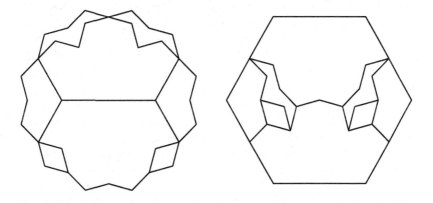

I133: Improved dodecagram to a hexagon [GNF]

ular hexagons. The geometry of these figures would seem to be more cooperative than we can perhaps expect for figures with more than twelve sides. In the next chapter we will sample selected dissections for regular polygons of 15, 16, 20, and 24 sides, some of which are quite ingenious.

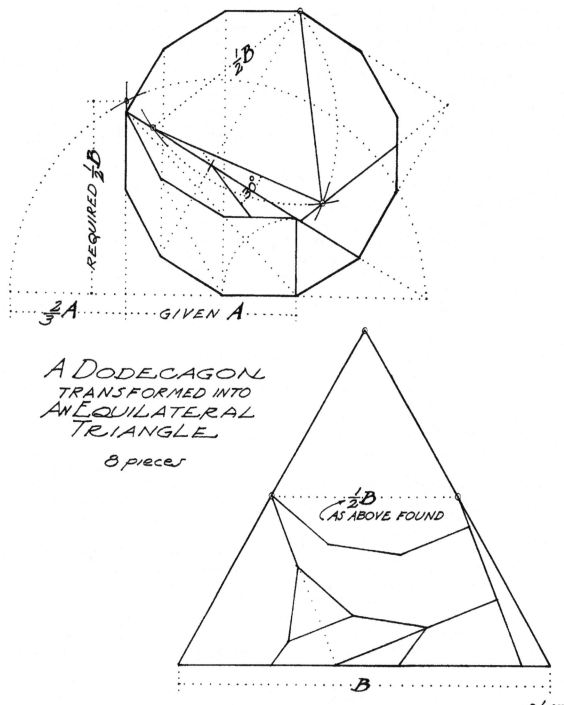

½B

REQUIRED ½B

⅔A   GIVEN A

A DODECAGON
TRANSFORMED INTO
An EQUILATERAL
TRIANGLE

8 pieces

½B
AS ABOVE FOUND

B

BI P28 D15
..FREESE..

PLATE
112

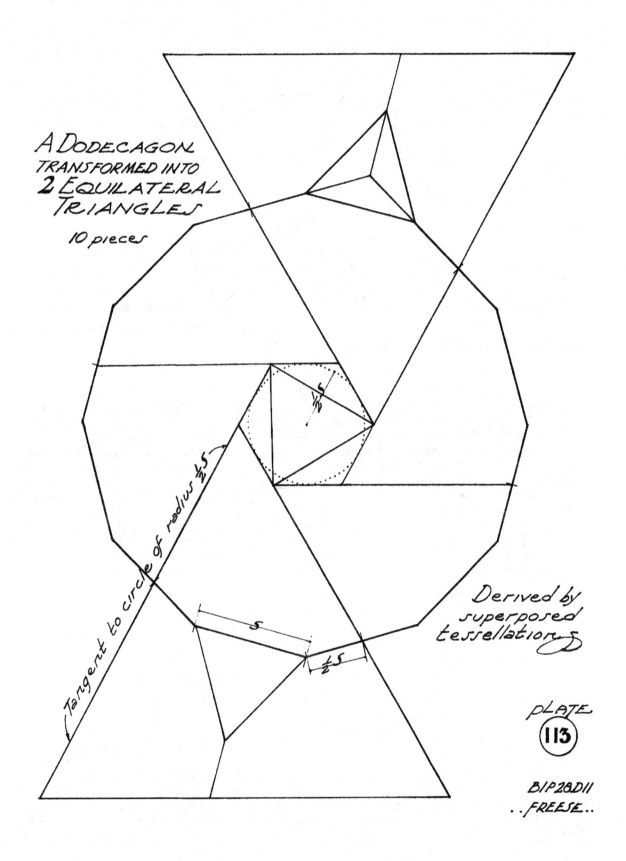

A DODECAGON
TRANSFORMED INTO
2 EQUILATERAL
TRIANGLES

10 pieces

Tangent to circle of radius $\frac{1}{2}s$

Derived by
superposed
tessellations

PLATE
113

BIP28DII
..FREESE..

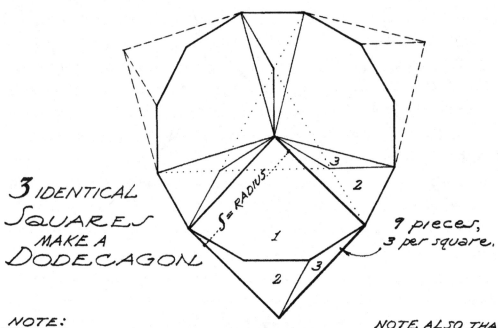

3 IDENTICAL
SQUARES
MAKE A
DODECAGON

S = RADIUS

3

2

1

2 3

9 pieces,
3 per square.

NOTE:
 DISSECTIONS OF THE DODECAGON,
TO TRANSFORM IT INTO ONE OR MORE
SQUARES, ARE SIMPLE & DIRECT;
EVERY DISSECTIVE ANGLE
BEING ONE THAT IS
GEOMETRICALLY
CONSTRUCTIBLE.
NO OTHER
POLYGON
HAS THIS
PROPERTY.

NOTE ALSO THAT, IN BOTH
DISSECTIONS, S IS GIVEN
BY THE DODECAGON
ITSELF, THUS ELIMINATING
ALL PRELIMINARY LINEAR
 CONSTRUCTION
 TO FIND SAME.

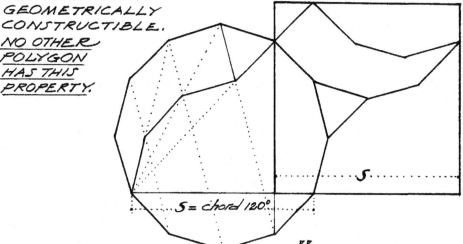

S = chord 120°

S

THE DODECAGON "SQUARED" with 6 pieces.
(min.)

THERE ARE OTHER HERETOFORE UNDISCOVERED 6-PIECE
SQUARINGS OF THE DODECAGON, BASED
ON THE UNIQUE PROPERTIES NOTED ABOVE.
A 5-PIECE SQUARING IS IMPOSSIBLE.

B1 P28 A2
B3 PS 23.94
..FREESE..

PLATE
114

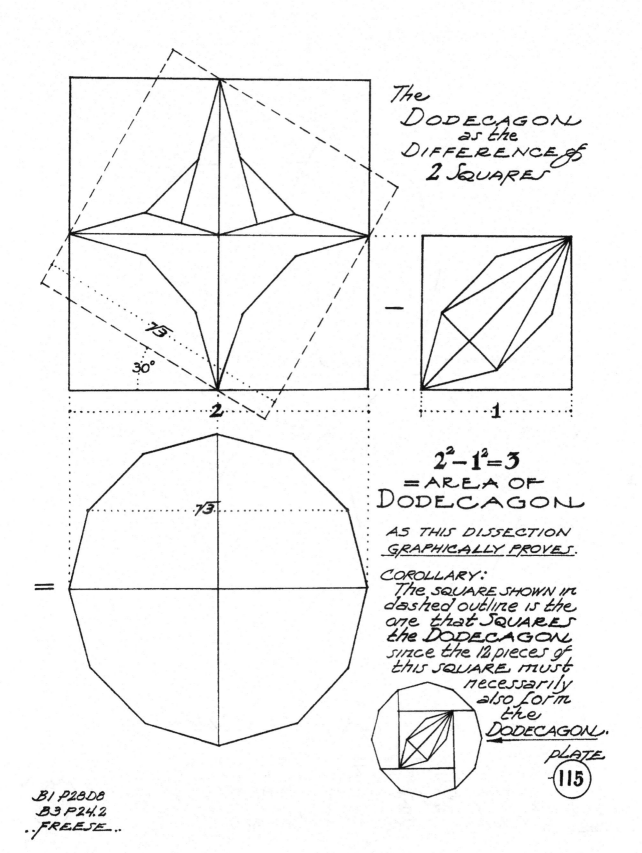

The
DODECAGON
as the
DIFFERENCE of
2 SQUARES

√3

30°

2

1

$$2^2 - 1^2 = 3$$
= AREA OF
DODECAGON

AS THIS DISSECTION
GRAPHICALLY PROVES.

COROLLARY:
The SQUARE SHOWN in
dashed outline is the
one that SQUARES
the DODECAGON
since the 12 pieces of
this SQUARE must
necessarily
also form
the
DODECAGON.

PLATE
115

√3

√3

B1 P28 D8
B3 P24.2
..FREESE..

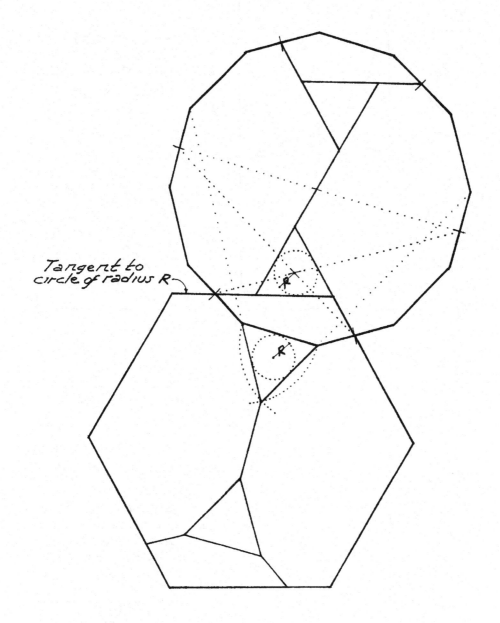

Tangent to circle of radius R

A DODECAGON transformed into A HEXAGON

with 6 pieces only.

This remarkably simple dissection was derived by superposing a tessellation of hexagons & equilateral triangles on a tessellation of dodecagons & equilateral triangles.

B1P28D13

..FREESE..

PLATE 116

A 2-in-1 DODECAGON
of 11 pieces; the minimum.

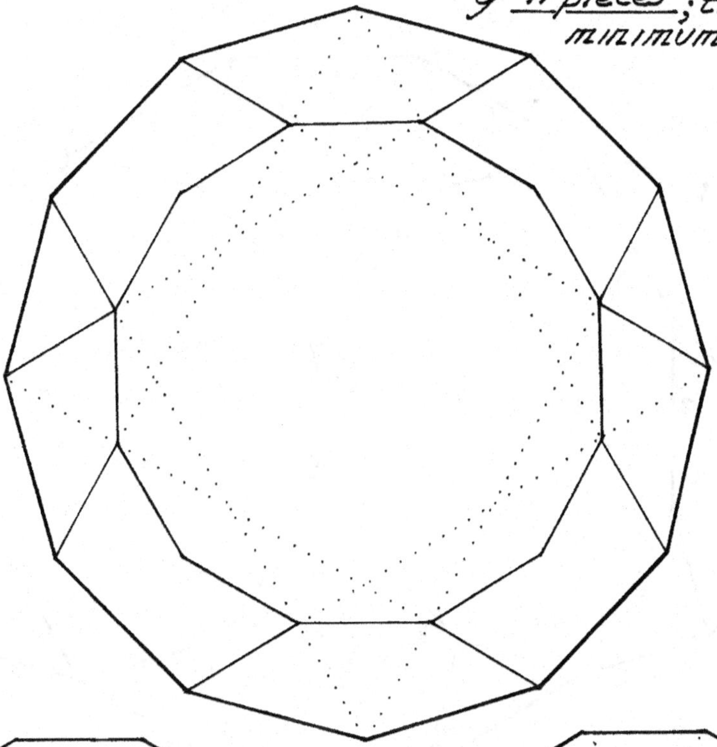

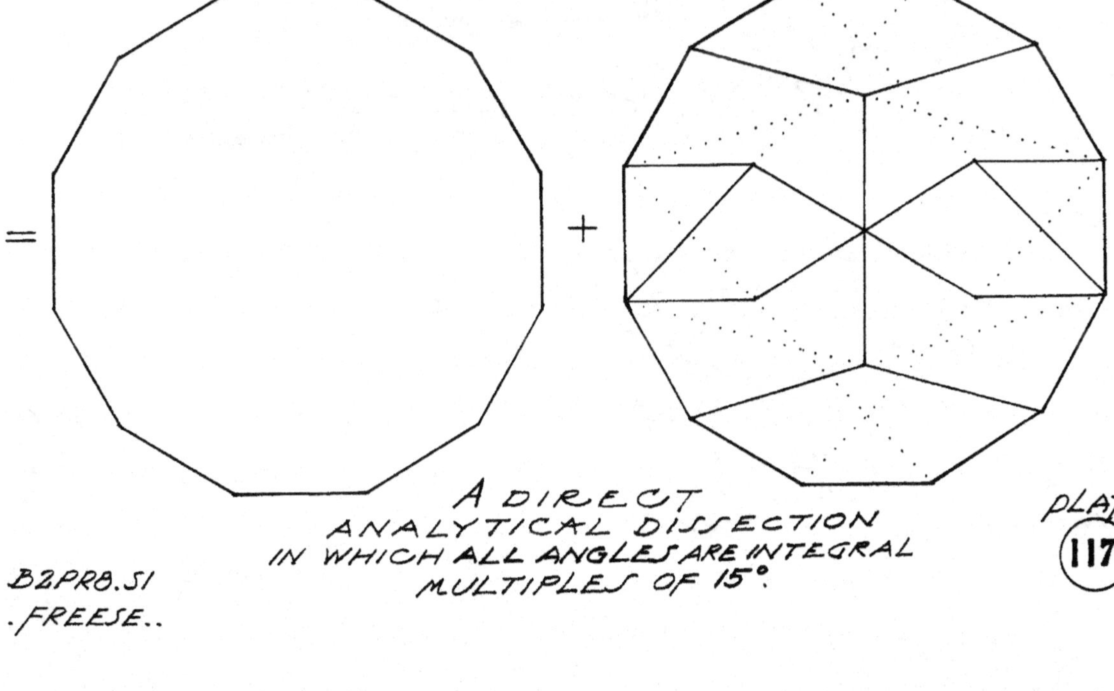

= +

A DIRECT
ANALYTICAL DISSECTION
IN WHICH ALL ANGLES ARE INTEGRAL
MULTIPLES OF 15°.

B2PR8.SI
..FREESE..

PLATE
(117)

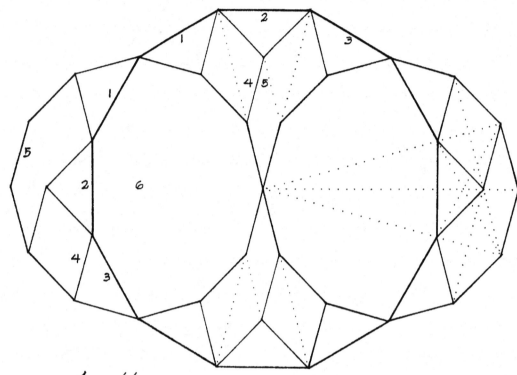

Another
2-IN-1 DODECAGON
12 Pieces

Based on the Rhombic Method
of Doubling any 4m-gon, hence
applicable to a SQUARE, an
OCTAGON, a DODECAGON,
a HEXADECAGON, etc., the
total number of pieces
always equalling
the number of
respective sides

B2 PR 10
..FREESE..

PLATE
(118)

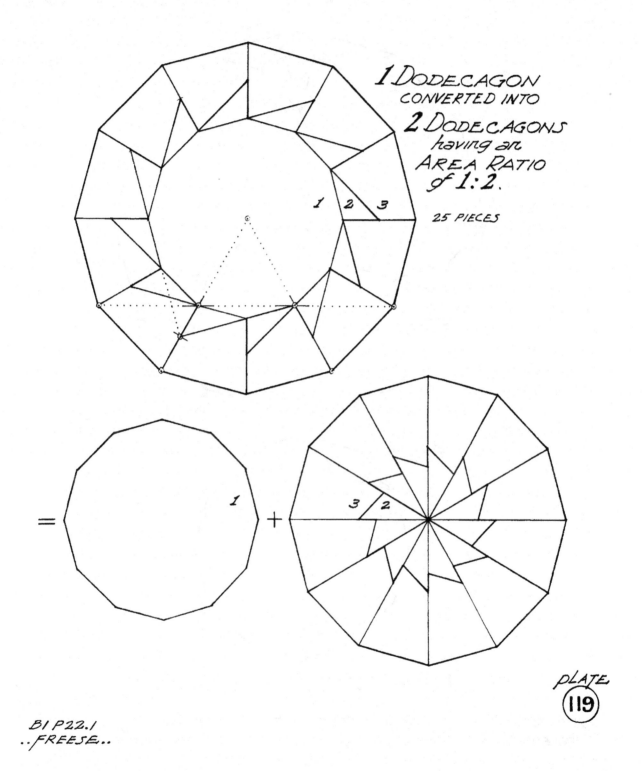

1 DODECAGON CONVERTED INTO 2 DODECAGONS having an AREA RATIO of 1:2.

1 2 3

25 PIECES

PLATE
(119)

B1 P22.1
..FREESE..

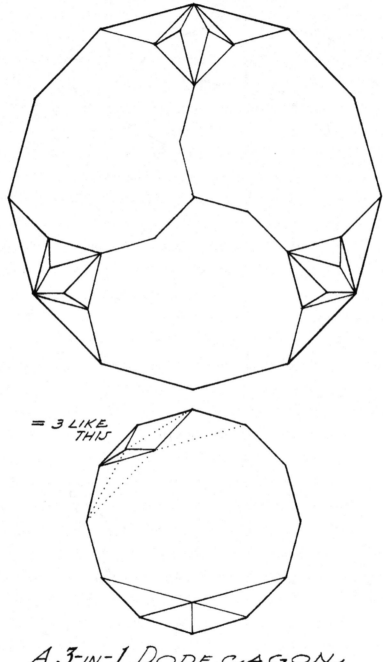

= 3 LIKE
THIS

A 3-IN-1 DODECAGON

24 pieces

BIP28DI
..FREESE..

PLATE
120

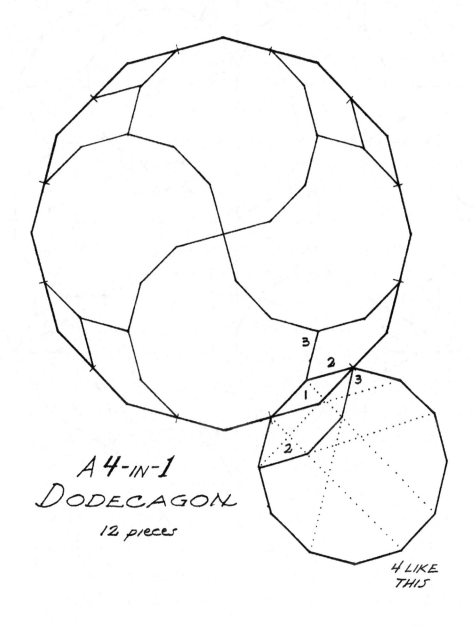

A 4-ɪɴ-1
Dodecagon

12 pieces

3
2
3
1
2

4 LIKE
THIS

PLATE
121

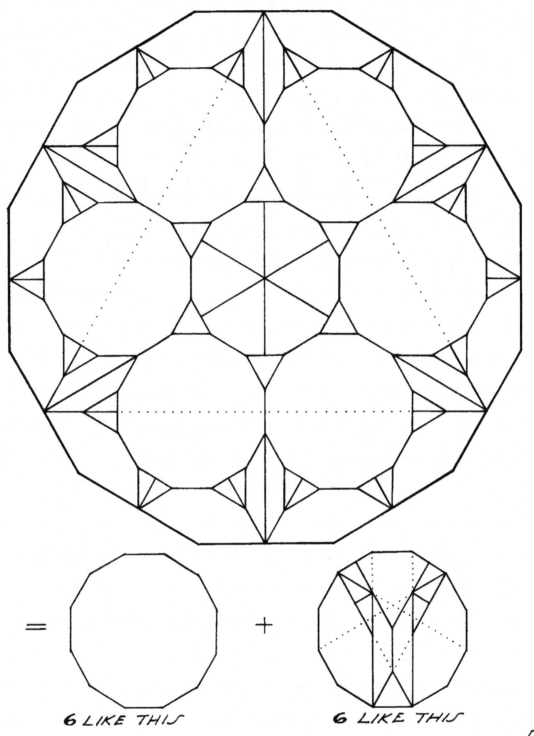

= 6 LIKE THIS   +   6 LIKE THIS

A *12-in-1* DODECAGON, 78 pieces.

B4PZ23
..FREESE..

PLATE
(122)

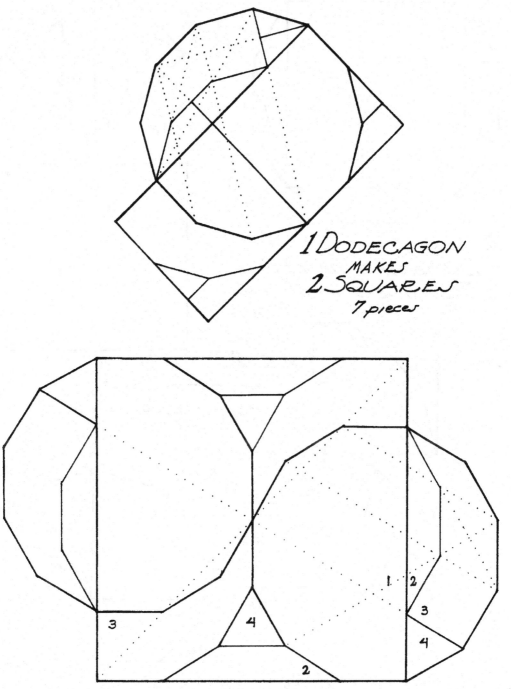

1 DODECAGON
MAKES
2 SQUARES
7 pieces

2 DODECAGONS MAKE 1 SQUARE
8 pieces

B2.PROD3
B2.PR8B.S3
..FREESE

PLATE
123

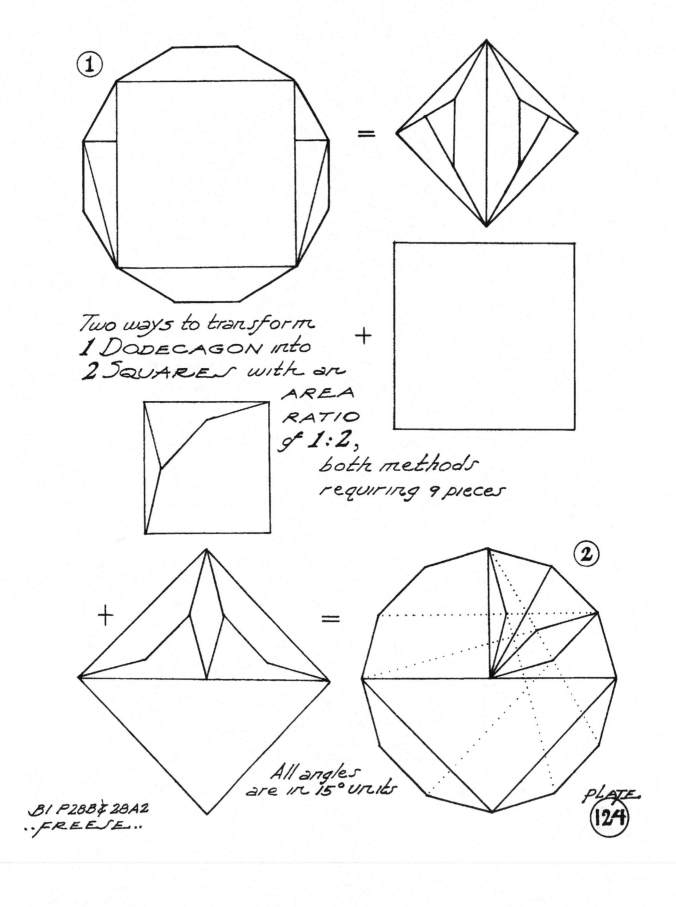

① 　 ＝

Two ways to transform
1 DODECAGON into
2 SQUARES with an
　　　AREA
　　　RATIO
　of **1:2**,
　　　both methods
　　requiring 9 pieces

+

+ ＝

All angles
are in 15° units

②

PLATE
124

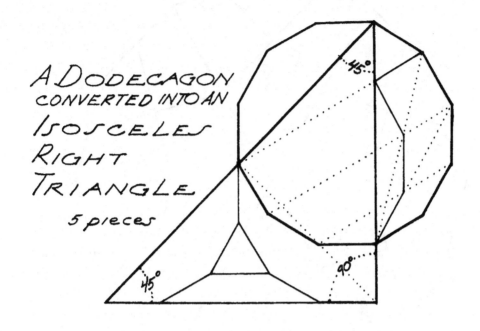

A DODECAGON
CONVERTED INTO AN
ISOSCELES
RIGHT
TRIANGLE
5 pieces

45°

90°

45°

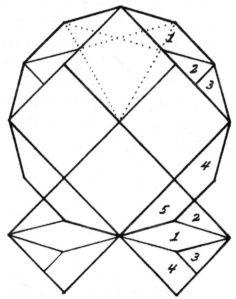

A DODECAGON
CONVERTED INTO 6 SQUARES
14 pieces
(min.)

PLATE
125

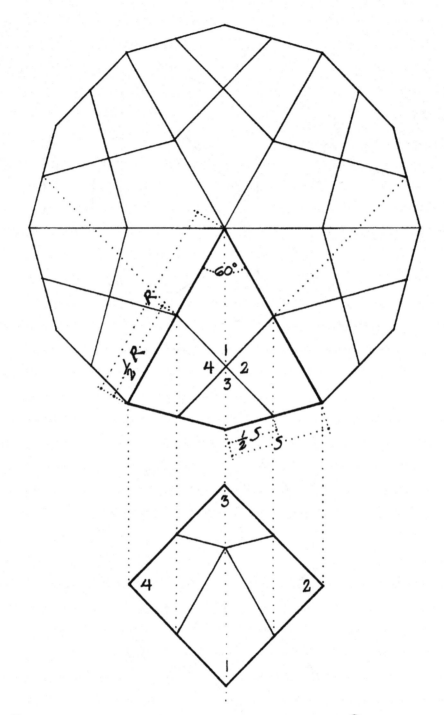

SQUARING A DODECAGON DELTOID,
HENCE THE
CONVERSION OF A DODECAGON INTO **6**
IDENTICAL SQUARES
OF 4 PIECES EACH

PLATE
(126)

B2 PR841
..FREESE..

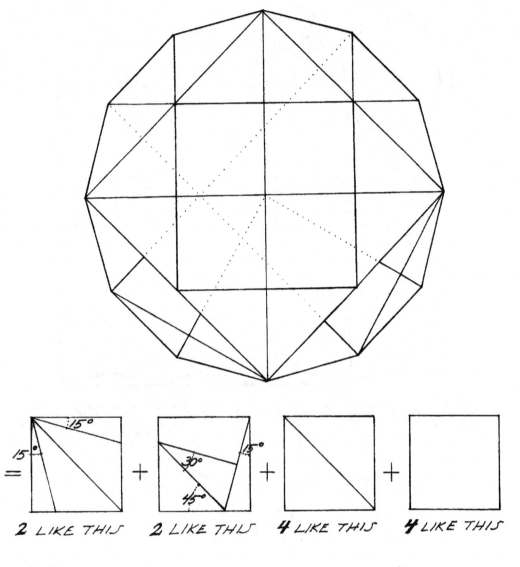

2 LIKE THIS   +   2 LIKE THIS   +   4 LIKE THIS   +   4 LIKE THIS

A DODECAGON CONVERTED INTO 12 SQUARES
28 pieces

B2 PR 8C6
..FREESE..

PLATE
127

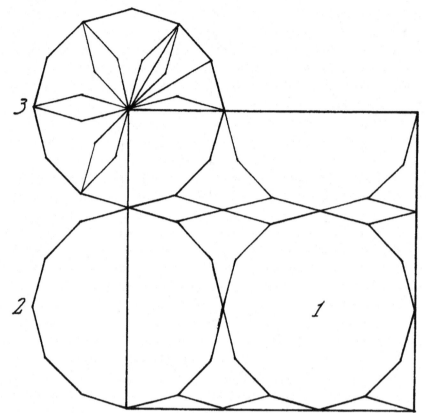

3 DODECAGONS MAKE 1 SQUARE

16 pieces

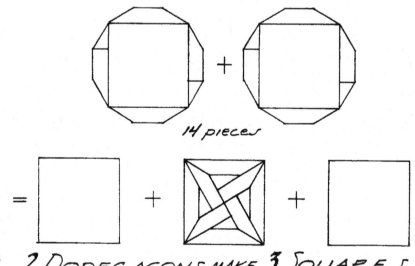

14 pieces

2 DODECAGONS MAKE 3 SQUARES

B2PR8E
B2PR8H3
..FREESE..

PLATE
(128)

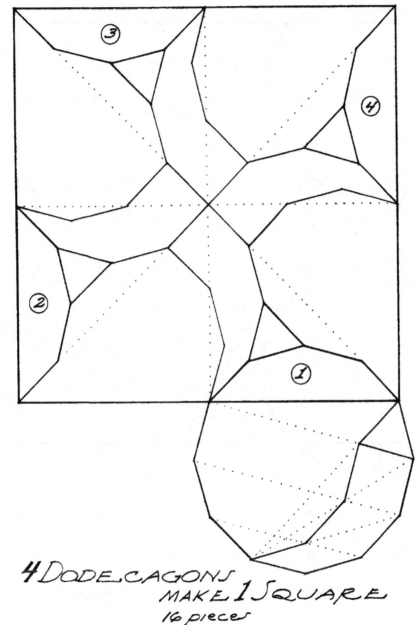

4 DODECAGONS
MAKE 1 SQUARE
16 pieces

PLATE
(129)

B2.P8D2
..FREESE..

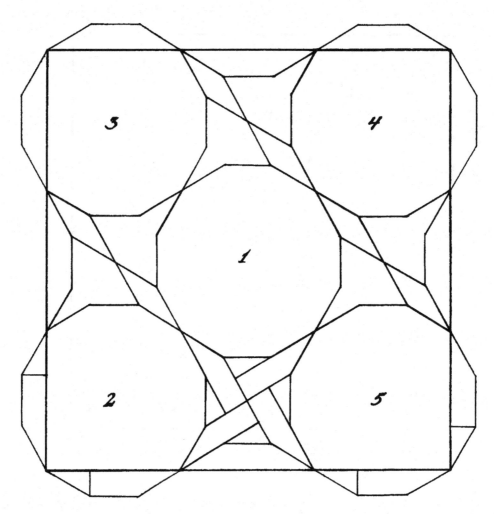

6 DODECAGONS
MAKE
1 SQUARE

29 pieces

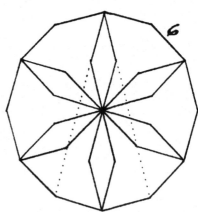

B2PR8D7
..FREESE..

PLATE
130

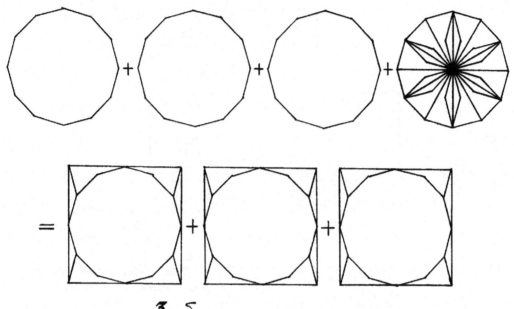

3 SQUARES
directly compounded from
4 DODECAGONS
27 pieces

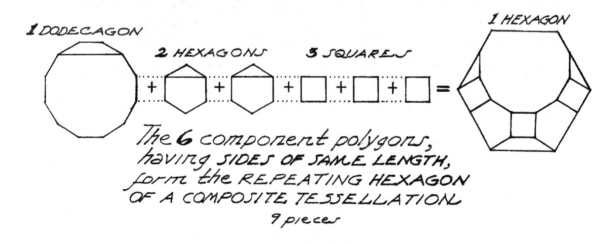

1 DODECAGON   2 HEXAGONS   3 SQUARES   1 HEXAGON

The 6 component polygons,
having SIDES OF SAME LENGTH,
form the REPEATING HEXAGON
OF A COMPOSITE TESSELLATION
9 pieces

B2 PR8H2
B1P28D9
..FREESE..

PLATE
131

A DODECAGON or a GREEK CROSS
formed from
the same
7 pieces

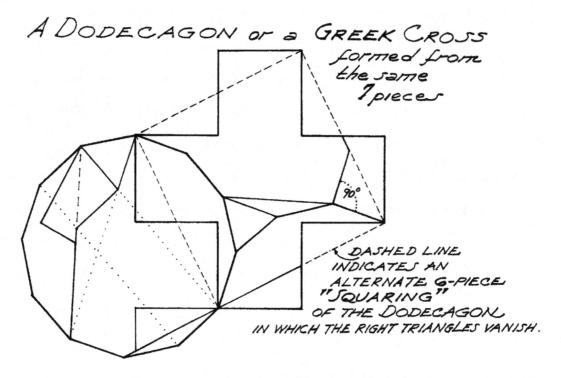

90°

DASHED LINE
INDICATES AN
ALTERNATE 6-PIECE
"SQUARING"
OF THE DODECAGON
IN WHICH THE RIGHT TRIANGLES VANISH.

A DODECAGON
dissected to form
a LATIN CROSS
9 pieces

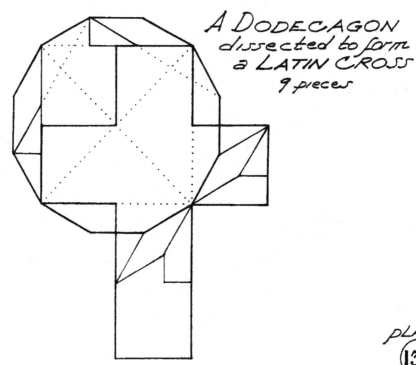

B2PR8D5
B2 PR8EI & 8C5
..FREESE..

PLATE
132

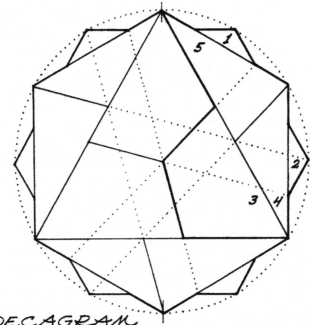

A Dodecagram
TRANSFORMED INTO
A Hexagon
15 pieces

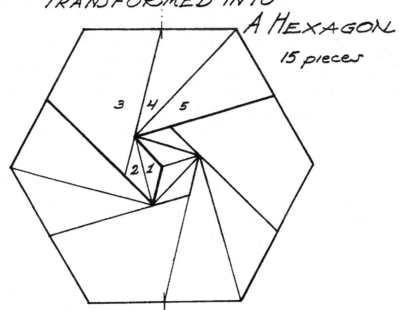

PLATE
133

Chapter 14

Many-sided Polygons (Plates 134–141)

As I wrote in my (2006) book, Freese was way ahead of his time in dissecting many-sided figures to squares. He did not include all of his work on this topic in his 1957 manuscript, but he had created at least one dissection for every regular polygon up to 21 sides, and also for a 24-sided polygon. In Plates 134 through 141 of his manuscript he included his dissections for polygons with 15, 16, 20, and 24 sides. In this same group of plates he also included the dissections of two congruent polygons to one for that same set of polygons. Eventually, Gavin Theobald handily bested each of Freese's dissections of those four polygons to squares. Yet even so, some of Freese's approaches were quite resourceful and foreshadowed some of the strategies that Theobald has employed.

Freese's dissections of two polygons to one are less inventive for polygons with a sufficiently large number of sides. The cases in which the number of sides is a multiple of 4 are special, and Freese used a simple approach in which the number of pieces equals the number of sides, as we see in Plates 137, 139, and 141. When the number of sides is sufficiently large and not a multiple of 4, Freese used a variant of Harry Hart's approach (1877) that creates the same number of pieces as Hart, as we see in Plate 135. Yet I was flabbergasted to discover a technique, which I described in my 2006b article, that improves on Hart's and Freese's approaches when the number of sides is sufficiently large and <u>not</u> a multiple of 4. My technique renders a reduction in the number of pieces over Freese's method for polygons with 13, 14, or 15, or 17, 18, or 19, or 21, 22, or 23, and so on, number of sides. As luck would have it, Freese showed his own dissection only for the case of 15 sides.

Thus I single out Freese's dissections of these many-sided figures to squares. In particular, I point the reader to his dissection of the (15-sided) pentadecagon to a square, in Plate 134, to his dissection of the (20-sided) icosagon to a square, in Plate 138, and to his dissection of the (24-sided) icositetragon to a square, in Plate 140. And yet, the real stars in this chapter are Gavin Theobald's amazing improvements over Freese's efforts. They are a joy over which to marvel!

Plate 134: Freese appears to have been the first person to attempt a dissection of a (15-sided) pentadecagon to a square. He approached this dissection piecemeal: First he sliced five thin trapezoids off the exterior of the pentadecagon, leaving a regular pentagon. Then he converted the regular pentagon via a P-strip dissection to a rectangle whose length is that of the desired square. Freese next converted the five trapezoids to a thin rectangle whose length is also that of the desired square, using a P-slide, as shown in his accompanying smaller figure labeled with a circled 2. This took eleven pieces, so that Freese completed the dissection in seventeen pieces.

Note that Freese used a somewhat similar technique for his Plate 166, in which he dissected an enneagon to an equilateral triangle. There he sliced three trapezoids off the exterior of the enneagon, leaving an equilateral triangle. He then converted those three trapezoids to a larger trapezoid, upon which he set the equilateral triangle. While Freese noticed that he could use four

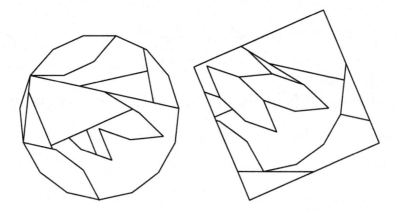

I134a: Improved pentadecagon to a square [Gavin Theobald]

hinges to connect some of the pieces in his enneagon dissection, he could have introduced five hinges into his pentadecagon dissection. Those hinges would connect pieces 5 and 6, pieces 12 and 13, pieces 14 and 17, pieces 15 and 16, and pieces 16 and 17.

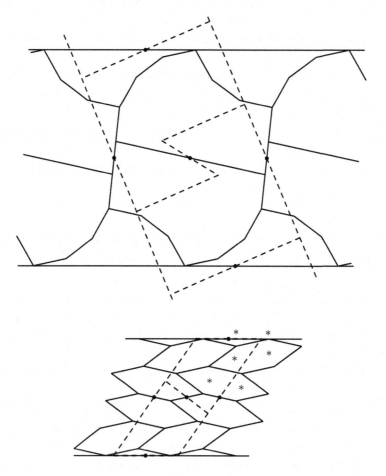

I134b: Crosspositions for improved pentadecagon to a square [Gavin Theobald]

In January 2001, Gavin Theobald first dissected the pentadecagon to the square, originally in thirteen pieces. In September 2003, he found a better dissection, with only twelve pieces. And within a couple of weeks, he had found an 11-piece dissection!

To produce his amazing 11-piece dissection (Figure I134a), Theobald first cut the pentadecagon into five pieces, which he pieced together into two T-strip elements. He included as much of the area as possible into two large pieces, as we see in Figure I134b. Again, the way that these two pieces fit together is most impressive! The three remaining pieces, namely two flattened hexagons and an isosceles triangle, form a T-strip element. Theobald could then have crossposed each with a rectangle whose long side would equal that of the side length of the square. Instead, he cut a trapezoid out of the square, choosing the shape of the trapezoid to fit conveniently in the crossposition of Figure I134b. The trapezoid has twice the height of a rectangle for the two flattened hexagons and the triangle.

By themselves, these are terrific ideas, which would lead to a 13-piece dissection. Yet Theobald had one more trick. To reduce the seven pieces resulting from the second crossposition, he cut a cavity into the largest piece of the square, into which he could insert the two flattened hexagons and the isosceles triangle. The positions of the hexagons and triangle conform to the positions of their smaller pieces when those smaller pieces fill the trapezoid. Gavin then replaced those seven pieces by the five pieces that extend out from the upper left corner of the square. Thus he avoided cutting the two flattened hexagons and the isosceles triangle, which are marked with asterisks in the lower crossposition in Figure I134b.

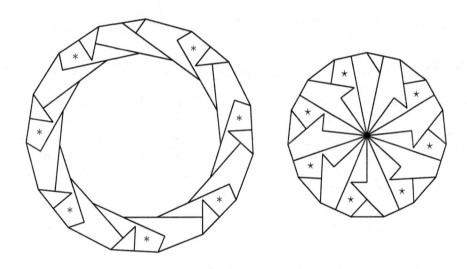

I135a: Improved two pentadecagons to one, with turnover [GNF]

Plate 135: In this plate Freese applied to regular pentadecagons his general technique (from Plate 50) for dissecting two congruent regular polygons to one. His dissection technique is a variation of Hart's method (Plate 55) and uses the same number of pieces, $2n+1$, where n is the number of sides in the polygon. This gives a total of 31 pieces for the pair of pentadecagons.

However, in (2006b) I found methods that are better whenever $n > 12$. The first method uses $2n - \lfloor n/2 \rfloor + 1$ pieces, with $\lfloor n/2 \rfloor$ pieces turned over. First, add a small isosceles right triangle to $\lfloor n/2 \rfloor$ nontriangular pieces, and then cut $\lfloor n/2 \rfloor$ nontriangular pieces to accommodate the expanded pieces. The cavities left in the small and the large polygons are then larger isosceles right triangles. The one drawback is that the accommodating nontriangular pieces are so accommodating that they

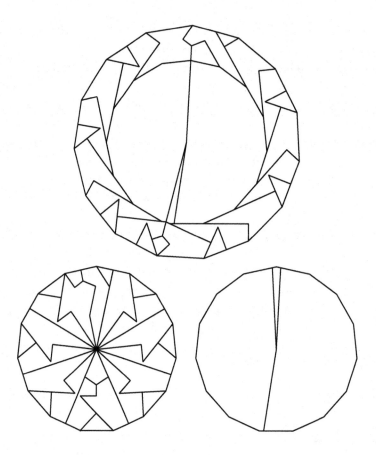

I135b: Improved two pentadecagons to one, with no turnover [GNF]

need to be turned over! Yet this approach reduces the number of pieces by approximately 25%, as we see in Figure I135a, where the large pentadecagon has just 24 pieces.

If we choose not to turn over pieces, then we can still do better than Freese's and Hart's methods. We can get a 28-piece dissection for pentadecagons, as we see in Figure I135b. The basic idea is to take half of the ring from Figure I135a, and then make the second half be a mirror image of it. We make several adjustments to get this idea to work. First, we slice up several pieces into two pieces each, to produce an even number of edges on the ring sections. Second, a slice that goes from a vertex of the larger pentadecagon to its center will not go through a vertex of the smaller, inner pentadecagon. This forces us to cut the inner pentadecagon and rearrange the pieces to form an irregular 17-sided figure. Third, we can save two pieces if we merge pairs of small triangles created by the initial slice and also cut holes to accommodate those merged pieces.

These latter methods work for every regular polygon with more than twelve sides. The idea behind Figure I135a also works for similar but not congruent regular polygons, as long as there are a sufficient number of sides. For example, it works for n-sided regular polygons of relative area $1 : 2$ to a larger n-sided regular polygon, as long as $n \geq \lceil 180° / \arcsin(\sqrt{2}/4) \rceil = 18$.

Plate 136: Freese was the first to attempt a dissection of the regular (16-sided) hexadecagon to a square. His approach was to convert the hexadecagon to a rectangle, which he then converted to a square by applying a P-slide. Starting the conversion to the rectangle, Freese first sliced pieces 1, 2, and 8 off of the top, and pieces 5, 6, and 7 off of the bottom. Then he cut pieces 3 and 4 off of the lefthand side, and cut the agglomeration of pieces 9, 10 , 11, and 12 also off of the left.

Freese then focused on the righthand side, placing triangles 1 and 2 on the top right, triangles 5 and 6 on the bottom right, and moving the collection of pieces 9, 10 , 11, and 12 over to the right. He thus produced a rectangle containing all pieces except for pieces 3, 4, 7, and 8. Freese noticed that pieces 3, 4, 7, and 8 formed a rectangle. He then took the rightmost portion of the rectangle containing all pieces except for pieces 3, 4, 7, and 8 and performed a P-slide on it, producing a rectangle of length equal to the length of pieces 7 and 8. Sliding piece 10 up and piece 9 down, so that piece 9 goes to the right of piece 10, he could then place the rectangle with pieces 3, 4, 7, and 8 above pieces 9 and 10. Freese had determined the width of piece 9 so that a rectangle of all of the pieces would result. Finally, he cut lines between pieces 13, 14, and 15 as shown and slid the pieces, carrying out the P-slide to give the square.

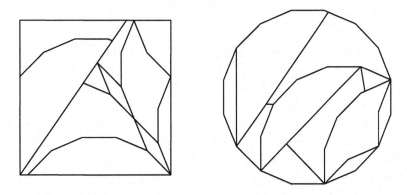

I136a: Improved hexadecagon to a square [Gavin Theobald]

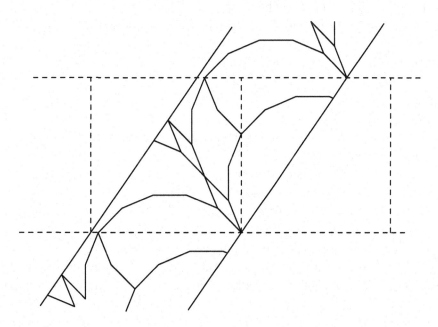

I136b: Crossposition for improved hexadecagon to a square [Gavin Theobald]

At the time when he produced it, Freese's dissection was impressive, and the count of 15 pieces was quite good. However since then, Gavin Theobald has generated a much more striking 11-piece dissection, in Figure I136a. Gavin cut the hexadecagon into seven pieces which he rearranged to form a P-strip element. He then crossposed this strip with a strip of squares, as shown in Figure I136b. The resulting dissection appears simpler than it actually is, because the strip of squares must be tilted very slightly clockwise. (If you're sharp you may spot the tilt of the square in Figure I136a.) What looks like a trapezoid cut off the lefthand side of the hexadecagon is actually a pentagonal piece, because it has a small side projecting off from what appears to be its top angle. Also, the piece in the upper right corner of the square appears to be a quadrilateral, but is actually an irregular hexagon, because it has two tiny sides, one projecting off to the right from its lowest angle, and the other projecting up from what appears to be its top angle. One more piece has an extra side that is hard to spot without magnification. Can you identify it?

Plate 137: Freese demonstrated his "rhombic method" from Plate 118, producing a 16-piece dissection of two hexadecagons to one in this plate. I gave a similar dissection in (1972a), long before I had seen Freese's manuscript.

Plate 138: Freese appears to have been the first to dissect a regular (24-sided) icosagon to a square. He adapted the completing the tessellation approach that was so successful in producing the 5-piece dissection of a regular octagon to the square (see Plate 89). He added eight right triangles of two different sizes, two congruent rectangles, and one fairly large rectangle, while at the same time cutting eight notches corresponding to the eight right triangles out of the exterior of the icosagon. Using essentially a double P-slide (see Plate 145), his dissection contains altogether 19 pieces.

Unfortunately, Freese missed a simple trick that would have reduced the number of pieces by one: Delete pieces 8 and 9 in the icosagon, and replace pieces 8 and 9 in the square by the isosceles triangle that is the merge of them. This action would have shaved a corner off piece 2 and another corner off piece 1 in the square. Then in the icosagon, replace pieces 1 and 2 with the shaved pieces 1 and 2, and fill in the gap between the shaved pieces 1 and 2 with the isosceles triangle that is the merge of pieces 8 and 9.

Recently Gavin Theobald discovered an incredible improvement on Freese's dissection, reducing the number of pieces down to just 13. Gavin cut the icosagon into five pieces, two of which he cut into a rectangle using a strip dissection, as we see in Figure I138a. The resulting rectangle has length equal to the side length of the desired square. The remaining three of the five pieces are two flattened hexagons and a flattened octagon. Gavin then performed a P-slide on the rectangle, which created two new triangles which he slid down and to the right to extend the right side of the rectangle to form the lower right corner of the desired square. This slide leaves a cleared-out corner cavity on the lower left for the desired square. Gavin cut a zigzag piece above the cavity and slid it into the cavity. He then shaved a thin rhombus off of the flattened octagon, and sliced the rhombus into two thin isosceles triangles. Finally he inserted the two isosceles triangles, the two flattened hexagons, and the remnant of the flattened octagon into the space above the zigzag piece. Hiding in plain sight is the crucial property, that the grouping of those five pieces have a uniform vertical thickness, which is just stunning! The resulting 13-piece dissection stands triumphantly in Figure I138b!

Plate 139: Freese once again demonstrated his "rhombic method" from Plate 118, this time producing a 20-piece dissection of two icosidecagons to one in this plate.

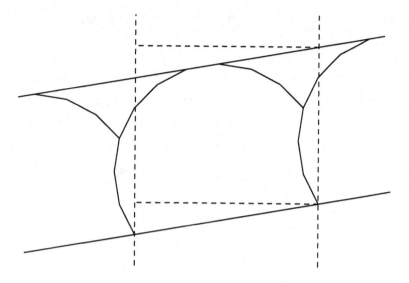

I138a: Crossposition for improved icosagon to a square [Gavin Theobald]

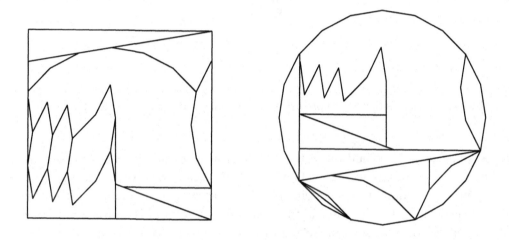

I138b: Improved icosagon to a square [Gavin Theobald]

Plate 140: Freese was the first to dissect the (24-sided) icositetragon to the square. His approach was to slice eight trapezoids off the exterior of the figure, leaving a regular octagon. In (1926) Henry Dudeney had published in the *Strand Magazine*, volume 71, page 522, a dissection of a regular octagon to the square. Dudeney sliced the octagon into four pieces that reassemble to make a rectangle. Freese laid this out by moving pieces 1 and 2 to beneath piece 10 and then moving pieces 3, 4, 5, and 6 to the right of pieces 8 and 10. He then cut the eight trapezoids into twelve pieces, which he reassembled into a rectangle. Finally, he performed a P-slide on that second rectangle and the righthand side of the first rectangle to get a square, using 22 pieces altogether.

Disappointingly, Freese missed two simple tricks that would have reduced the number of pieces by four. He could have avoided cutting the two trapezoids off piece 7, if he had cut flattened hexagons off the corresponding portions of pieces 4 and 8. (A flattened hexagon corresponds to a pair of trapezoids merged together.) He could also have avoided cutting one trapezoid off piece 2 if he had cut a flattened hexagon off piece 10 and then chopped it in half. Slicing the two remaining trapezoids across their widths would then complete the improvement.

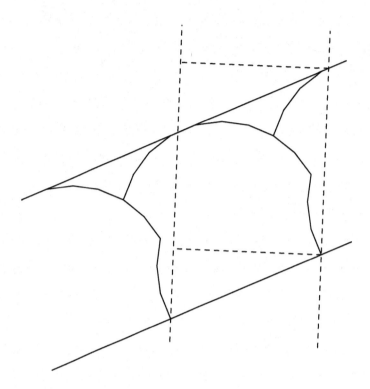

I140a: Crossposition for improved icositetragon to a square [Gavin Theobald]

But yet again, Gavin Theobald came through with a tremendous improvement. In a fashion similar to his dissection of the icosagon, Gavin first cut the icositetragon into six pieces, two of which he cut into a rectangle using a strip dissection, as we see in Figure I140a. The resulting rectangle has length equal to the side length of the desired square. The remaining pieces are two flattened hexagons, a flattened octagon, and a thin decagon that we would get if we cut a flattened hexagon from a flattened decagon. Gavin then performed a P-slide on the rectangle, which used a triangle and a pentagon that slid to form the left edge of the square. The slide creates a cavity that is open on the lower left. As with his dissection of the icosagon, Gavin cut a zigzag piece to the right of the cavity and slid it into the cavity. He then shaved a thin rhombus off of one of the flattened hexagons, and cut a thin rhombus off of the flattened octagon. Finally he inserted the isosceles triangles, the irregular decagon, the thin rhombus, the irregular octagon, a flattened hexagons, and the remaining piece from the flattened hexagon. Once again, Gavin designed his dissection with a crucial property, namely that the stack of those six pieces have a uniform horizontal thickness, so that they neatly fit into the space created by sliding the zigzag piece. And once again, Gavin produced a stunning 13-piece dissection, which you can admire in Figure I140b!

A word of caution: There is one piece in this dissection that has over 20 edges bounding it. It has a tiny edge on its top left side that without magnification looks like a vertex. If that were actually a vertex rather than a tiny edge, then a 12-piece dissection would be possible.

It is probably difficult for most readers to absorb quickly all aspects of the phenomenal dissections in Figures I138b and I140b. For over two decades Gavin has conjured up an amazing insight into how to attack these challenging dissection problems. He has revisited many of these problems several times, slowly unlocking the special techniques that allow him to achieve ever more efficient dissections. Indeed, a couple of years ago I had included less efficient, but still impressive, examples of Gavin's work on these two problems. Then, with less than two months to go before this book was due, he improved these two dissections again, and I had to scramble to include diagrams and discussion of them into the final version of the book. It is a bit nerve-racking to labor under this time pressure, but I think that readers will understand how very special the results are. And this is just a taste of all that Gavin has found: He has crafted smart dissections of many other regular polygons to a square, enough to keep the heads of people like me, and Ernest Freese, and numerous dissectionists, spinning well into the future!

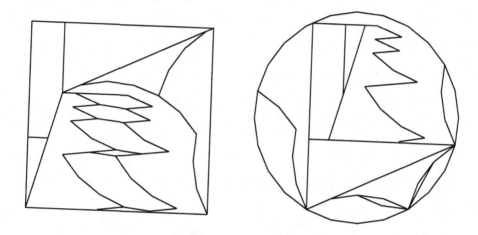

I140b: Improved icositetragon to a square [Gavin Theobald]

Plate 141: In this plate Freese demonstrated his "rhombic method" from Plate 118, producing a 24-piece dissection of two regular icositetragons to one in this plate. Since one type of rhombus in these polygons is a 60°-rhombus, as there is in a regular dodecagon, could there be a dissection of fewer than 24 pieces, in analogy to the dissections in Plate 117 and Figure I117?

For each of the four many-sided regular polygons that Freese dissected in this chapter, he focused on dissecting it to a square and also dissecting two congruent figures to a larger similar polygon. Thus he used two common dissection types for each figure to provide a seeming coherence to the chapter, even if the dissections to squares did not share much real commonality. Yet Freese's decision to dissect many-sided polygons to squares was truly daring. During his era, no one else came anywhere close to attempting such challenges. His results were certainly insightful and remarkable. Then a half-century later, a similarly daring dissectionist, Gavin Theobald, embarked on an even more ambitious program, dissecting regular polygons with any number of sides, not just a multiple of 4. I could fill many more pages showing all of them, rather than just showing the ones that Freese had attempted. But perhaps that treat should wait for Gavin to write his own book!

Let's move forward to the next chapter, in which Freese grouped a variety of different dissection problems together, perhaps deliberately showing off the diverse nature of dissections in general.

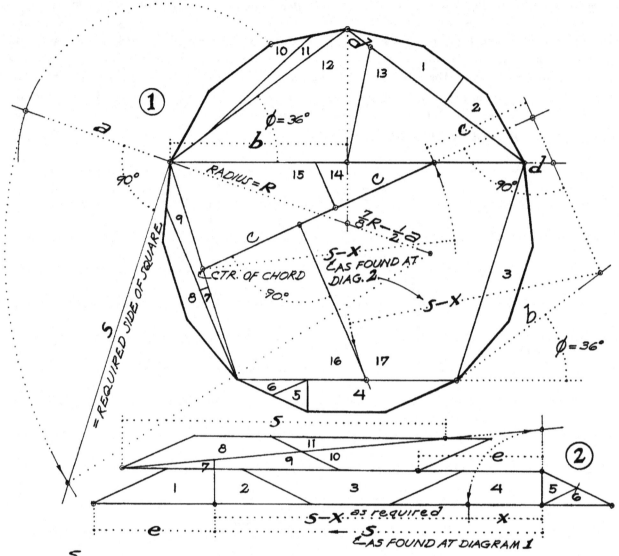

① a 90° RADIUS = R 15 14 c ∅ = 36° b c b d 90°

$2R - \frac{1}{2}a$

S-X (AS FOUND AT DIAG. 2) S-X

CTR. OF CHORD 90°

10 11 12 13 1 2 c

9 8 7 16 17 6 5 4 3

S = REQUIRED SIDE OF SQUARE

∅ = 36°

② s 8 11 e 7 9 10 1 2 3 4 5 6

e S-X as required X

← S
(AS FOUND AT DIAGRAM 1)

SQUARING THE PENTADECAGON
By taking it as the sum of two RECTANGLES.

17 pieces... 15 SIDES

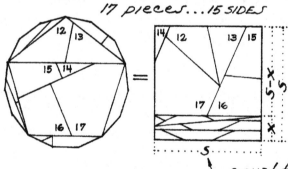

This method is general, being applicable to ANY polygon.

In this case the given polygon is resolved into the inscribed pentagon & the five circumsegments, then each of these elements, the segments as a GROUP, converted into a RECTANGLE whose longer side is made equal to the required side of the SQUARE. Hence, the pieces forming the RECTANGLES must consequently form either the PENTADECAGON or its SQUARE.

B3 P.S 25.98
..FREESE..

PLATE
134

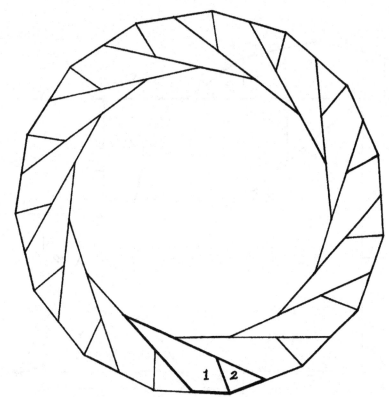

A 2-in-1 Pentadecagon
by RING EXPANSION
31 pieces

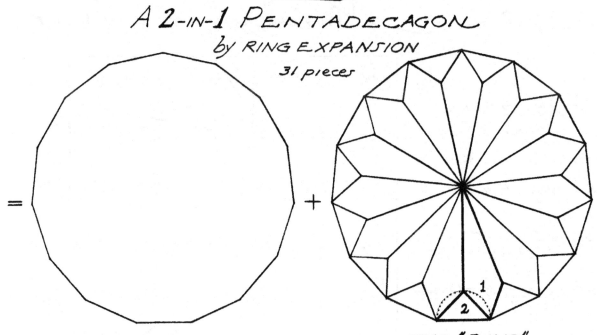

$=$

THE NUCLEUS
1 piece

$+$

THE "RING"
30 pieces,
2 to a side.

PLATE
135

B4 PZ58
..FREESE..

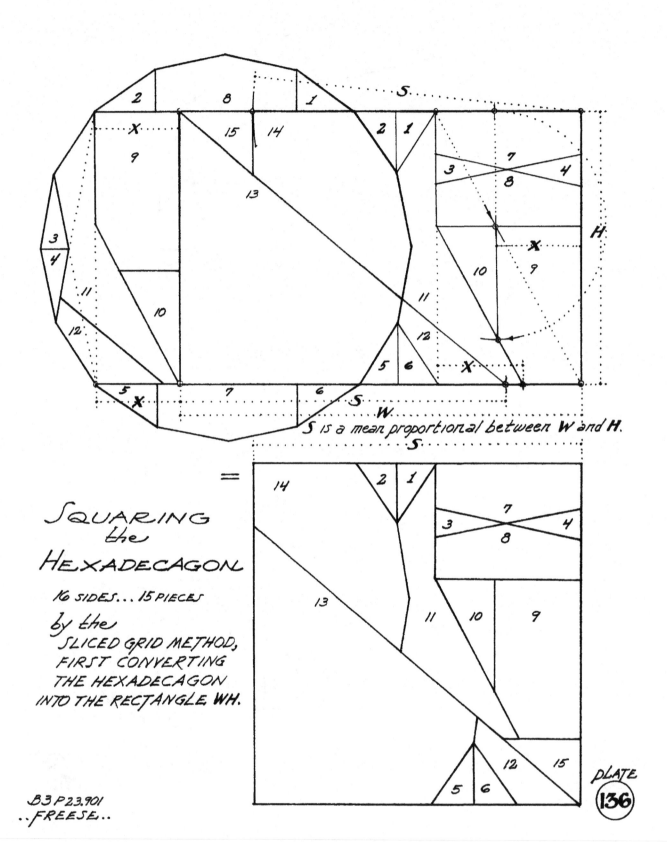

S is a mean proportional between W and H.

SQUARING
the
HEXADECAGON

16 SIDES... 15 PIECES

by the
SLICED GRID METHOD,
FIRST CONVERTING
THE HEXADECAGON
INTO THE RECTANGLE WH.

B3 P23.901
..FREESE..

PLATE
136

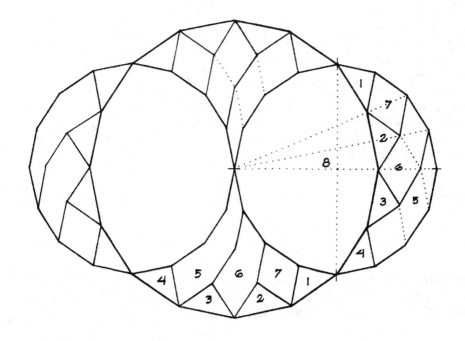

A 2-IN-1 HEXADECAGON
by the Rhombic Method.

16 PIECES

PLATE
137

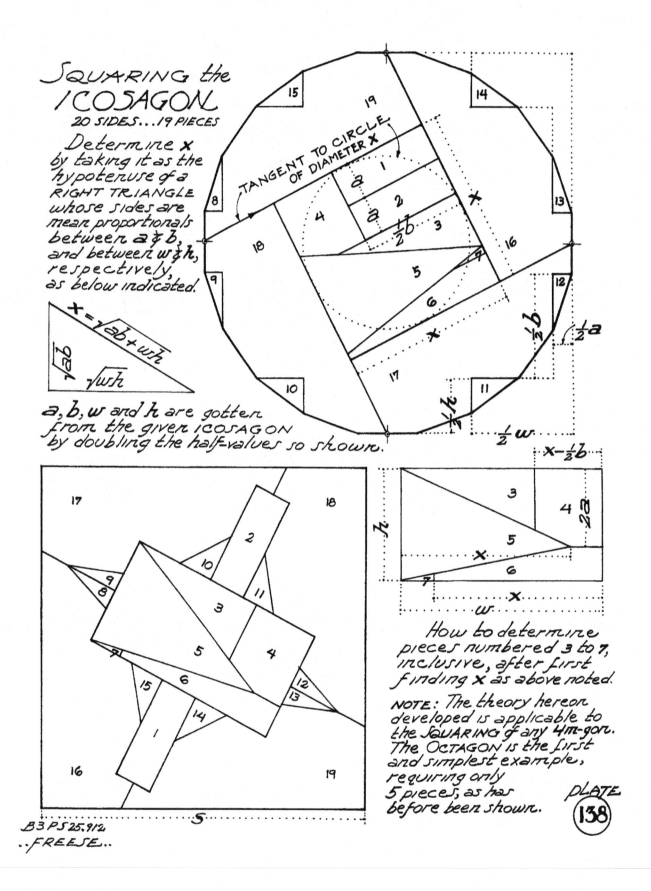

Squaring the ICOSAGON

20 SIDES ... 19 PIECES

Determine **x** by taking it as the hypotenuse of a RIGHT TRIANGLE whose sides are mean proportionals between a & b, and between w & h, respectively, as below indicated.

$$x = \sqrt{ab + wh}$$

$\sqrt{ab}$ $\sqrt{wh}$

a, b, w and h are gotten from the given ICOSAGON by doubling the half-values so shown.

TANGENT TO CIRCLE OF DIAMETER X

$\frac{1}{2}a$ $\frac{1}{2}b$ $\frac{1}{2}w$ $\frac{1}{2}h$ $\frac{1}{2}D$

How to determine pieces numbered **3 to 7**, inclusive, after first finding **x** as above noted.

NOTE: The theory hereon developed is applicable to the SQUARING of any 4m-gon. The OCTAGON is the first and simplest example, requiring only 5 pieces, as has before been shown.

$x - \frac{1}{2}b$ $2a$ h w

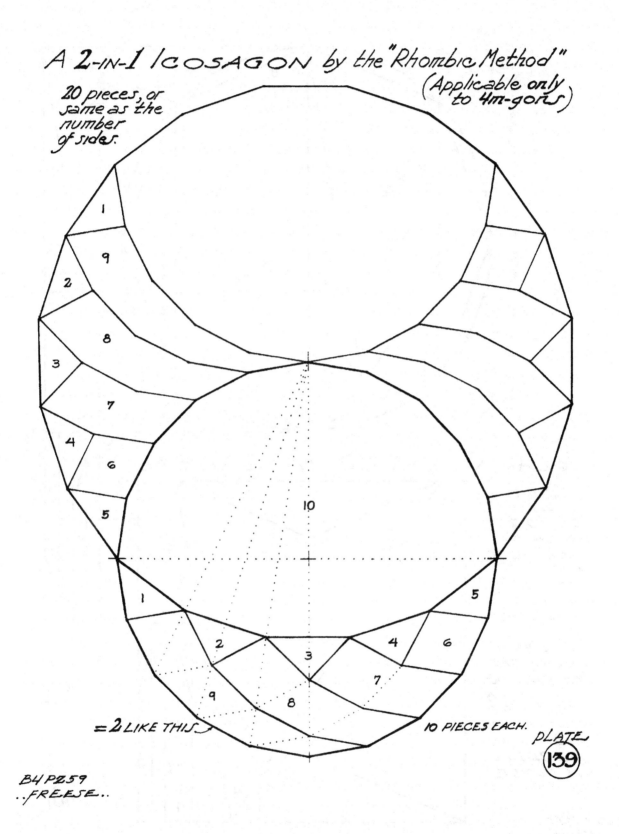

A 2-IN-1 ICOSAGON by the "Rhombic Method"
(Applicable only to 4m-gons)

20 pieces, or same as the number of sides.

=2 LIKE THIS

10 PIECES EACH.

B4 P259
..FREESE..

PLATE
139

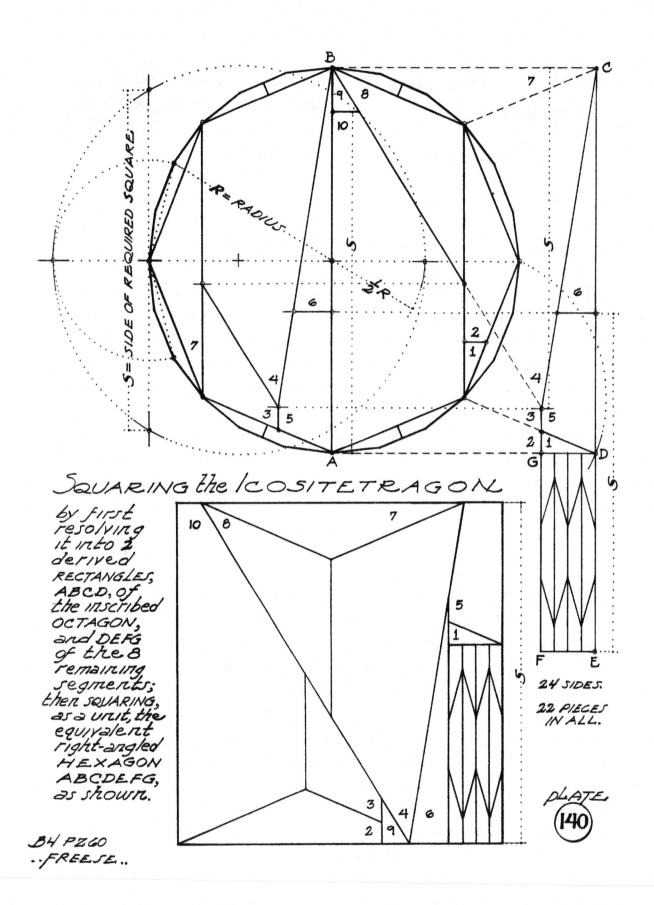

SQUARING the ICOSITETRAGON

by first
resolving
it into 2
derived
RECTANGLES,
ABCD, of
the inscribed
OCTAGON,
and DEFG
of the 8
remaining
segments;
then SQUARING,
as a unit, the
equivalent
right-angled
HEXAGON
ABCDEFG,
as shown.

S = SIDE OF REQUIRED SQUARE

R = RADIUS

2 R

24 SIDES.

22 PIECES
IN ALL.

PLATE
140

B4 PZ60
..FREESE..

A **2**-in-**1** ICOSITETRAGON, by the
(24 SIDES)
Rhombic Method.
24 PIECES

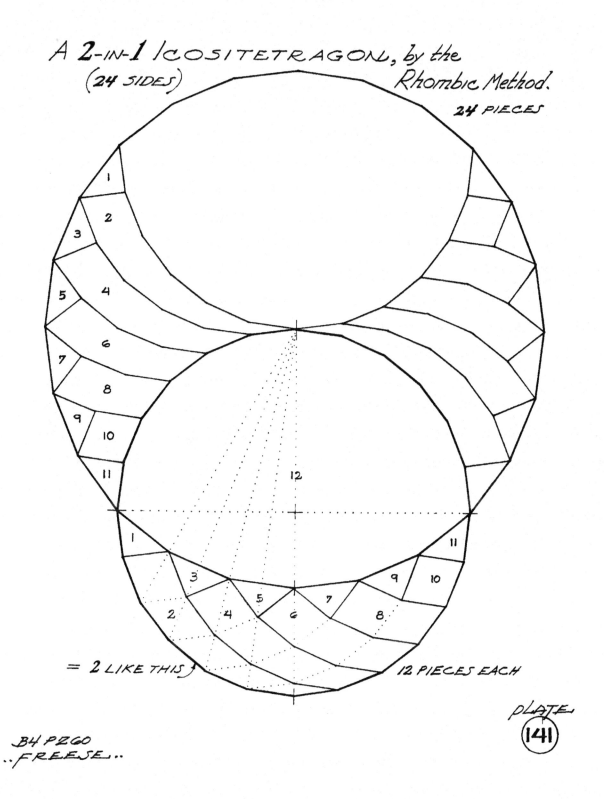

= 2 LIKE THIS }

12 PIECES EACH

PLATE
141

B4 P260
..FREESE..

Chapter 15

Miscellaneous Figures   (Plates 142–153)

The contents of this chapter are a grab-bag of offbeat dissections. From locating small polygons in the heart of larger polygons in Plate 142, to dissections with integral edge lengths in Plates 143-144, to dissections that have an artistic flair in Plates 146-147, the dissections in the first half of this chapter are decidedly different in concept from what we have seen so far.

Furthermore, the dissections in the second half are even more innovative. Plate 148 opens up the possibility of generalizations that Freese did not imagine. Plates 149-152 pose intriguingly novel problems, such as squares whose sidelengths come from the chords of regular octagons, squares that have been perforated, and various curvilinear figures. Finally, we see in Plate 153 a cut-corner square that is an eye-opener.

Enjoy the dizzying effect of the whirling squares in Plate 147, and snatch inspiration from the perforation in Plate 150!

Plate 142: For a given equilateral triangle, Freese showed how to identify an equilateral triangle of one thirteenth its area in the center of the given triangle, and similarly for a square and a hexagon. Freese based these "fractional dissections" on dissections of thirteen congruent figures to a larger figure. For the triangle, see Freese's Plate 18. For the hexagon, see Freese's Plate 73. For the square, see Emile Fourrey's 1907 book, which described the work of Abū'l-Wafā. Why did Freese not show Abū'l-Wafā's 21-piece dissection? Perhaps Freese did not know of that dissection.

Freese's contribution here was to use straight line segments to bound each of the small figures. For the square, each line segment runs from a corner of the given square to a point two thirds of the way along a nonadjacent side. (Such line segments are part of Abu'l-Wafa's dissection.) For the triangle, each line segment is from a midpoint of a side to two thirds of the way along an adjacent side. (Again, such line segments are part of the relevant dissection, in this case in Plate 18.) For the hexagon, each line segment is from a corner to the midpoint of a side almost directly opposite the corner. (Such line segments are not strictly part of the relevant dissection in Plate 73. However, Freese indicates such a line segment by dotted lines extending certain straight line segments in that figure.)

Plate 143: Freese discussed how to cut an equilateral triangle with integral side length into two scalene triangles whose side lengths are all integral. Starting with a 60° angle, he noted that $c^2 = a^2 - ab + b^2$, where a is the side length of the equilateral triangle, c is the length of an edge from the apex down to the base, and b is the length of that portion of the base from the 60° angle to the point on the base where the edge from the apex meets it. He had applied the Law of Cosines.

Freese noted that the scalene triangles come in pairs, (a, b, c) and (a, b_1, c), where $b_1 = a - b$. There is an interesting relationship between this problem and the one in Plate 163. To underscore this relationship, we do not require $b < b_1$ as Freese did, and we relabel b_1 as $\bar{a}$. We then generate quadruples of values a, b, c, and $\bar{a}$ with the following formulas. For whole numbers m and n, where $m > n$ and $m - n$ is not a multiple of 3, let $a = m^2 + 2mn$, $b = m^2 - n^2$, $c = m^2 + mn + n^2$, and $\bar{a} = a - b$.

Plate 144: Freese seems to have been the first to have given a 4-piece dissection of an $((m^2 + n^2) \times (m+n)^2)$-rectangle to an $m(m+n)$-square and an $n(m+n)$-square. He identified his technique for the case in which m and n are positive integers, but it works just fine when m and n are any pair of positive real numbers. If the two squares are attached, then we can merge pieces 2 and 4 so that three pieces suffice.

Plate 145: Freese defined an *L-shape*, or a "right-angled hexagon," as the merge of two rectangles such that they are adjacent along one side and they each have a side that lines up with the other. He noted that the L-shape has a 5-piece dissection to a square whenever the following two conditions hold: First, the side length of the square is at least half the length of the longer of the two side lengths that are adjacent to each other. Second, the side length of the square is at most the length of the shorter of the two sides that are adjacent to each other.

Freese essentially applied a P-slide to each of the two rectangles, but he was not the first to identify this technique. In his monthly puzzle column, "Perplexities," in the *Strand Magazine*, volume 73 (1927), pages 420 and 526, Henry Dudeney reproduced a dissection by George Wotherspoon that used precisely this technique. My 1997 book describes Wotherspoon's technique and calls it a "double P-slide." This dissection is translational.

Freese also noted that when both rectangles are squares, there is a 3-piece dissection. Earlier, George Biddle Airy had discovered that same dissection, which Augustus DeMorgan mailed to W. Rowan Hamilton in 1855, and which Robert Graves reported in (1889). The 3-piece dissection is both hingeable and translational, as noted by Philip Kelland in (1864). Freese indicated the position of the two hinges in his drawing at the bottom of the plate. Two other pairs of hinges are possible.

Plate 146: Freese dissected twenty squares to one, using the method of Abū'l-Wafā, which Emile Fourrey described in his 1907 book. Freese suggested the dissection's hingeability by illustrating the rotation. The dissection is also translational.

Plate 147: Freese was the first to dissect 25 congruent squares such that they could form a 5-square with a $\sqrt{17}$-square centered within it, such that the $\sqrt{17}$-square contains a 3-square centered within it, such that the 3-square contains a $\sqrt{5}$-square centered within it, such that the $\sqrt{5}$-square contains a 1-square centered with it. He called this multiple dissection the "Whirling Squares," because he tilted each successively formed square further in the same direction. He formed the $\sqrt{5}$-square by using the method of Abū'l-Wafā. His $\sqrt{17}$-square derives using the method of Abū'l-Wafā, but with his 3-square replacing nine 1-squares.

Plate 148: Freese displayed a 5-piece dissection that Henry Perigal (1873) had published earlier. Freese noted that for squares whose side lengths are rational, all of the cuts are of rational length and all areas of the resulting pieces are rational. When suitably arranged, the dissection is translational.

Freese erroneously claimed that there is just one case in which two 4-piece conversions are possible, namely for squares whose side lengths are in the ratio of $3:4:5$. However, Henry Dudeney and Sam Loyd had published many years earlier 4-piece dissections for not only $3:4:5$, but also for $5:12:13$ and $7:24:25$, and each of several different dissection methods applied in each case. In my 1997 book I gave a 4-piece method for all squares in the so-called Pythagoras class. The class is infinite, and it is easy to extend the different methods of Dudeney and Loyd to produce 4-piece dissections for each member of the class. In my book I also gave a 4-piece method for all squares in the so-called Plato class, and again it is easy to produce a second method with the same result.

A member of the Plato class is $8:15:17$, and I show two 4-piece dissections for it in Figure I148. The lower dissection differs from the upper one by having had two squares of side length 6 removed from the largest piece and merged onto the two smaller pieces. The dotted lines indicate how each square is merged onto its piece. A similar operation applies for each dissection in my (1997) book that I gave for the Plato class. Edo Timmermans, in (2005) and (2007), described many more variants of various 4-piece dissections of squares, some quite complicated.

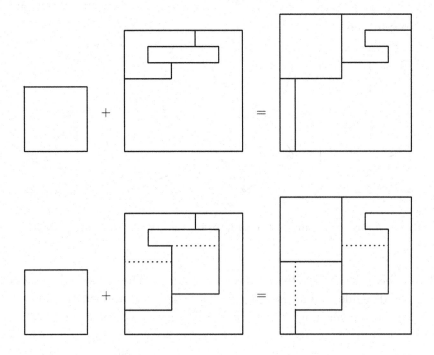

I148: Two 4-piece dissections of squares with side lengths of $8:15:17$ [GNF]

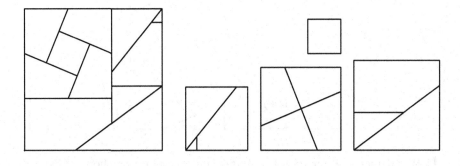

I149a: Improved square of the squares on the octagon's chords [GNF]

Plate 149: Freese was the first to dissect the squares of side lengths equal to the side and three chords of an octagon to a larger square. He repeatedly applied Perigal's dissection to get his 17-piece dissection. When suitably arranged, the dissection is translational.

However, Freese missed a simple relationship that helps reduce the number of pieces substantially: The area of the smallest square (with side length equal to the side of the octagon) plus the area of the square whose side length is the next-to-the-longest chord equals the area of the square whose side length is the longest chord. Indeed, this is easy to see by examining the diagram in the upper righthand corner of Plate 149, which contains a right triangle with dimensions equal to those three lengths.

Taking advantage of this relationship, we can find the 11-piece dissection in Figure I149a as follows: Use Perigal's dissection on the smallest and the third-smallest squares, producing a square of the same area as the largest square. Then use the P-slide on the largest square and also on the second-smallest square, so we can pack them into a resulting large square. When suitably arranged, the dissection is translational.

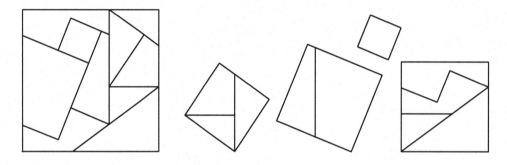

I149b: Further improved square of the squares on the octagon's chords [Gavin Theobald]

When this book was ready to go to the publisher, Gavin Theobald found a way to reduce the number of pieces to 10, while maintaining translationality. He applied a trick from Figures 4.15 and 4.16 of my 1997 book to get more mileage out of Perigal's dissection, resulting in Figure I149b.

There is another way to take advantage of the special relationship, which leads to a 12-piece hingeable dissection, as we see in Figures I149c and I149d. Note that pieces 11 and 12 form a quadrilateral congruent to piece 6, as do pieces 9 and 10. Also, pieces 1 through 5 form a quadrilateral similar to piece 6, as do pieces 7 through 10.

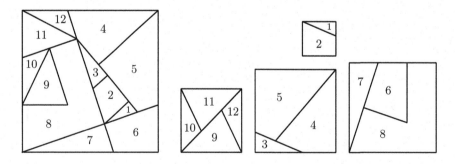

I149c: Hingeable square of the squares on the octagon's chords [GNF]

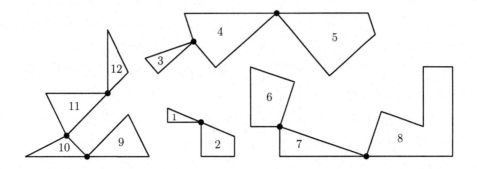

I149d: Hinging of previous square of the squares on the octagon's chords [GNF]

Plate 150: The dissections on this plate show an interesting application of Perigal's dissection. Freese fills the holes for the perforated squares on the left with pieces from a square cut from the middle of the perforated figures in the center. The middle disappears by the rearrangement of pieces as in Perigal's dissection (See Plate 35). Note that version III is translational, even though Freese did not display it as such.

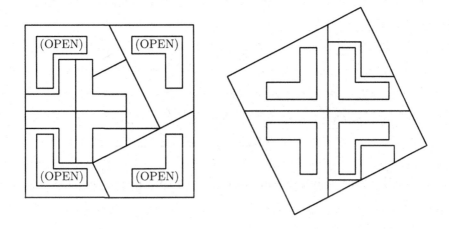

I150: Improved perforated square, version III, to a square [Gavin Theobald]

Gavin Theobald found a nifty way to improve Freese's dissection of the third version of a perforated square to a normal square. His 11-piece dissection in Figure I150 uses one fewer piece. Gavin employed what he calls a "hole dissection" to convert a small square in Freese's central panel into the four figures that fill in Freese's four perforations. Lovely! And still translational!

Plate 151: Freese showed how to dissect several curved figures to a square. Necessarily these curved figures have boundaries such that the concave boundaries exactly match the convex boundaries. Freese considered what has elsewhere been called a pendulum (the shape at the top left) and also an axe head (the shape in the middle). These shapes appeared in dissections contained in Leonardo da Vinci's manuscript *Codex Atlanticus*, which I discussed in my 2002 book. In that book I showed a hinged dissection of the same two figures as in Freese's dissection number 2.

Plate 152: This plate continues the curvilinear dissections from the previous plate and makes heavy use of the pendulum and the axehead. In particular, in his Figure II, Freese gave a 3-way

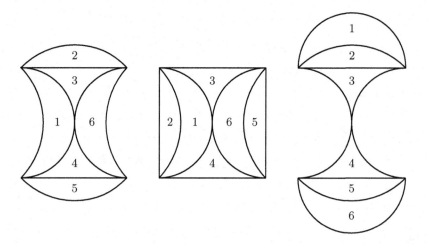

I152a: Hingeable dissection of axehead to square to double pendulum

dissection of an axehead to a square to a double pendulum. I label the pieces of this hingeable dissection in Figure I152a and specify a hinging of this dissection in Figure I152b.

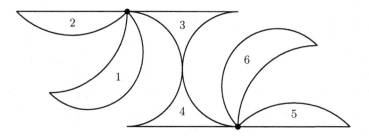

I152b: Hinging of axehead to square to double pendulum [GNF]

Plate 153: Freese considered a figure obtained by cutting an isosceles right triangle out from the corner of a square, and dissected the resulting figure to a square. As he pointed out, one quadrant of an octagon is such a figure. Yet following through directly with this observation would have led to a 5-piece dissection.

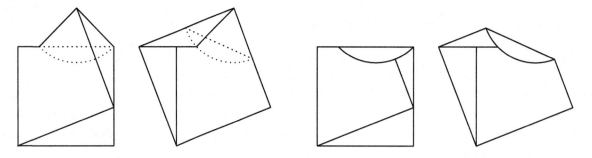

I153a: Square plus isosceles right triangle to square [Sam Loyd] with (dotted) alterations and its corresponding improved square to a "cut-corner square" [Gavin Theobald]

However Gavin Theobald discovered a 4-piece dissection, which we can derive from a 3-piece dissection of a square with an attached isosceles right triangle to a square. Sam Loyd published that dissection, shown with the solid lines on the left of Figure I153a, in his puzzle column in the *Philadelphia Press* on March 16, 1902. If we remove the isosceles right triangle at the top and cut a curved piece beneath it, as as shown with the dotted lines, we get Gavin's dissection, as we see on the right of that figure. Then we can slide the curved piece around to bound the diagonal cut for the cut-corner square. Note that Gavin's nifty dissection works over a wide range of ratios of side length of the isosceles right triangle to that of the square. What a great use of a curved cut!

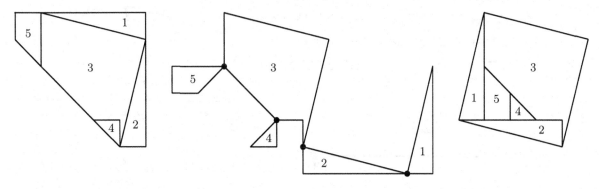

I153b: A hinged dissection of a different "cut-corner square" to a square [GNF]

If we try for a hinged dissection of a cut-corner square, then it seems that we need more pieces. For the case that $A = 5$ and $B = 4$, we can achieve a 5-piece hingeable dissection (Figure I153b).

In this chapter Freese broadened his range of dissections. The double P-slide in Plate 145 and even his having missed a 5-piece dissection of a cut-corner square embedded within Plate 92 are also worthy of appreciation. The sense of awe will continue in the next chapter, in which we will encounter some stunning designs and dissections of crosses.

3 EXAMPLES OF FRACTIONAL DISSECTION

TO CUT FROM THE CENTER OF
A GIVEN EQUILATERAL TRIANGLE,
SQUARE OR HEXAGON, A
SIMILAR POLYGON EQUAL
IN AREA TO $\frac{1}{13}$th THE AREA
OF THE GIVEN POLYGON.

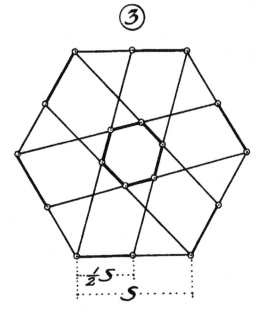

①

The central triangle
is $\frac{1}{13}$th the area of the
outer triangle.

②

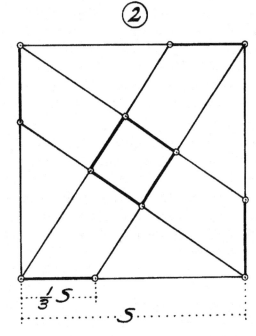

The central square
is $\frac{1}{13}$th the area of the
outer square.

③

The central hexagon
is $\frac{1}{13}$th the area of the
outer hexagon.

PLATE
142

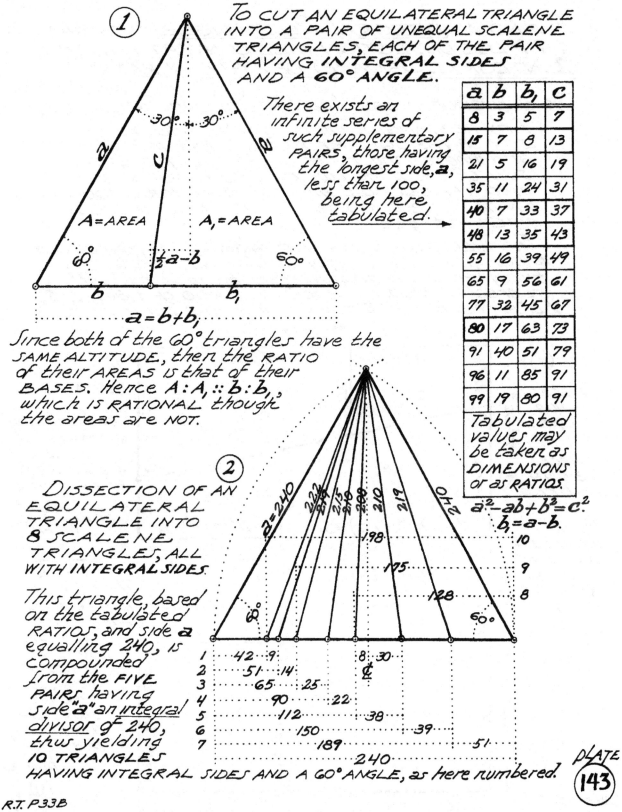

1 TO CUT AN EQUILATERAL TRIANGLE INTO A PAIR OF UNEQUAL SCALENE TRIANGLES, EACH OF THE PAIR HAVING *INTEGRAL SIDES* AND A *60° ANGLE.*

There exists an infinite series of such supplementary PAIRS, those having the longest side, **a**, less than 100, being here tabulated. ⟶

30° + 30°

a c a

A=AREA   A₁=AREA

60° $\frac{1}{2}a-b$ 60°

b b_1

$$a = b + b_1$$

Since both of the 60° triangles have the SAME ALTITUDE, then the RATIO of their AREAS is that of their BASES. Hence $A : A_1 :: b : b_1$, which IS RATIONAL though the areas are NOT.

a	b	b_1	c
8	3	5	7
15	7	8	13
21	5	16	19
35	11	24	31
40	7	33	37
48	13	35	43
55	16	39	49
65	9	56	61
77	32	45	67
80	17	63	73
91	40	51	79
96	11	85	91
99	19	80	91

Tabulated values may be taken as DIMENSIONS or as RATIOS.

$$a^2 - ab + b^2 = c^2$$
$$b_1 = a - b.$$

2 DISSECTION OF AN EQUILATERAL TRIANGLE INTO 8 SCALENE TRIANGLES, ALL WITH *INTEGRAL SIDES.*

This triangle, based on the tabulated RATIOS, and side **a** equalling 240, is compounded from the FIVE PAIRS having side "**a**" an *integral divisor* of 240, thus yielding 10 TRIANGLES

a=240 222 217 215 216 209 210 219 240

198   10

175   9

128   8

60°   60°

1	42	9		8	30
2	51	14		¢	
3	65	25			
4	90	22			
5	112			38	
6	150			39	
7	189			51	
	240				

HAVING INTEGRAL SIDES AND A 60° ANGLE, as here numbered.

PLATE 143

TO FIND AN _INTEGRAL RECTANGLE_, SUCH THAT A **4**-PIECE DISSECTION WILL CONVERT IT INTO **2** INTEGRAL SQUARES OF A GIVEN RATIO.

The derivation is based on the following THEOREM:

Any RECTANGLE whose shorter side is the SUM OF THE SQUARES of any two integers, and whose longer side is the SQUARE OF THE SUM of the same two integers, can be cut into 4 PIECES that will form 2 INTEGRAL SQUARES having respective sides in the RATIO of the two given integers.

The THEOREM is demonstrated by the formulated values given on the typical Diagrams, 1 & 2, the two given or assumed integers being m & n, with m greater than n; hence the RATIO of sides $S_1 : S_2$ of the resultant 2 INTEGRAL SQUARES being $m : n$.

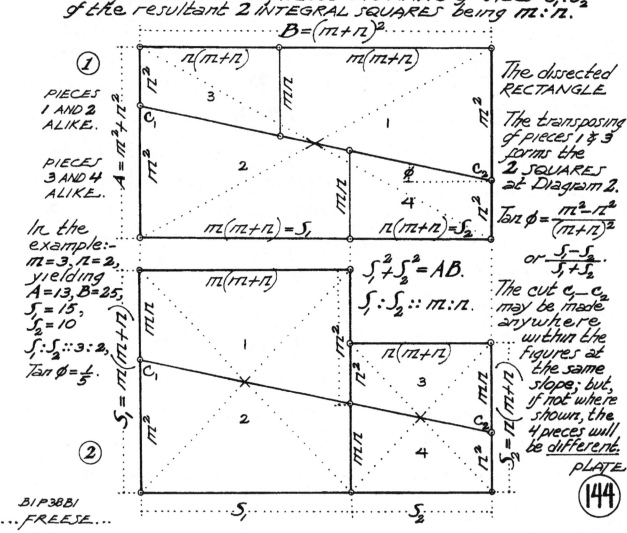

PIECES 1 AND 2 ALIKE.

PIECES 3 AND 4 ALIKE.

In the example:-
$m = 3$, $n = 2$,
yielding
$A = 13$, $B = 25$,
$S_1 = 15$,
$S_2 = 10$,
$S_1 : S_2 :: 3 : 2$,
$\mathrm{Tan}\,\phi = \frac{1}{5}$.

The dissected RECTANGLE

The transposing of pieces 1 & 3 forms the **2** SQUARES at Diagram 2.

$$\mathrm{Tan}\,\phi = \frac{m^2 - n^2}{(m+n)^2}$$

or $\frac{S_1 - S_2}{S_1 + S_2}$.

The cut $c_1 - c_2$ may be made anywhere within the figures at the same slope; but, if not where shown, the 4 pieces will be different.

$$S_1^2 + S_2^2 = AB.$$

$$S_1 : S_2 :: m : n.$$

pLATE

(144)

A SPECIAL 5-PIECE TRANSFORMATION OF A RIGHT-ANGLED HEXAGON, OR L-SHAPE, INTO A SQUARE. ✦

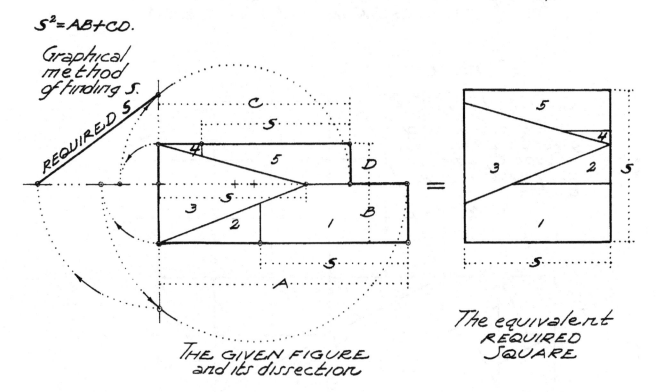

$S^2 = AB + CD.$

Graphical method of finding S.

REQUIRED S

THE GIVEN FIGURE and its dissection

=

The equivalent REQUIRED SQUARE

LIMITATION:
When S (as found by the given graphical method) is greater than C, or less than ½A, then a 5-piece dissection is impossible.

✦ When two squares form the L-shape, then a 3-piece dissection will suffice, as here shown. $S^2 = B^2 + C^2$

PLATE 145

B3 PS 25.911
...FREESE...

A *20-in-1* Square
and a graphic demonstration
of the GENERALITY of
The PYTHAGOREAN THEOREM

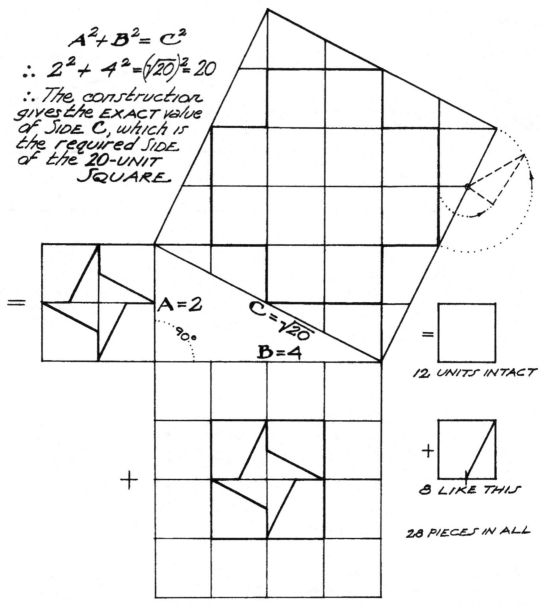

$$A^2 + B^2 = C^2$$

$$\therefore \ 2^2 + 4^2 = (\sqrt{20})^2 = 20$$

∴ The construction gives the EXACT value of SIDE C, which is the required SIDE of the 20-UNIT SQUARE

A = 2

C = √20

B = 4

90°

=

+

= 12 UNITS INTACT

+ 8 LIKE THIS

28 PIECES IN ALL

PLATE
146

B4 PA3
...FREESE...

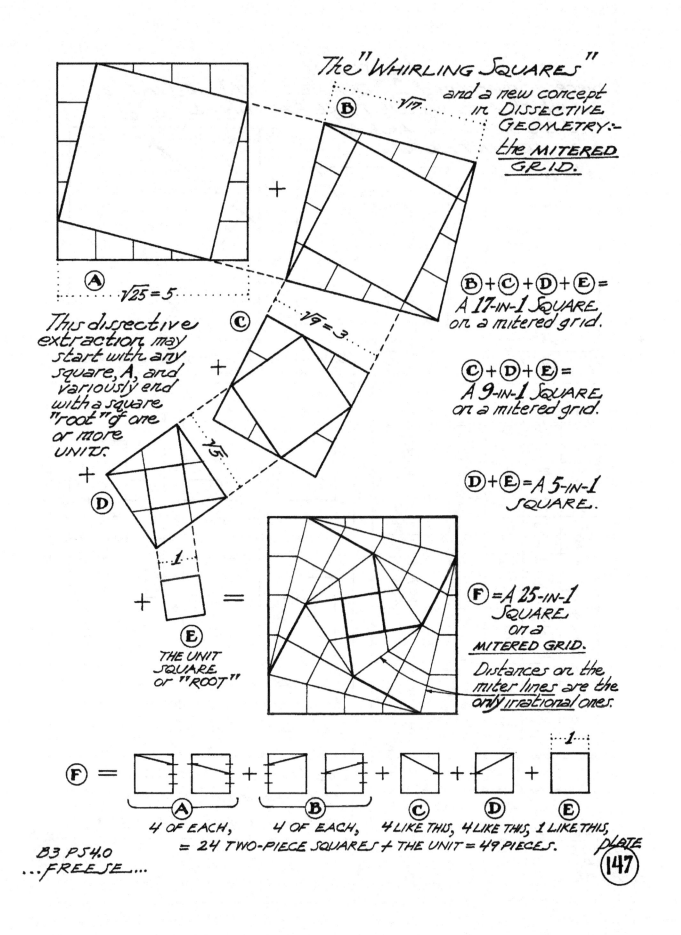

The "WHIRLING SQUARES"
and a new concept in DISSECTIVE GEOMETRY:— the MITERED GRID.

$\sqrt{17}$

Ⓑ

Ⓐ $\sqrt{25} = 5$.

This dissective extraction may start with any square, A, and variously end with a square "root" of one or more UNITS.

Ⓒ $\sqrt{9} = 3$.

Ⓓ $\sqrt{5}$

1

Ⓔ
THE UNIT SQUARE or "ROOT"

Ⓑ + Ⓒ + Ⓓ + Ⓔ = A 17-IN-1 SQUARE on a mitered grid.

Ⓒ + Ⓓ + Ⓔ = A 9-IN-1 SQUARE on a mitered grid.

Ⓓ + Ⓔ = A 5-IN-1 SQUARE.

Ⓕ = A 25-IN-1 SQUARE on a MITERED GRID.
Distances on the miter lines are the only irrational ones.

Ⓕ = [Ⓐ 4 OF EACH,] + [Ⓑ 4 OF EACH,] + Ⓒ 4 LIKE THIS, + Ⓓ 4 LIKE THIS, + Ⓔ 1 LIKE THIS, 1
= 24 TWO-PIECE SQUARES + THE UNIT = 49 PIECES.

B3 PS4.0
...FREESE....

PLATE
147

THE SERIES OF
5-piece conversions of 2 SQUARES into 1 SQUARE
for which ALL VALUES ARE RATIONAL
and the SIDES AND AREAS ARE INTEGRAL.

$A^2 + B^2 = C^2$

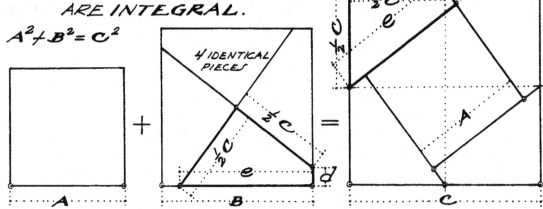

4 IDENTICAL PIECES

A + B = C

Numerical values may be taken either as actual DIMENSIONS or as RATIOS.

SIDES			d	e
A	B	C		
3	4	5	$\frac{1}{2}$	$3\frac{1}{2}$
5	12	13	$3\frac{1}{2}$	$8\frac{1}{2}$
7	24	25	$8\frac{1}{2}$	$15\frac{1}{2}$
8	15	17	$3\frac{1}{2}$	$11\frac{1}{2}$
9	40	41	$15\frac{1}{2}$	$24\frac{1}{2}$
11	60	61	$24\frac{1}{2}$	$35\frac{1}{2}$
12	35	37	$11\frac{1}{2}$	$23\frac{1}{2}$
13	84	85	$35\frac{1}{2}$	$48\frac{1}{2}$
15	112	113	$48\frac{1}{2}$	$63\frac{1}{2}$
16	63	65	$23\frac{1}{2}$	$39\frac{1}{2}$
17	144	145	$63\frac{1}{2}$	$80\frac{1}{2}$
19	180	181	$80\frac{1}{2}$	$99\frac{1}{2}$
20	21	29	$\frac{1}{2}$	$20\frac{1}{2}$
20	99	101	$39\frac{1}{2}$	$59\frac{1}{2}$
21	220	221	$99\frac{1}{2}$	$120\frac{1}{2}$
23	264	265	$120\frac{1}{2}$	$143\frac{1}{2}$
24	143	145	$59\frac{1}{2}$	$83\frac{1}{2}$
25	312	313	$143\frac{1}{2}$	$168\frac{1}{2}$

Tabulated values up to A=25

In all such cases, the sides are also those of INTEGRAL right triangles, and $d = \frac{1}{2}(B-A)$, $e = \frac{1}{2}(B+A) = B-d$.

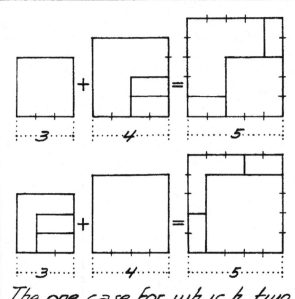

3 + 4 = 5

3 + 4 = 5

The one case for which two 4-piece conversions are also possible is, as above shown, when A=3, B=4, C=5, the General 5-piece Conversion thus requiring one piece more in this unique instance.

PLATE
148

B4 PA3.3
T.R.B.R1
..FREESE..

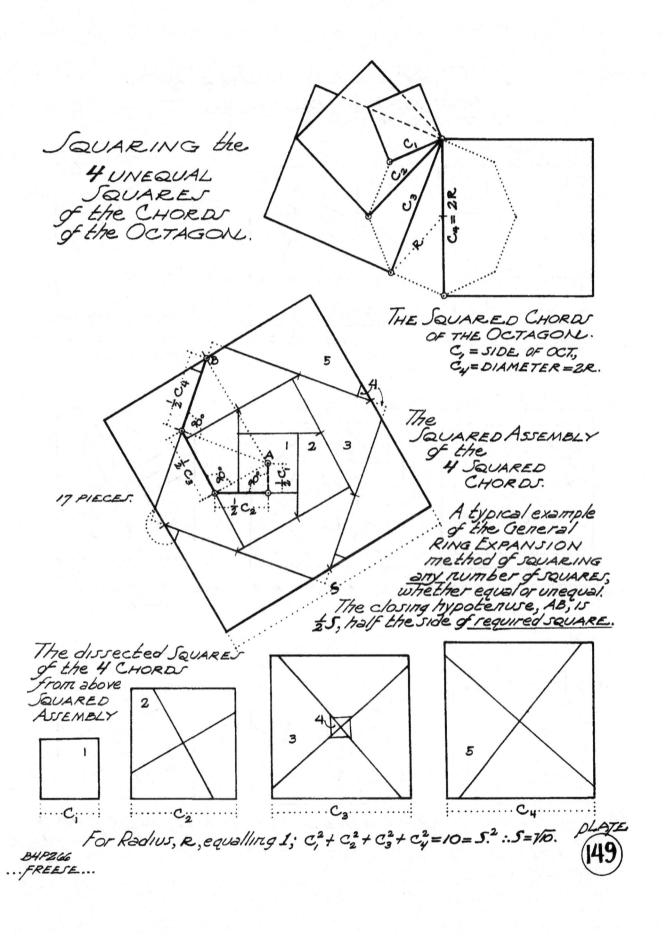

SQUARING the
4 UNEQUAL
SQUARES
of the CHORDS
of the OCTAGON.

$C_4 = 2R$

THE SQUARED CHORDS
OF THE OCTAGON.
C_1 = SIDE OF OCT.,
C_4 = DIAMETER = 2R.

The
SQUARED ASSEMBLY
of the
4 SQUARED
CHORDS.

17 PIECES.

A typical example
of the General
RING EXPANSION
method of SQUARING
any number of SQUARES,
whether equal or unequal.
The closing hypotenuse, AB, is
½S, half the side of required SQUARE.

The dissected SQUARES
of the 4 CHORDS
from above
SQUARED
ASSEMBLY

For Radius, R, equalling 1: $C_1^2 + C_2^2 + C_3^2 + C_4^2 = 10 = S^2$ ∴ $S = \sqrt{10}$.

PLATE 149

B4P266
...FREESE...

.... CONVERTIBLE SQUARES ....
from PERFORATED to SOLID
or vice versa

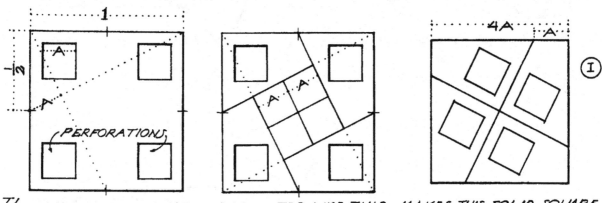

This PERFORATED SQUARE, DISSECTED LIKE THIS, MAKES THIS SOLID SQUARE.
8 PIECES.

①

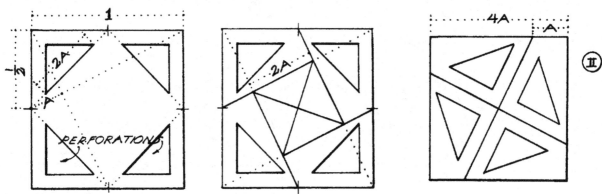

This PERFORATED SQUARE, DISSECTED LIKE THIS, MAKES THIS SOLID SQUARE.
8 PIECES.

②

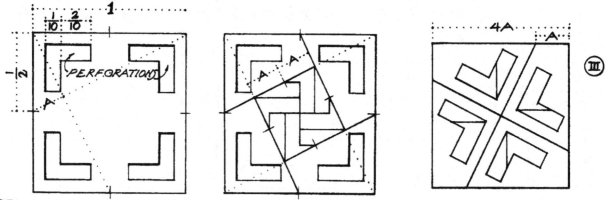

This PERFORATED SQUARE, DISSECTED LIKE THIS, MAKES THIS SOLID SQUARE.
12 PIECES

③

B3PS33.01
...FREESE...

PLATE
150

DISSECTIVE TRANSFORMATIONS OF CURVILINEAR ELEMENTS INTO SQUARES HAVING INTEGRAL AREAS

THE 1-PIECE UNIT

① R=1 R=1 √2

THE CURVILINEAR ELEMENT OF THE CIRCLE HAVING THE MAXIMUM INTEGRAL AREA EQUAL TO 2R². ∴ FOR R=1, AREA = 2.

② 3 PIECES

= √2 √2

AREA = (√2)² = 2.

③ 4 PIECES

= AREA = 2.

= + 1 1

AREA OF EACH SQUARE = 1.

④ 4 PIECES

AREA = 2.

= 3 2 1 + 4

LUNE.
AREA = 1.   AREA = 1.

THOUGH EACH OF THESE HAS AN INTEGRAL AREA, NEITHER CAN BE DISSECTED TO FORM A SEPARATE SQUARE.

= 2 4 1 3 √2 √2

AREA = 2.

PLATE 151

B4 P267
...FREESE...

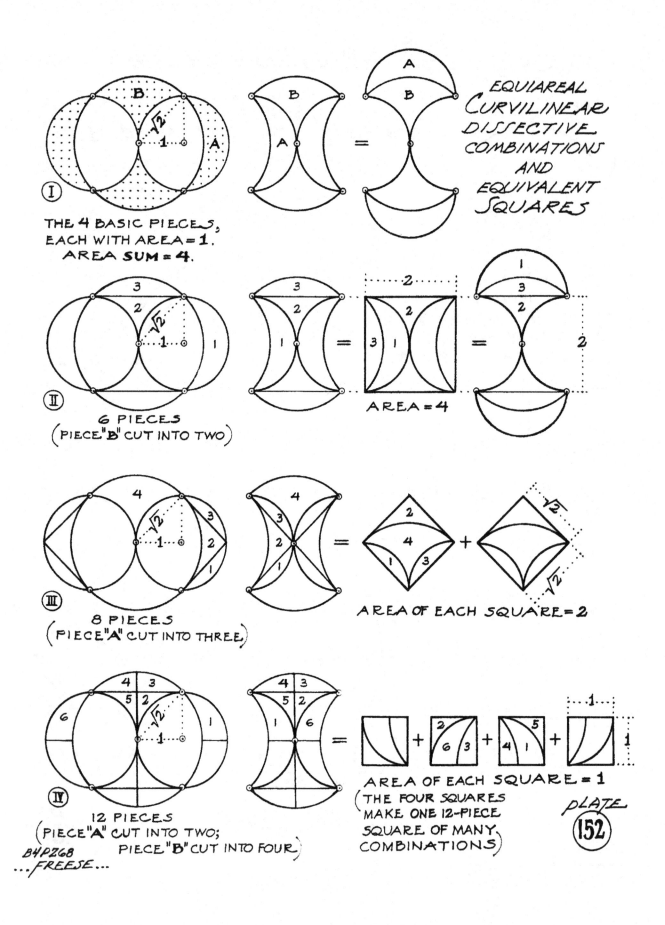

EQUIAREAL
CURVILINEAR
DISSECTIVE
COMBINATIONS
AND
EQUIVALENT
SQUARES

Ⓘ THE 4 BASIC PIECES,
EACH WITH AREA = 1.
AREA SUM = 4.

Ⓘ 6 PIECES
(PIECE "B" CUT INTO TWO)

AREA = 4

Ⓘ 8 PIECES
(PIECE "A" CUT INTO THREE)

AREA OF EACH SQUARE = 2

Ⓘ 12 PIECES
(PIECE "A" CUT INTO TWO;
PIECE "B" CUT INTO FOUR)

AREA OF EACH SQUARE = 1
(THE FOUR SQUARES
MAKE ONE 12-PIECE
SQUARE OF MANY
COMBINATIONS)

B4P268
...FREESE...

PLATE
152

$$S^2 = A^2 - \left(\tfrac{C}{2}\right)^2$$

Graphical
determination
of S.

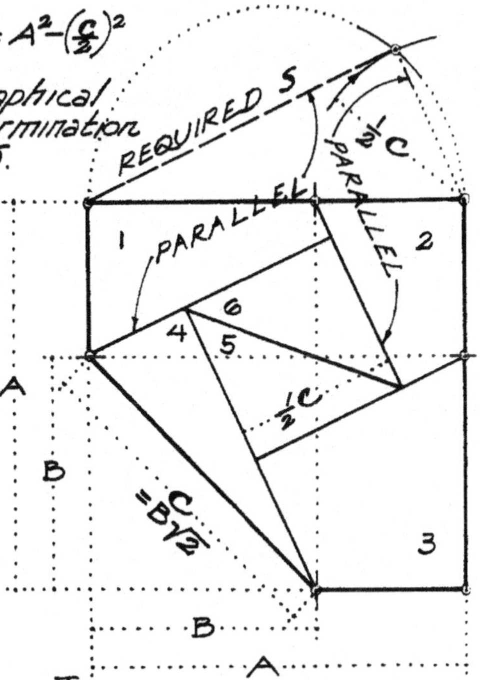

SQUARING A
CUT-CORNER SQUARE
by a direct 6-piece dissection.

REQUIRED S

PARALLEL

PARALLEL

½C

PARALLEL

1   2

4   6
5

½C

=C½
=B½

A

B

3

B

A

The GIVEN FIGURE
and its dissection.

=

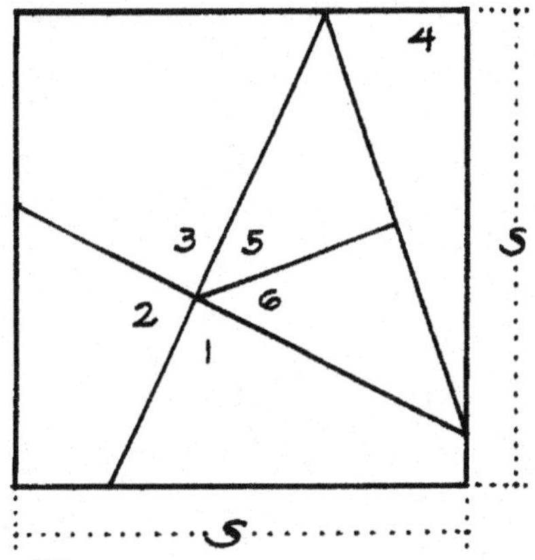

4

3   5

2   6

1

S

S

The equivalent
REQUIRED SQUARE
(S is always irrational)

LIMITATION: When B exceeds .656A, then a 6-piece
dissection is impossible.

The above direct method
applied to the SQUARING
of an OCTAGON QUADRANT,
hence transforming the
OCTAGON into 4 SQUARES,
or into 1 SQUARE by
combining the four.

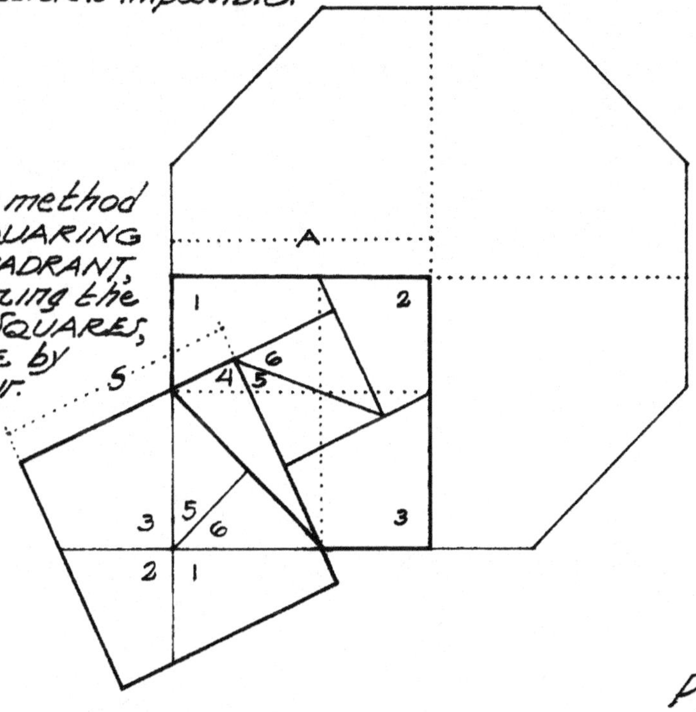

A

S

1   2

4   6
5

3   5
6   3

2   1

PLATE
153

B3 PS31.0
B3 PS31.1
...FREESE...

Chapter 16

More Crosses   (Plates 154–160)

This is a short but stupendous chapter. At first, Plates 154-157 feature some straightforward dissections involving Greek Crosses, a quadrate cross, a St. Andrew's Cross, a "stepped cross," and a crosslet. But then Plates 158-160 feature remarkable dissections based on the technique of completing the tessellation for four different more complicated crosses. For each of these latter four dissections, Freese indicated how to place some hinges, and we complete this task by making additional cuts so that the dissections are fully hingeable.

Not only are Freese's last four dissections thrilling, but the way to extend them to be fully hingeable is jaw-dropping! And if that isn't enough of a shock, Gavin Theobald found a way to reduce both dissections on Plate 160 by one piece.

Plate 154: Freese's dissections of the two square frames seem to have been original. We can hinge both of them, the top one with seven hinges, and the bottom one with six hinges. The latter has six rather than seven, because each of the two squares uses just three hinges. Both dissections are translational.

Plate 155: Freese was the first to pose and then solve a 4-way dissection of a swastika to a quadrate cross to a square to a pair consisting of a St. Andrews cross and a "tetraskelion." This multi-way dissection is translational. His 12-piece dissection is surprisingly symmetrical, but a bit of a letdown, since eight of the pieces are just unit squares.

After some experimentation, I found the 11-piece solution in Figure I155. Note that I simplified the tetraskelion, at the expense of the St. Andrews cross. On balance though, we have one fewer piece.

Plate 156: Freese gave 4-piece dissections of the two stepped crosses in Plates 30 and 31. In this plate he considered the stepped cross with yet another step per side and gave two different dissections of it to a square. The dissection at the top of this plate follows the same method as the dissections of both stepped crosses in the earlier plates. It is a mystery why he presented the second dissection, since it has many more pieces. Perhaps he was impressed because it can be partially hinged, as he indicates. Yet the dissection at the top of this plate can be fully hinged, as can the dissections of the stepped crosses in Plates 30 and 31. When suitably arranged, the dissection at the top is also translational.

Plate 157: Freese seems to have been the first to dissect a "crosslet" to a square. His 6-piece dissection would seem to be a minimal dissection, proved perhaps by an analysis of cases.

The dissection of a tetraskelion to a square can be derived by superposing a tessellation of tetraskelions with a tessellation of squares. It is hingeable, and when suitably arranged is also translational.

I155: Improved "equivalent figures" [GNF]

Plate 158: Freese defined a family of crosses that are generalizations of the Greek Cross, which he displayed in Plate 19. The generalized cross results from taking a square whose side length is $A + 2B$ and cutting from each corner a square whose side length is B, for values A and B with A sufficiently larger than B. Freese then noted that a completing the tessellation approach yields a 5-piece dissection. He also noted that the four identical pieces can be hinged.

In my 2002 book, I described exactly the same dissection, but noted that we can find a fully hinged rather than partially hinged dissection, if we use seven pieces rather than five. The fully hinged dissection, in Figure I158a, relies on the same sort of construction that I employed in Figure I89a. We see the pieces and hinges swung out in a line in Figure I158b. If you swing the pieces counter-clockwise around piece 1, you will obtain the cross, while if you swing the pieces clockwise around piece 1, you will obtain the square.

Plate 159: When it comes to dissecting geometric figures such as crosses, there is some latitude on exactly how to define those figures. In this plate Freese defined a family of Maltese Crosses, in which the angle between two adjacent arms of the cross is precisely 45°. Using a variation of the completing the tessellation approach, Freese then described a 9-piece dissection for each such cross to a square, applicable whenever the cross is sufficiently "fat".

Curiously, Freese did not mention a competing definition of the Maltese Cross, which Henry Dudeney described in (1926). Using a 5×5 grid, Dudeney laid out a Maltese Cross so that the "elbows" between the arms are located at the vertices of the small square at the center of the grid, and the outer corners of each arm are at vertices of the small squares at the corners of the grid. Readers should recognize this version of the Maltese Cross as the one in Figure 1.11 in our introduction, where we have already savored A. E. Hill's marvelous 7-piece dissection. In my 1997 book I generalized that definition to grids that are 7×7, 9×9, and so on, and identified a 7-piece

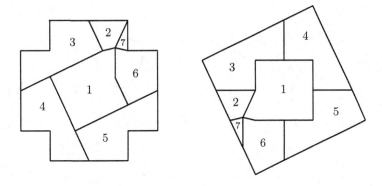

I158a: Hingeable generalized Greek Cross to a square [GNF]

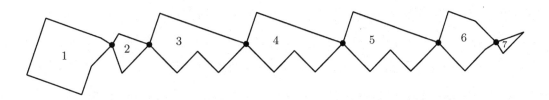

I158b: Hinged pieces for generalized Greek Cross to a square [GNF]

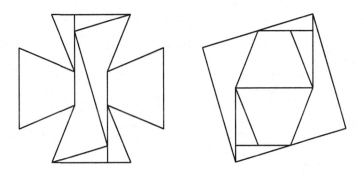

I159a: Certain 45° Maltese cross to a square [Harry Lindgren]

dissection to a square for each such Maltese Cross!

The 7-piece dissections that I described are for grid-based crosses that are "well-proportioned," where the diagonal distance between bases of the arms is equal to the distance between adjacent tips of the arms. As it turns out, there is exactly one 45 degree angled cross that is well-proportioned. For this particular cross there is an 8-piece dissection, shown in Figure I159a, that was discovered by Harry Lindgren (1961).

Freese gave a "limiting case" dissection in the lower right corner of his plate, in which he indicated two hinges on the outer pieces in the dissection. These hinges will also work in his general dissection too, and we can actually place one more hinge to connect all four outer pieces together. This suggests finding a fully hingeable dissection of this cross to a square. In Freese's dissection, the isosceles right triangles cause some problems, but we can split each into two smaller isosceles triangles that we hinge together and also hinge to the small central square. We then split

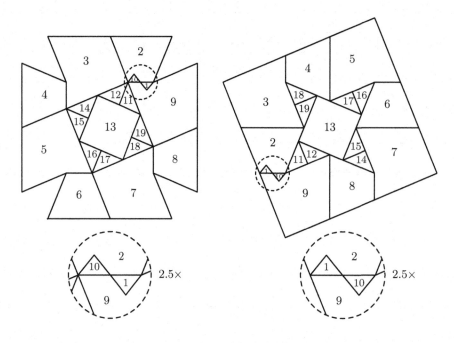

I159b: Hingeable dissection of 45° Maltese cross to a square [GNF]

each outer piece into two pieces, and cut out a pair of small pieces that interchange their positions, producing the hingeable dissection in Figure I159b.

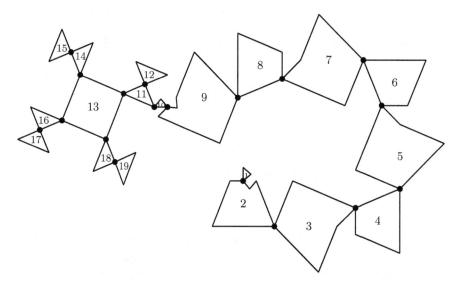

I159c: Hinged pieces for 45° Maltese cross to a square [GNF]

We hinge these nineteen pieces as in Figure I159c. Notice that the pieces appear in counterclockwise order around the cross, starting with piece 1, while pieces 1-10 appear in clockwise order around the (large) square, starting with piece 1, and pieces 11-19 appear (more or less) in counterclockwise order from piece 11 onward.

To unfurl the pieces from the cross, proceed as follows: Swing pieces 1-4 as a unit around the hinge connecting pieces 4 and 5. Continue by swinging pieces 1-6 as a unit around the hinge connecting pieces 6 and 7. Then swing pieces 1-8 around the hinge connecting pieces 8 and 9, and finally unfurl pieces 9-19. To swing out the pieces from the square: Swing pieces 1-3 as a unit around the hinge connecting pieces 3 and 4. Next swing pieces 1-5 as a unit around the hinge connecting pieces 5 and 6. Then swing pieces 1-7 around the hinge connecting pieces 7 and 8. Then swing pieces 1-9 around the hinge connecting pieces 9 and 10, and finally unfurl pieces 11-19.

Plate 160: The two dissections in this plate are real gems! They are elaborate applications of the completing the tessellation technique. To describe his geometric constructions, Freese used the architectural term "spandrel." A spandrel is the space between two arches or between an arch and a rectangular enclosure. In the examples that Freese displayed, he called the four spaces between a cross and the enclosing square the four spandrels.

Freese considered two figures, which we shall call F_1 and F_2, each with 4-fold rotational symmetry, such that F_1 and F_2 together tile the plane and F_2 has smaller area than F_1. He then considered two squares, square S_1 of area equal to that of F_1 and square S_2 of area equal to that of F_2. He superposed the tiling of figures F_1 and F_2 with a tiling of the two different-sized squares, and found a dissection of F_2 to square S_2. This then gave him a dissection of F_1 to S_1.

In the top half of this plate, Freese illustrated his technique by having F_2 be a stepped cross. (See Plates 30 and 31 for examples of a stepped cross.) Then F_1 is an unusual sort of cross, and an 8-piece dissection of F_1 to S_1 results. In the lower half of the plate, Freese chose F_2 to be a regular octagon, which makes F_1 a sort of cross. If we use the 5-piece dissection of a regular octagon to a square, the resulting dissection of F_1 to square S_1 takes nine pieces. Forty years later, in my 1997 book on page 155, I gave two other dissections that are of the same general type, namely an 8-pointed star $\{8/3\}$ to a square, and a 12-pointed star $\{12/4\}$ to a square. However those dissections do not involve what might be called spandrels.

An interesting feature of both dissections in this plate is that we can partially hinge them. Freese identified two of four possible hinges for the outer four pieces from F_1 in both dissections, yet did not note that we can also hinge the dissection of the stepped cross (Figure F_2).

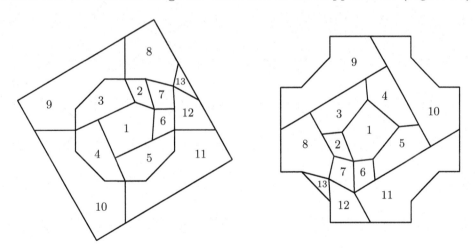

I160a: Hingeable dissection of octagon-based cross to a square [GNF]

We can convert Freese's dissection of an octagon-based cross to a square into a hingeable dissection in a manner similar to my conversion of the dissection of an octagon to a square as shown in Figure I89a. Just as I dissected one of the four congruent pieces in the octagon-to-square dissection,

I dissect one of Freese's four congruent large pieces, so that two of the resulting pieces are congruent triangles. Just as I glued a congruent triangle to the small square in the octagon-to-square hingeable dissection, I glue a congruent triangle to the remaining triangle from the octagon-to-square dissection, giving piece 7. This gives the 13-piece hingeable dissection in Figure I160a.

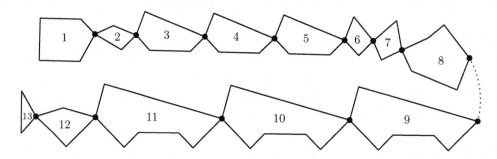

I160b: Hinged pieces for octagon-based cross to a square [GNF]

The thirteen pieces hinge together in one long chain as shown in Figure I160b. (Since the chain is too long to show in line, I have split it into two chains, where the rightmost hinges in both chains should actually be the same hinge, as indicated by the dotted line connecting them.) Winding the whole chain clockwise around piece 1 gives the cross, while winding the chain counterclockwise around piece 1 gives the square. This is perhaps the most artful of the hinged dissections in this book, with Freese's double application of the completing tessellation technique, adapted to hinging, and employing two compound pieces (namely, pieces 1 and 7).

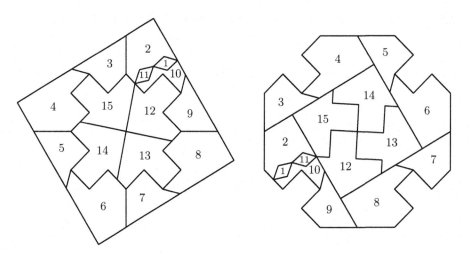

I160c: Hingeable stepped-spandrel-based cross to a square [GNF]

Let's return to Freese's dissection in the top half of the plate. The dissection is partially hingeable, and Freese identified the locations of two hinges for the stepped-spandrel-based cross. We can also hinge the stepped cross itself and then drop it into the hole in the center of the square. But to be fully hinged, we must hinge the pieces of the stepped cross to the remaining pieces.

We could cut one of the outer four pieces of Freese's square to accomplish the hinging, but I suspect that Freese would not approve, because we would need "wobbly hinges," which I defined

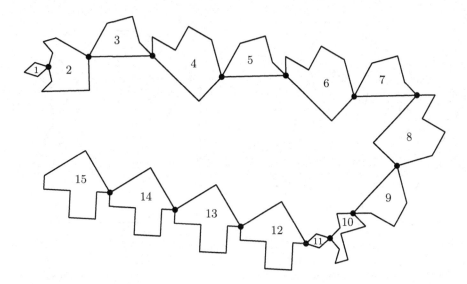

I160d: Hinged pieces for stepped-spandrel-based cross to a square [GNF]

on page 15 of my 2002 book. Instead, make four additional cuts so that we can swing the four blunt ends of the stepped cross into the appropriate cavities of the outer pieces. Thus we cut the four pieces that form the outside of Freese's square into pieces 3 and 4, pieces 5 and 6, pieces 7 and 8, and pieces 1, 2, 9, 10, and 11, giving the 15-piece hingeable dissection in Figure I160c, with its hinged pieces in Figure I160d.

Starting with piece 1 in the square, the pieces wind counterclockwise around the exterior of the square until piece 10, after which the pieces wind clockwise around in the center. The directions are reversed as we form the stepped-spandrel-based cross. However, we must take care to unfurl the pieces around without obstruction, as I described on pages 14 and 15 of my 2002 book. Here are the reverse operations for opening up the figures. Starting with the square, we first rotate pieces 1, 2, and 3 as a unit around the hinge between pieces 3 and 4. With piece 2 unblocked from piece 11, rotate pieces 1 and 2 as a unit around the hinge between pieces 2 and 3. With pieces 2 and 3 unblocked from piece 15, rotate pieces 1 through 5 as a unit around the hinge between pieces 5 and 6. Next we rotate pieces 1 through 4 as a unit around the hinge between pieces 4 and 5. Persisting with this pattern unlocks pieces 1 through 11, after which we release pieces 12 through 15.

On the other hand, if we unfurl the stepped-spandrel-based cross, we rotate pieces 1 through 4 as a unit around the hinge between pieces 4 and 5. Then we rotate pieces 1 through 6 as a unit around the hinge between pieces 6 and 7. This sequence eventually frees pieces 1 through 8. Finally we rotate pieces 1 through 13 around the hinge between pieces 13 and 14.

Because the prospect of fully hinging the two special crosses had been irresistible, that prospect played a role in diverting attention from a greater goal, namely reducing the number of pieces in Freese's original two dissections of crosses. It fell to Gavin Theobald to question if Freese's remarkable dissections used the fewest number of pieces. And Gavin delivered a stunner by showing that each of Freese's two dissections could be performed with one fewer piece.

For the stepped-spandrel-based cross, Gavin discovered that instead of dissecting a stepped cross into four pieces to form a square, he could slice the cross into just two pieces. He could then fit the two pieces together and cut a third piece out of a neighboring piece to form an indentation that would accommodate the pair of fitted-together pieces. The third piece would then accommodate the other end of the pair, as we see in Figure I160e.

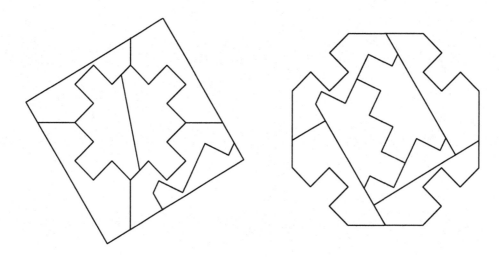

I160e: Improved stepped-spandrel-based cross to a square [Gavin Theobald]

The same sort of game plan works for the dissection of the octagon-based cross to a square. The shapes are a bit trickier in Figure 160f, where Gavin expanded the larger square hidden within the octagon-based cross to include an irregular shape, for which Gavin cut an accommodating notch out of a piece from the largest square. Gavin took care to design the pieces so that they would fit into the now-cramped confines of both the cross and the square. Lost now is any hint of the symmetry of the original dissection, but what remains is the amazement that Gavin could have pulled off this engineering feat!

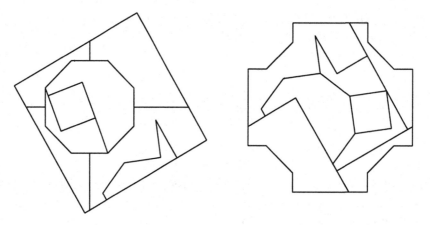

I160f: Improved octagon-based cross to a square [Gavin Theobald]

What a treat to close out a chapter with a flurry of wild techniques! First, Freese applied the completing the tessellation technique to produce four stunning dissections in the last three plates. Second, we gulped down Freese's bait and converted those four dissections from being partially hingeable to being fully hingeable. And then Gavin Theobald knocked our socks off by knocking a piece off each of Freese's last two dissections! What awaits us in the next chapter are more offbeat dissection problems, including concave polygons, and star-bursting polygrams with matching co-polygrams—wild stuff!

MULTIPLE TRANSFORMATIONS OF A SQUARE FRAME.

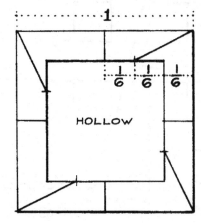

1

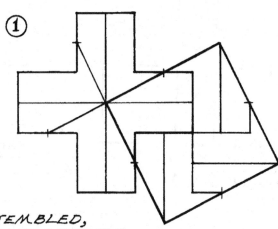

An 8-PIECE CUT-UP, RE-ASSEMBLED,
MAKES EITHER
A GREEK CROSS OR
A SOLID SQUARE.

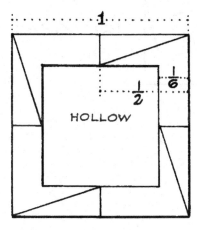

2

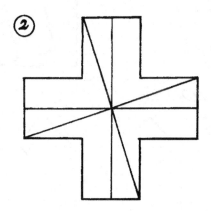

A SLIGHTLY DIFFERENT CUT-UP THAT FORMS THE SAME CROSS
OR 2 SOLID SQUARES.

8 PIECES.

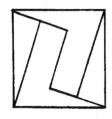

NOTE:
IN BOTH EXAMPLES, IF THE
DIAGONAL LINES WERE
OMITTED, THEN NO SQUARES,
BUT ONLY THE CROSS, COULD
BE FORMED FROM THE FOUR
REMAINING RIGHT-ANGULAR
PIECES.

B3PS33.02
...FREESE...

PLATE
154

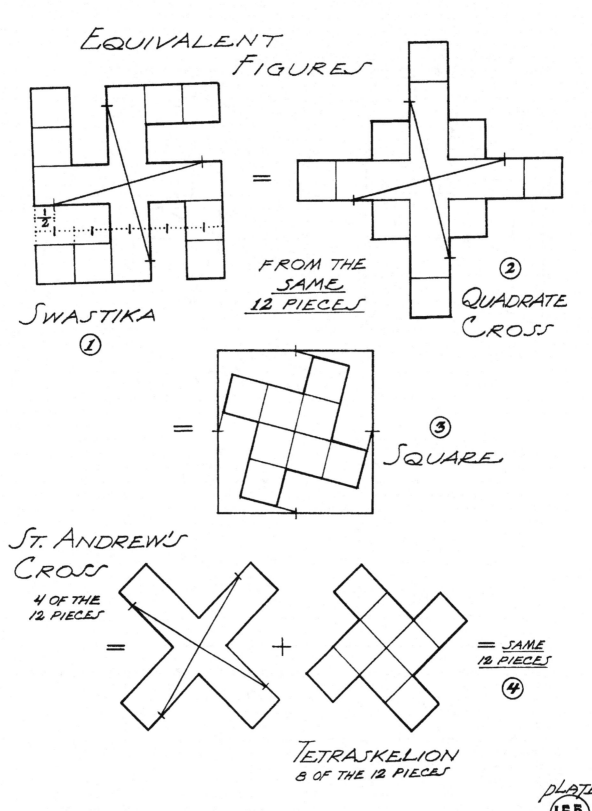

EQUIVALENT FIGURES

SWASTIKA
①

= FROM THE SAME 12 PIECES

QUADRATE CROSS
②

= SQUARE
③

ST. ANDREW'S CROSS
4 OF THE 12 PIECES

= X + ⊕ = SAME 12 PIECES
④

TETRASKELION
8 OF THE 12 PIECES

B4 PZ63
...FREESE...

PLATE
155

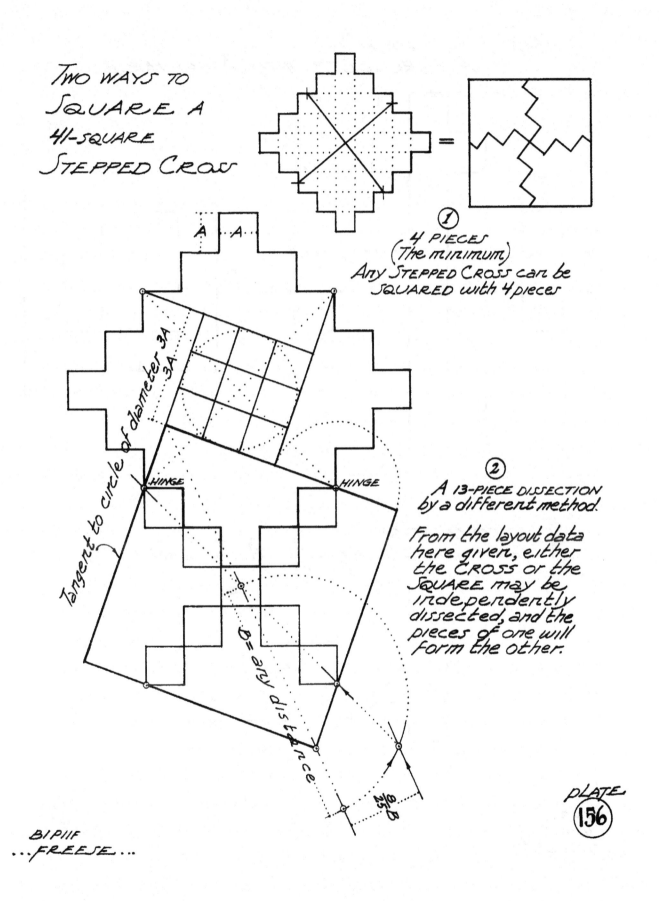

TWO WAYS TO
SQUARE A
41-SQUARE
STEPPED CROSS

=

①

4 PIECES
(The minimum)
Any STEPPED CROSS can be
SQUARED with 4 pieces

A A

Tangent to circle of diameter 3A
3A

HINGE   HINGE

B = any distance

②

A 13-PIECE DISSECTION
by a different method.

From the layout data
here given, either
the CROSS or the
SQUARE may be
independently
dissected, and the
pieces of one will
form the other.

$\frac{9}{25}$ B

PLATE
156

B1 P11F
...FREESE...

TRANSFORMATION OF
A CROSSLET INTO A SQUARE

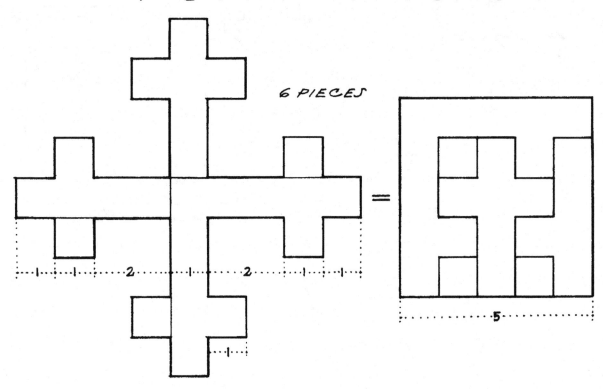

6 PIECES

2   2

=

5

TRANSFORMATION OF
A TETRASKELION INTO A SQUARE
4 PIECES

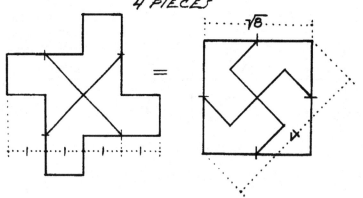

√8

=

4

PLATE
157

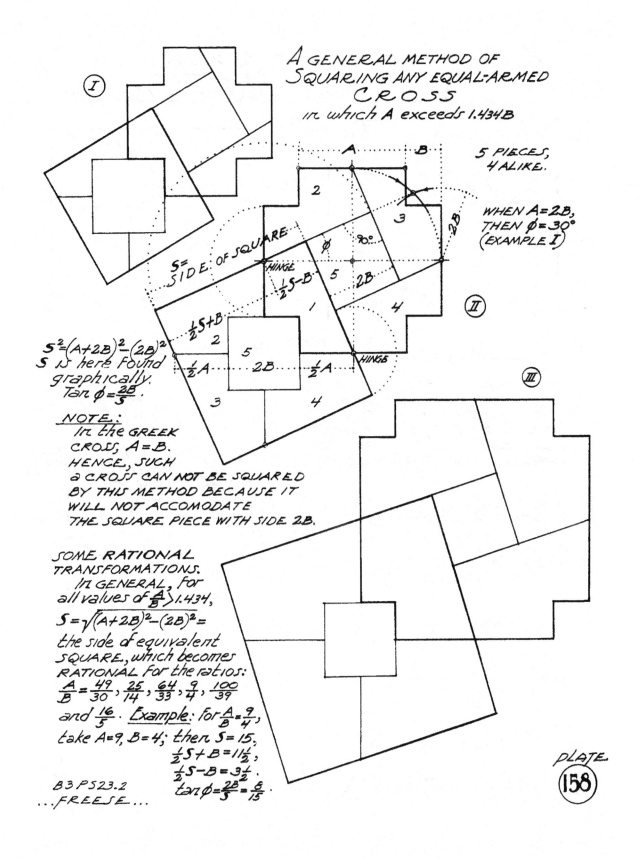

A GENERAL METHOD OF
SQUARING ANY EQUAL-ARMED
CROSS
in which A exceeds 1.434B

5 PIECES,
4 ALIKE.

WHEN A=2B,
THEN ϕ=30°
(EXAMPLE I)

S=
SIDE OF SQUARE

HINGE

90°

ϕ

$\frac{1}{2}$S-B

5   2B

1

4

2

3

2B

HINGE

$\frac{1}{2}$S+B

2

5

$\frac{1}{2}$A   2B $\frac{1}{2}$A   HINGE

3   4

I

II

III

$S^2=(A+2B)^2-(2B)^2$
S is here found
graphically.
Tan $\phi=\frac{2B}{5}$.

NOTE:
In the GREEK
CROSS, A=B.
HENCE, SUCH
a CROSS CAN NOT BE SQUARED
BY THIS METHOD BECAUSE IT
WILL NOT ACCOMODATE
THE SQUARE PIECE WITH SIDE 2B.

SOME RATIONAL
TRANSFORMATIONS.
In GENERAL, for
all values of $\frac{A}{B}$>1.434,
$S=\sqrt{(A+2B)^2-(2B)^2}=$
the side of equivalent
SQUARE, which becomes
RATIONAL for the ratios:
$\frac{A}{B}=\frac{49}{30},\frac{25}{14},\frac{64}{33},\frac{9}{4},\frac{100}{39}$
and $\frac{16}{5}$. Example: for $\frac{A}{B}=\frac{9}{4}$,
take A=9, B=4; then S=15,
$\frac{1}{2}$S+B=11$\frac{1}{2}$,
$\frac{1}{2}$S-B=3$\frac{1}{2}$,
tan $\phi=\frac{2B}{5}=\frac{8}{15}$.

B3 PS23.2
...FREESE...

PLATE
158

A GENERAL METHOD OF SQUARING
the variable
MALTESE CROSS

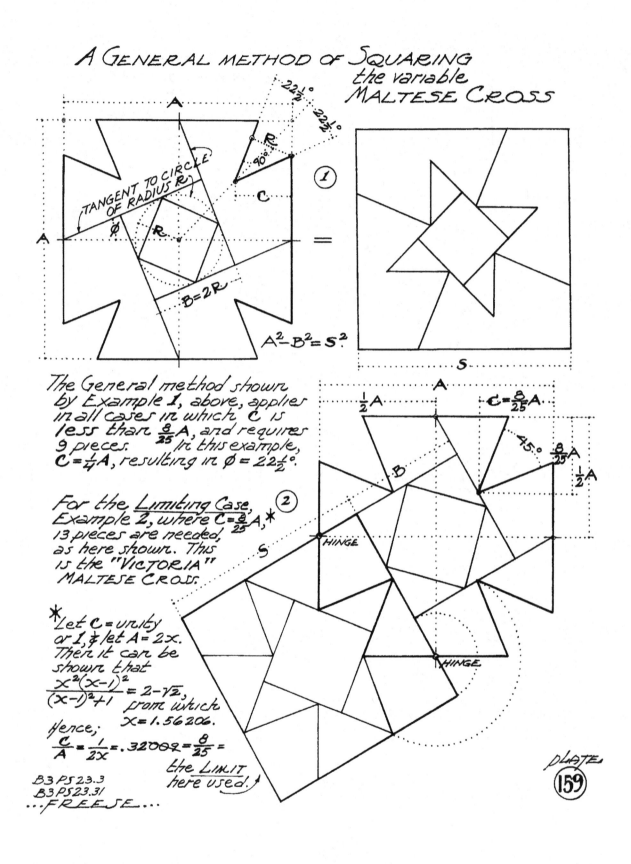

TANGENT TO CIRCLE OF RADIUS R

$22\frac{1}{2}°$

$22\frac{1}{2}°$

90°

$\textcircled{1}$

R

C

Ø

R

A

A

B = 2R

$A^2 - B^2 = S^2$

=

S

The General method shown by Example 1, above, applies in all cases in which C is less than $\frac{8}{25}A$, and requires 9 pieces. In this example, $C = \frac{1}{4}A$, resulting in Ø = $22\frac{1}{2}°$.

For the *Limiting Case*, Example 2, where $C = \frac{8}{25}A$,* 13 pieces are needed, as here shown. This is the "VICTORIA" MALTESE CROSS.

$\textcircled{2}$

$\frac{1}{2}A$

A

$C = \frac{8}{25}A$

B

45°

$\frac{8}{25}A$

$\frac{1}{2}A$

S

HINGE

HINGE

*Let C = unity or 1, & let A = 2x. Then it can be shown that
$$\frac{x^2(x-1)^2}{(x-1)^2+1} = 2-\sqrt{2},$$
from which x = 1.56206.

Hence;
$$\frac{C}{A} = \frac{1}{2x} = .32008 = \frac{8}{25} =$$
the LIMIT here used!

B 3 PS 23.3
B 3 PS 23.31
...FREESE...

PLATE
159

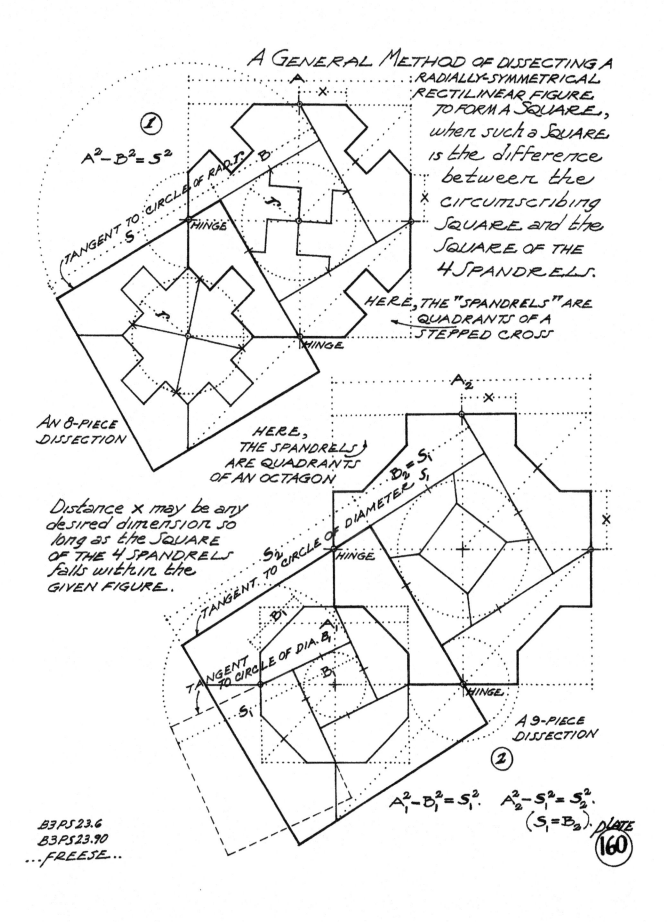

A GENERAL METHOD OF DISSECTING A RADIALLY-SYMMETRICAL RECTILINEAR FIGURE TO FORM A SQUARE, when such a SQUARE is the difference between the circumscribing SQUARE and the SQUARE OF THE 4 SPANDRELS.

①

$A^2 - B^2 = S^2$

A

X

TANGENT TO CIRCLE OF RAD.T.

B

S

HINGE

HINGE

X

HERE, THE "SPANDRELS" ARE QUADRANTS OF A STEPPED CROSS

AN 8-PIECE DISSECTION

HERE, THE SPANDRELS ARE QUADRANTS OF AN OCTAGON

Distance x may be any desired dimension so long as the SQUARE OF THE 4 SPANDRELS falls within the GIVEN FIGURE.

A_2

X

$B_2 = S_1$
S_1

TANGENT TO CIRCLE OF DIAMETER S_1

S_1

HINGE

X

B_1

A_1

TANGENT TO CIRCLE OF DIA. B_1

B_1

TANGENT TO CIRCLE OF DIA. B_1

S_1

HINGE

A 9-PIECE DISSECTION

②

$A_1^2 - B_1^2 = S_1^2$. $A_2^2 - S_1^2 = S_2^2$.
$(S_1 = B_2)$. PLATE

Chapter 17

More Miscellaneous Figures   (Plates 161–178)

This chapter features dissections of several distinct varieties. First come dissections in which a figure has a central symmetric core surrounded by a symmetric arrangement of pieces. These include Plates 161, 165, 167, 168, and 169 and involve a pentagon, an octagon, an enneagon, a decagon, and a dodecagon as the central cores. A second theme is combining figures that have sides of the same length, as in Plates 171, 172, and 173. Finally are Plates 174-178, which all involve what Freese calls "concave polygons."

As beautiful as many of Freese's dissections are in this chapter, they have achieved more impact by attracting some gorgeous improvements, such as the unexpected designs based on Plates 171 and 176. If only Freese could have lived to see those designs!

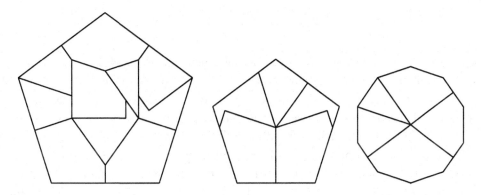

I161: Improved pentagon and decagon of same inradius to pentagon [GNF]

Plate 161: Freese discovered that there is a convenient relationship between a regular pentagon and a regular decagon of the same inradius, which makes it easy to dissect them to a larger pentagon. His dissection is easy to understand, even though it uses sixteen pieces.

We can substantially improve the dissection by identifying three triangles that we can combine together to make various pieces. Two of the triangles are what appear in Freese's dissection. The third triangle is congruent to each of the five isosceles triangles that we must cut away from the smaller pentagon to give the decagon. Each piece in the 10-piece dissection in Figure I161 is the merge of one or more of these triangles. Compare the underlying structure of the large pentagon with the implied structure of a pentagon in Plate 177.

Plate 162: Freese showed how to cut a regular pentagon into five pieces of equal area with line segments coming from the apex of the pentagon. He indicated that the line segments ended at points where a side of the pentagon crossed a side of a pentagon rotated 36° about a circumscribing circle, but he supplied no proof or citation for such a result. Having found no such result in various sources, I construct a proof myself, based on a rhombic structure of the pentagon as shown in Figure I162. The rhombuses, indicated by dotted lines, are of two different shapes: one with angles of 36° and 144°, and the other with angles of 72° and 108°.

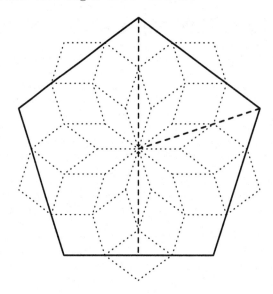

I162: Pentagon structure for cutting into five equal areas [GNF]

The pentagon contains ten of the $(36°, 144°)$-rhombuses and 17.5 of the $(72°, 108°)$-rhombuses. Note that one fifth of the area of the pentagon will contain the combined area of two $(36°, 144°)$-rhombuses and 3.5 of the $(72°, 108°)$-rhombuses, as indicated by the dashed lines demarcating the upper right portion of the pentagon. If we let the side length of each rhombus be 1, then the length of the longer diagonal of the $(72°, 108°)$-rhombus will be $2\cos 36° = (1 + \sqrt{5})/2 = \phi$ (the so-called "golden ratio"). Thus one fifth of the area of the pentagon is $2\sin 36° + 3.5\sin 72°$.

As indicated by the vertical dashed line in the figure, the height of piece 3 is twice the length of the long diagonal of a $(36°, 144°)$-rhombus plus $3/2$ times the length of a short diagonal of a $(72°, 108°)$-rhombus, or $4\sin 72° + 3\sin 36°$. The base of piece 3 has a length of ϕ, so that the area of piece 3 is one half times its base times its height, namely $(\phi/2)(4\sin 72° + 3\sin 36°)$. We note that $\sin 72° = (\sqrt{10+2\sqrt{5}})/4$ and $\sin 36° = (\sqrt{10-2\sqrt{5}})/4$. Then piece 3 has area equal to one fifth of the pentagon if $(7-4\phi)\sin 72° = (3\phi-4)\sin 36°$, which follows by a direct proof.

We can also show that piece 1 has area equal to one fifth of the area of the pentagon: View as the base of this piece the side on its lower left. Extend the base (of length $1 + \phi$) and then drop a perpendicular from the apex of the pentagon down to the base. The altitude will be of length $(2 + \phi)\sin 72°$, and the area of piece 1 will thus be $(1/2)(1 + \phi)(2 + \phi)\sin 72°$. Again a direct proof will show that this area equals one fifth the area of the pentagon. By symmetry piece 5 will have the same area, and since pieces 2 and 4 are mirror images of each other, each must have again that same area.

Plate 163: Freese noted that distances between certain points in a hexagonal grid are integral multiples of the shortest distances. An example at the bottom of this plate shows a distance of 7, which results from a triangle with an angle of 120° bracketed by sides of length 3 and 5. He observed that this leads to a different dissection of 49 identical hexagons to a large hexagon than the two dissections at the top of the plate. Yet the lengths of the cuts in the dissections are all rational numbers.

Dissections such as the one at the bottom of the plate follow from the Law of Cosines, in a fashion similar to that described in Plate 143. Using the formulas from the commentary for that plate, we note that $\bar{a}^2 + \bar{a}b + b^2 = c^2$. We get all such instances by using the formulas from the commentary of Plate 143. The dissection in the bottom middle of Plate 163 is hingeable.

Plate 164: Freese defined a concave hexagon to be a concave figure with three angles of 30° alternating with three angles of 210°. Another description of this figure is the pseudo-star $\{3/(12/5)\}$. Freese seems to have been the first person to dissect three such figures to one. His 18-piece dissection may well be minimal.

Plate 165: The manner in which Freese cut the large octagon is essentially the same as the way Boris Kordemskii (1956) cut an octagonal ring into eight isosceles right triangles. (See Puzzle 158 in Kordemsky (1972).) Suitably arranged, the dissection is translational.

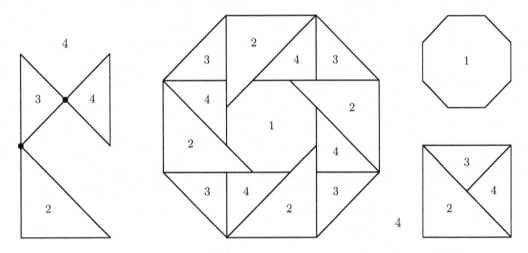

I165: Hinged four squares plus octagon to a large octagon [GNF]

The dissection is not hingeable, and yet by allowing four additional pieces, we can modify it to get a hingeable dissection. For each square, just split one of its two isosceles triangles into two smaller isosceles triangles, as in the middle and on the right of Figure I165. The hinged pieces are on the left of that figure.

Plate 166: Freese appears to have been the first to dissect a regular enneagon to an equilateral triangle. His approach is somewhat similar to what he employed for his dissection of a regular (15-sided) pentadecagon to a square in Plate 134. He sliced three trapezoids off of the enneagon, leaving a smaller equilateral triangle. He then cut an isosceles right triangle off one of the trapezoids so that the two resulting pieces, plus the two remaining trapezoids, would assemble to form a parallelogram. Finally, he used the strip method to dissect the parallelogram to a larger trapezoid, on which he could stack the smaller equilateral triangle. The result is a 9-piece dissection, of which Harry Lindgren 1964 wrote that if you could beat Freese and find an 8-piece dissection, you would have found a needle in a haystack.

Decades later Gavin Theobald found that needle, namely the 8-piece dissection in Figure I166a. Gavin's insight was to cut off two of the trapezoids, plus a flattened hexagon formed from gluing two trapezoids base to base. He then placed one of the two cut-off trapezoids in the cavity left from removing the hexagon. Next he formed an element for a strip by placing the hexagon atop the remaining trapezoid. Finally he crossed that strip with a strip of the desired larger trapezoids, creating just six additional pieces.

There is one feature worth noting in Freese's and Theobald's dissections. Freese identified the locations of four hinges with which some pieces in his dissection could be hinged together. Actually, Freese missed the possibility of a fifth hinge, between pieces 3 and 4. So Freese could have constructed a model of his dissection with a total of four assemblages: piece 1, piece 2, pieces 3,4,5,6, and pieces 7,8,9. Although Theobald did not note hinge locations, it is quite clear that his dissection also accommodates five hinges. Thus he could have constructed a model with just three hinged assemblages, as we see in Figure I166b.

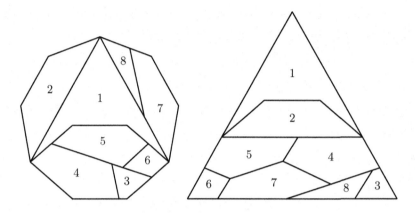

I166a: Regular enneagon to an equilateral triangle [Gavin Theobald]

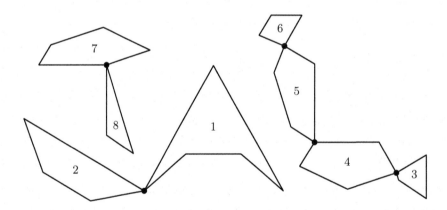

I166b: Partial hinging for Theobald's enneagon to equilateral triangle

Plate 167: The dissection in this plate is one of the more unusual dissections in Freese's manuscript. It seems to be an attempt to find geometrical structure in an enneagon. It is mildly interesting in that it hints at how to dissect nine enneagons to one. A more interesting structure for the enneagon is comprised of three equilateral triangles, three 80°-rhombuses, and three 40°-rhombuses. Cundy and Langford first used that structure in their 1960 paper, and I discussed the structure further in my 1997 book. Also see my discussion for Plate 101.

Plate 168: Freese was pleased by the attractive tilings of a regular decagon by pairs of $(36°, 108°, 36°)$ and $(72°, 36°, 72°)$ isosceles triangles. As he noted in the lower righthand corner, a pair of them together make a larger $(72°, 36°, 72°)$ isosceles triangle. We can reflect five such triangles to convert decagon 1 to decagon 3, and decagon 4 to decagon 5. We can also take the pair of small triangles to form a bilaterally symmetric concave figure, and reflect five such figures to convert decagon 2 to decagon 5, and also decagon 1 to decagon 5. However, Freese's statement that "The possible number of different PATTERNS IS UNLIMITED" seems either unjustified or unclear, or both. It is noteworthy that the dissection between any two of the five arrangements is translational.

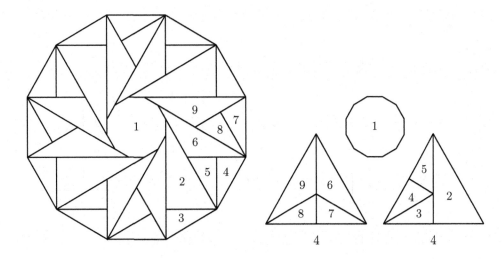

I169a: Hingeable eight triangles and {12} to large {12} [GNF]

Plate 169: Suitably arranged, this dissection is translational. Although it is not hingeable, a related 33-piece dissection (Figure I169a) is hingeable, as we can see in Figure I169b.

The dissection of eight triangles and small dodecagon to large dodecagon is related to the dissection of a hexagon to a dodecagon ring in Plate 88 and the dissection of a hexagon to eight triangles in Plate 76. As with the hexagon in Figure I188a, we can cut the eight triangles into fewer pieces, some of which we turn over. This gives us the 25-piece dissection in Figure I169c. Freese probably knew of this dissection but did not present it because he did not allow the turning over of pieces.

In an approach similar to that in Figure I188b, we can arrange some of the right triangles to radiate around on the left side of the dodecagon and an equal number of mirror-imaged right triangles to radiate around on the right side. We then must fill in the remaining area to achieve the 29-piece dissection in Figure I169d. Unfortunately, there seems to be no way to save even more pieces as we did in Figure I188c, because the equilateral triangles are not bunched together into a hexagon, as they are in Plate 88.

Plate 170: This plate displays some curious dissections. In the first, Freese showed how to dissect a dodecagon into nine pieces that form three isosceles triangles with apex angles of 30°. In the second, Freese showed that three particular triangles form either a dodecagon sector, or

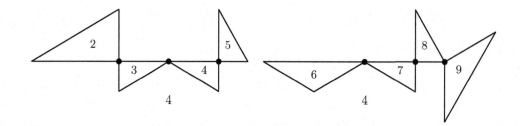

I169b: Hinged pieces for eight triangles and {12} to large {12} [GNF]

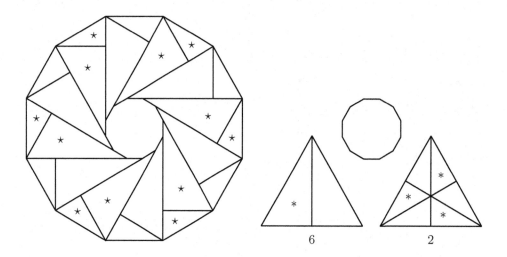

I169c: Improved eight triangles and {12} to large {12}, with turn over [GNF]

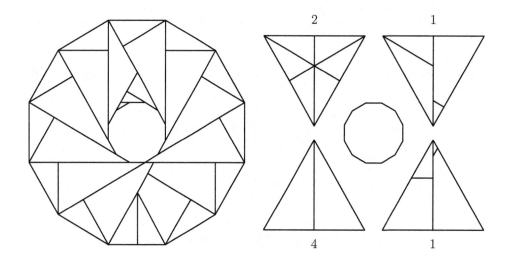

I169d: Improved eight triangles and {12} to large {12} with no turn over [GNF]

a dodecagon segment, or a dodecagon spandrel. In the third, Freese showed how to dissect an isosceles triangle with an apex angle of 36° into six pieces that will enclose a regular pentagon to form a rectangle. In the fourth, Freese showed how to dissect a regular octagon into five pieces that will form a tetraskelion. Unfortunately, Freese's use of the phrase "turns out" does not imply a hingeable dissection. Note that the resulting tetraskelion differs from the one in Plate 155.

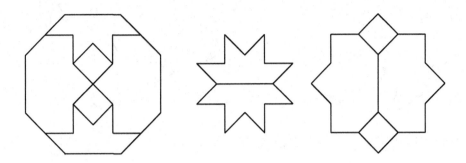

I171a: Improved octagram plus "co-octagram" to octagon [Robert Reid]

Plate 171: Freese noticed a natural relationship between any star $\{n/2\}$ and its related star or pseudo-star $\{n/((n-2)/2)\}$. This relationship allowed him to dissect such a pair into $n+1$ pieces to form a regular $\{n\}$ of twice the side length. If n is an even number, then there are two such dissections, one which leaves one star uncut, and the other which leaves the other star uncut. Suitably arranged, these dissections are all translational.

When n is even, it appears that Freese's dissections use more pieces than necessary. Lindgren 1964 gave a 6-piece dissection for the case of $n = 6$. In unpublished work, Robert Reid found a crafty 6-piece dissection (Figure I171a) for the case of $n = 8$. Both dissections are translational.

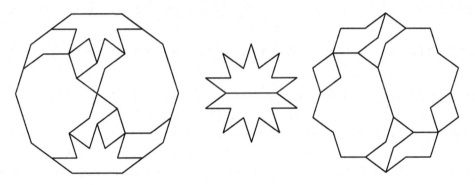

I171b: Improved decagram plus "co-decagram" to decagon [GNF]

We can pull off the same trick for $n = 10$, beating Freese's implied 11-piece dissection. Figure I171b shows a 10-piece dissection that is loosely inspired by Reid's dissection in Figure I171a. It too is translational!

To make sure that $n = 10$ is not an anomaly, I have found a 10-piece dissection in the case of $n = 12$. It's yet again a "star-burst" and still translational, in Figure 171c.

Plate 172: Freese noted that we can dissect a regular dodecagon to six equilateral triangles, six squares and one regular hexagon such that all side lengths are equal.

He also noted that we can dissect a regular decagon to two pentagrams and two regular pentagons. If we rearrange the pieces in the decagon, then we have achieved a translational dissection, as in Figure I172.

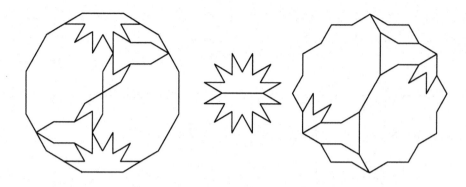

I171c: Improved dodecagram plus "co-dodecagram" to dodecagon [GNF]

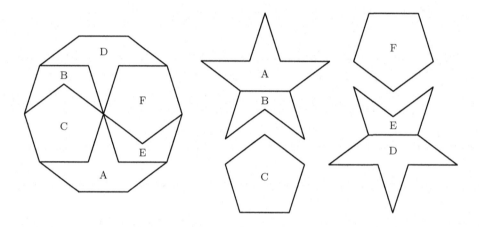

I172: Translational decagon to 2 pentagrams and 2 pentagons [GNF]

Plate 173: Freese may well have been the first to recognize the rule that every regular polygon with an even number of sides can be dissected to rhombuses whose side lengths equal the side length of the polygon. We have seen the application of this rule in Plates 118, 137, 139, and 141, in which Freese dissected two copies of such polygons to a larger such polygon. No one had previously discovered those dissections. Freese gave the formula for the number of such rhombuses in a regular polygon of n sides. It turns out to be the $(n/2 - 1)$st triangular number.

Plate 174: The dissections in this plate are somewhat confusing, because the third dissection is different in nature from the first two. So let's assume that the definition of concave polygon differs between the first two examples on the one hand and the third on the other hand. The next question is why Freese cut a regular hexagon to produce a concave hexagon, while he cut a square to produce a concave octagon.

To impose uniformity on the first two dissections, we interpret the first polygon in the first dissection as an equilateral triangle with a particular isosceles triangles attached to each of its three sides. The isosceles triangle will have an apex angle such that that angle plus the angle of the equilateral triangle be a straight angle. We view the first polygon in the second dissection as a square with an isosceles triangles attached to each of its sides. Thus each of the first two dissections is a dissection of a "capped regular n-gon" to a concave regular polygon of $2n$ sides.

The third dissection in this sequence will be of a capped regular pentagon to a concave decagon,

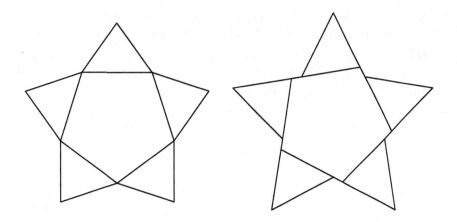

I174: Capped pentagon to concave decagon [GNF]

which we see in Figure I174. Since each angle in a regular pentagon is 108°, the apex angle of the isosceles triangle is 72°. The fourth dissection in the series, of a capped regular hexagon will have an isosceles triangle with all angles of 60°. Both the capped regular hexagon and its corresponding concave dodecagon will be hexagrams.

Plate 175: Freese appears to have been the first to identify the dissection of a regular dodecagon to three concave dodecagons. This dissection is somewhat predictable, since there is a straightforward dissection of a regular dodecagon to three congruent squares, and a simple dissection of a square to a concave dodecagon which we have seen in Plate 174. When suitably oriented, the dissection is translational.

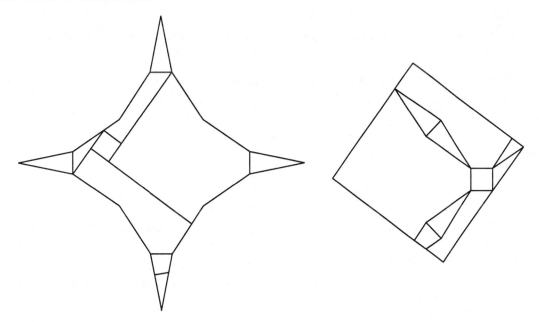

I176a: Improved concave hexadecagon to a square [Gavin Theobald]

Plate 176: Freese introduced a "concave hexadecagon" and gave a 13-piece dissection of it to a square. A concave hexadecagon is a 16-sided polygon in which four sequences of four edges bow in rather than bow out, giving a star-like object. We can tessellate the plane with the pair consisting of a regular hexadecagon and a concave hexadecagon. Freese found that if you clip off the outermost of the four points of the concave hexadecagon and slice each into two right triangles, then the right triangles fill the remaining cavities to produce an octagon with four long sides and four short sides. He then used the completing the tessellation technique to convert the octagon to a square.

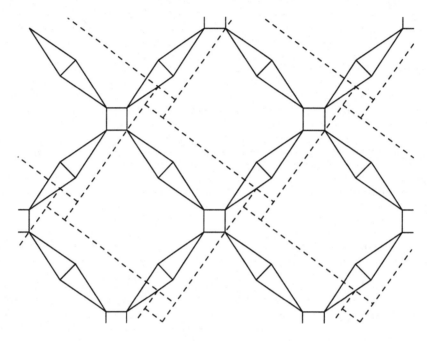

I176b: Tessellation for improved convex hexadecagon to a square [Gavin Theobald]

In my (2006) book I showed how to avoid slicing the four points into right triangles, but still used the completing the tessellation technique, reducing the number of pieces to eleven. Soon after, Gavin Theobald found a way to reduce the number of pieces to ten, as we can see in Figure I176a. He also cut off the points, but used a different tessellation in completing the tessellation. Figure I176b shows the superposition of the two tessellations.

Plate 177: Freese seems to have been the first to dissect three decagons and two concave decagons to two regular pentagons. Note that a concave pentagon is identical to the co-pentagram that Freese displayed in Plate 171. We can also identify it as a pseudo-star {5/1.5}. A variation of the 14-piece dissection is translational, as we see in Figure I177a.

The variation is also hingeable. We show the hingeable dissection in Figure I177b and the hinged pieces in Figure 177c. How placid would be the movement from the three decagons and the two concave pentagons to the two pentagons!

Plate 178: Freese appears to have been the first to dissect 10 congruent regular pentagons to a pair of concave pentagons and a regular decagon.

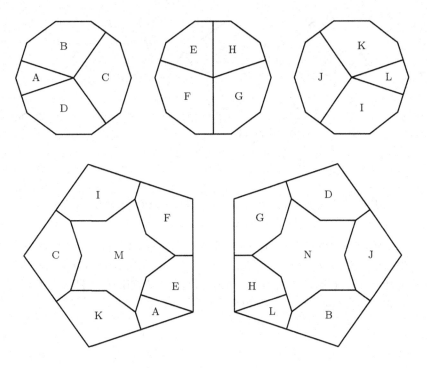

I177a: Translational 3 decagons plus 2 concave pentagons to 2 pentagons [GNF]

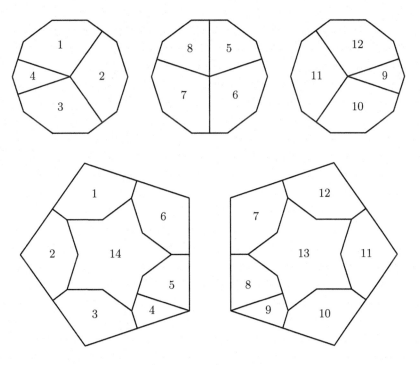

I177b: Hingeable 3 decagons plus 2 concave pentagons to 2 pentagons [GNF]

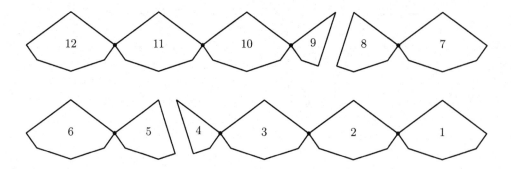

I177c: Hinged 3 decagons plus 2 concave pentagons to 2 pentagons [GNF]

In hindsight, characterizing the dissections in this chapter as miscellaneous is a bit unfair. This characterization fails to acknowledge the admirable accomplishments of Figures I161 and I169b, as well as the super properties conjured up in Figures I177a, I177b, and I177c. Or perhaps they end up giving the word miscellaneous a special luster!

EXPANDING A PENTAGON
BY ADDING A
PENTAGON RING FORMED FROM A DECAGON

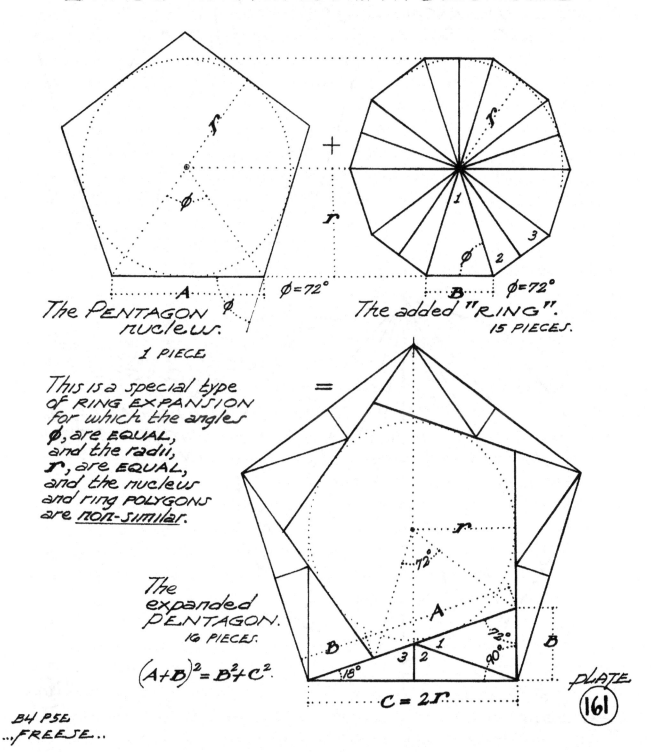

The PENTAGON
nucleus.

1 PIECE

$\phi = 72°$

The added "RING".

15 PIECES.

$\phi = 72°$

This is a special type
of RING EXPANSION
for which the angles
ϕ, are EQUAL,
and the radii,
r, are EQUAL,
and the nucleus
and ring POLYGONS
are _non-similar_.

The
expanded
PENTAGON.
16 PIECES.

$$(A+B)^2 = B^2 + C^2.$$

$C = 2r$

$72°$

$18°$ $90°$ $72°$

PLATE
161

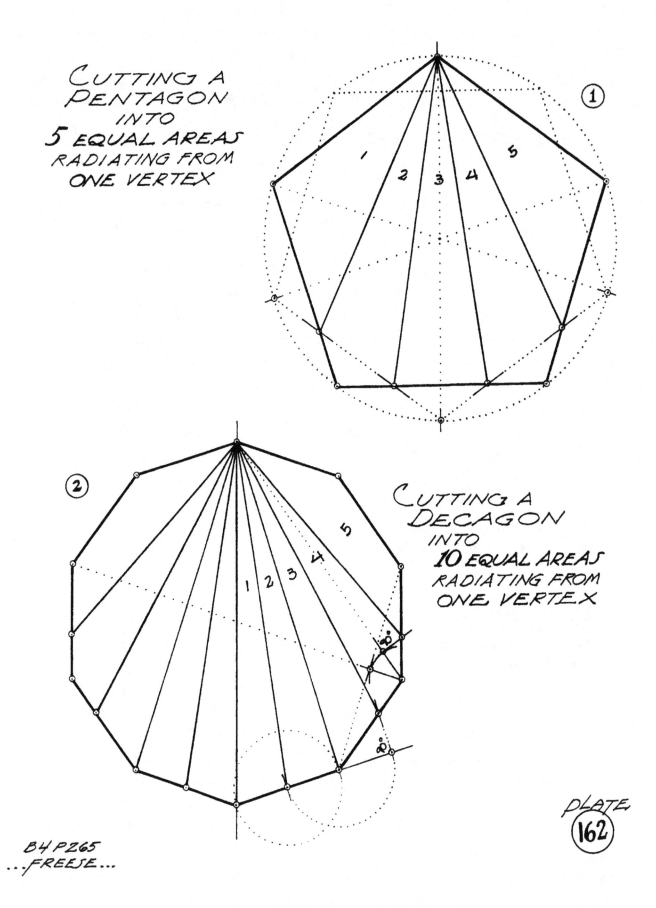

CUTTING A
PENTAGON
INTO
5 EQUAL AREAS
RADIATING FROM
ONE VERTEX

①

1 2 3 4 5

②

CUTTING A
DECAGON
INTO
10 EQUAL AREAS
RADIATING FROM
ONE VERTEX

1 2 3 4 5

PLATE
162

B4 PZ65
...FREESE...

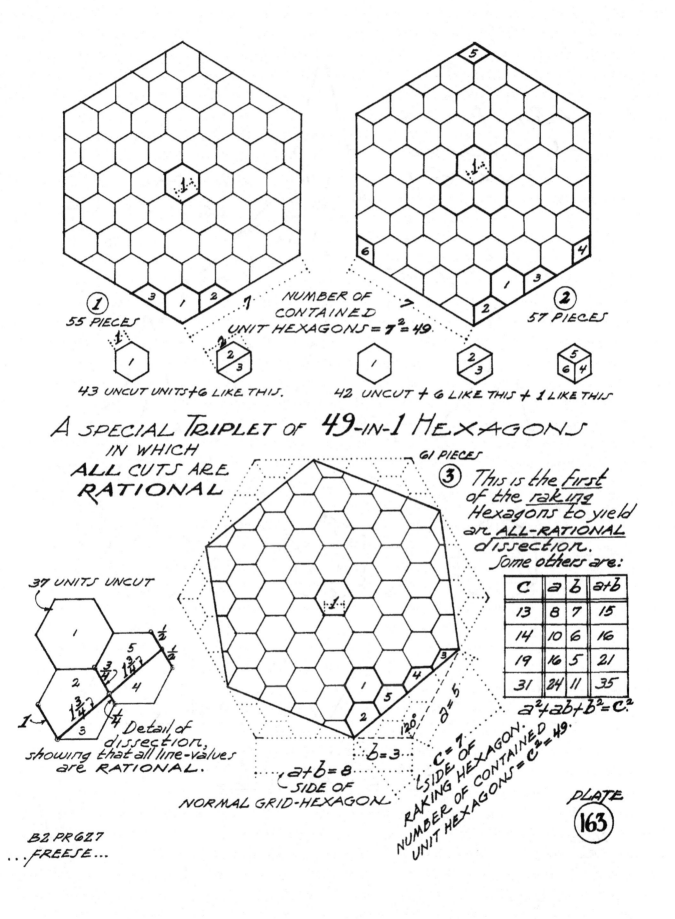

NUMBER OF
CONTAINED
UNIT HEXAGONS = $7^2 = 49$

① 55 PIECES

② 57 PIECES

43 UNCUT UNITS+6 LIKE THIS.

42 UNCUT + 6 LIKE THIS + 1 LIKE THIS

A SPECIAL TRIPLET OF 49-IN-1 HEXAGONS

IN WHICH
ALL CUTS ARE
RATIONAL

61 PIECES

③ This is the first of the raking Hexagons to yield an ALL-RATIONAL dissection. Some others are:

C	a	b	a+b
13	8	7	15
14	10	6	16
19	16	5	21
31	24	11	35

$a^2 + ab + b^2 = c^2$

37 UNITS UNCUT

Detail of dissection, showing that all line-values are RATIONAL.

$a+b = 8$
SIDE OF
NORMAL GRID-HEXAGON

$b = 3$
$a = 5$
$120°$

$c = 7$
SIDE OF
RAKING HEXAGON.
NUMBER OF CONTAINED
UNIT HEXAGONS = $c^2 = 49$.

PLATE
163

B2. PR627
...FREESE...

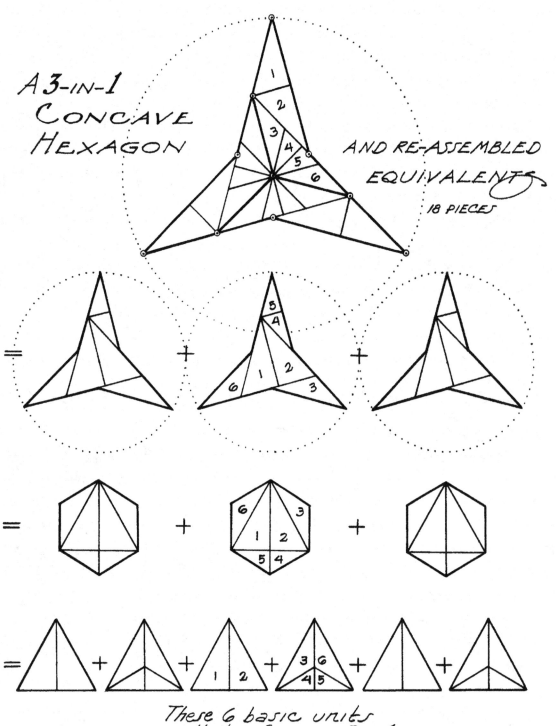

A 3-IN-1 CONCAVE HEXAGON

AND RE-ASSEMBLED EQUIVALENTS.

18 PIECES

These 6 basic units will also form a 3-IN-1 CONVEX HEXAGON.

B1P28D3
...FREESE...

PLATE 164

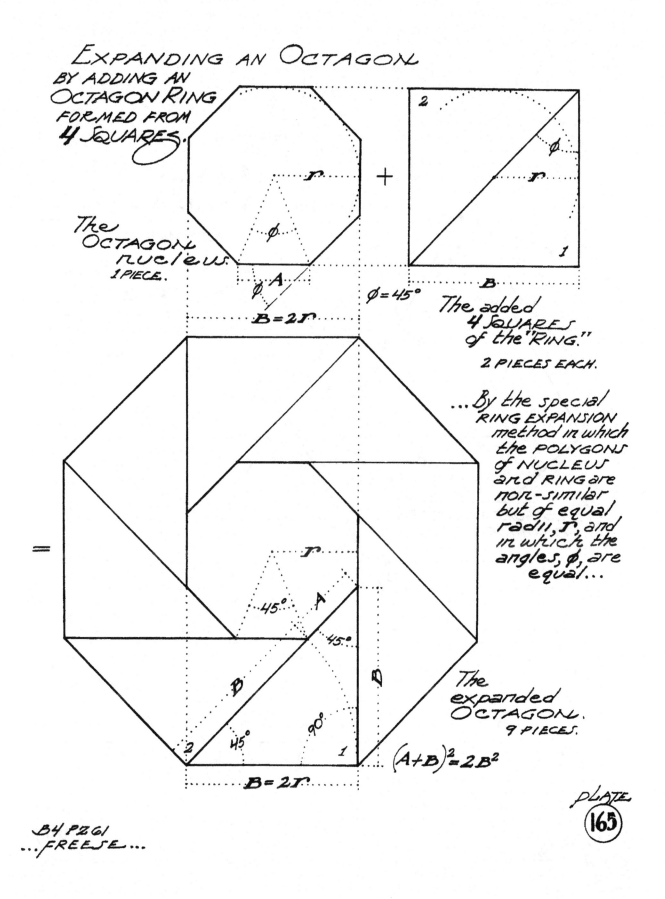

EXPANDING AN OCTAGON
BY ADDING AN OCTAGON RING FORMED FROM 4 SQUARES.

The OCTAGON nucleus. 1 PIECE.

$B = 2r$

$\phi = 45°$

The added 4 SQUARES of the "RING."

2 PIECES EACH.

... By the special RING EXPANSION method in which the POLYGONS of NUCLEUS and RING are non-similar but of equal radii, r, and in which the angles, ϕ, are equal...

The expanded OCTAGON. 9 PIECES.

$(A+B)^2 = 2B^2$

$B = 2r$

B4 PZ 61 ...FREESE...

PLATE
165

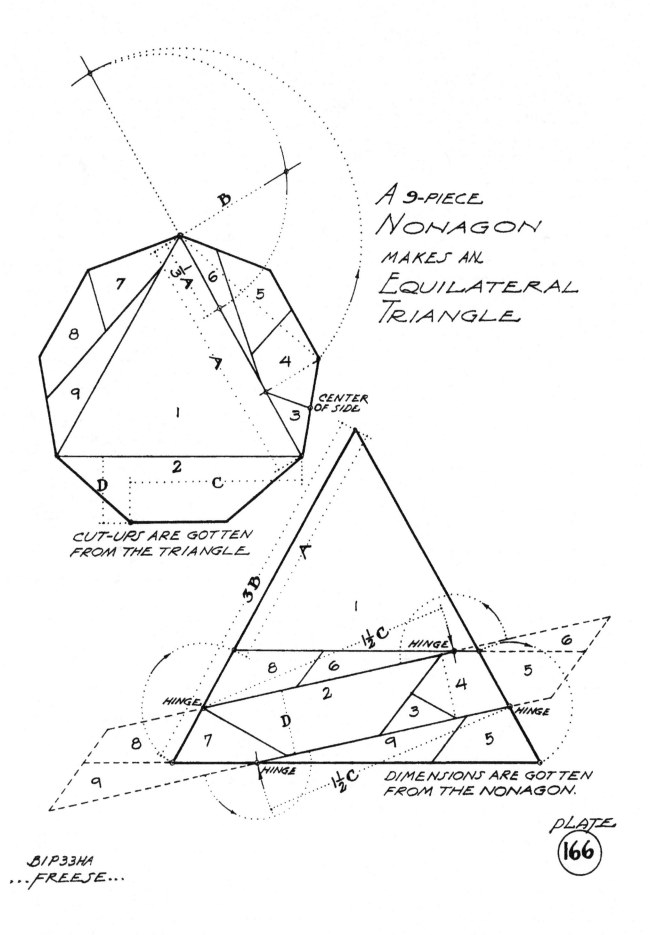

A 9-PIECE
NONAGON
MAKES AN
EQUILATERAL
TRIANGLE

B

7 3 A 6 5
4A
8
A
9 4
1 3
CENTER
OF SIDE
2
D C

CUT-UPS ARE GOTTEN
FROM THE TRIANGLE.

3B A
1
1½C HINGE 6
8 6 5
2 4
HINGE D 3 HINGE
8 7 9 5
HINGE
9 1½C DIMENSIONS ARE GOTTEN
FROM THE NONAGON.

PLATE
166

B1P33HA
...FREESE...

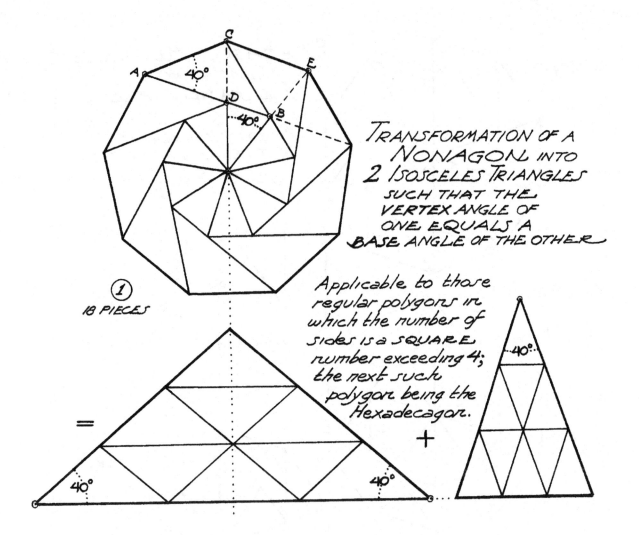

TRANSFORMATION OF A
NONAGON INTO
2 ISOSCELES TRIANGLES
SUCH THAT THE
VERTEX ANGLE OF
ONE EQUALS A
BASE ANGLE OF THE OTHER

① 18 PIECES

Applicable to those
regular polygons in
which the number of
sides is a SQUARE
number exceeding 4;
the next such
polygon being the
Hexadecagon.

40°

+

40°

=

40°   40°

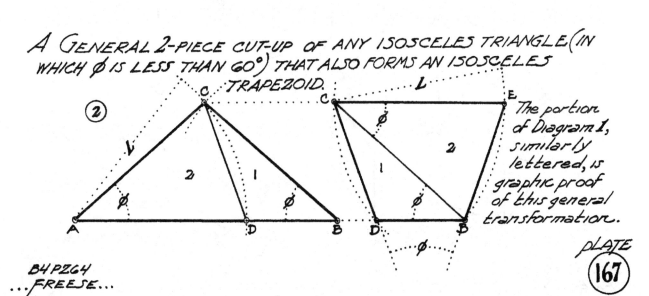

A GENERAL 2-PIECE CUT-UP OF ANY ISOSCELES TRIANGLE (IN
WHICH ∅ IS LESS THAN 60°) THAT ALSO FORMS AN ISOSCELES
TRAPEZOID.

②

L   C   C   L   E

2   1   ∅   1   2

∅   ∅   ∅   ∅

A   D   B   D   B

∅

The portion
of Diagram 1,
similarly
lettered, is
graphic proof
of this general
transformation.

PLATE
167

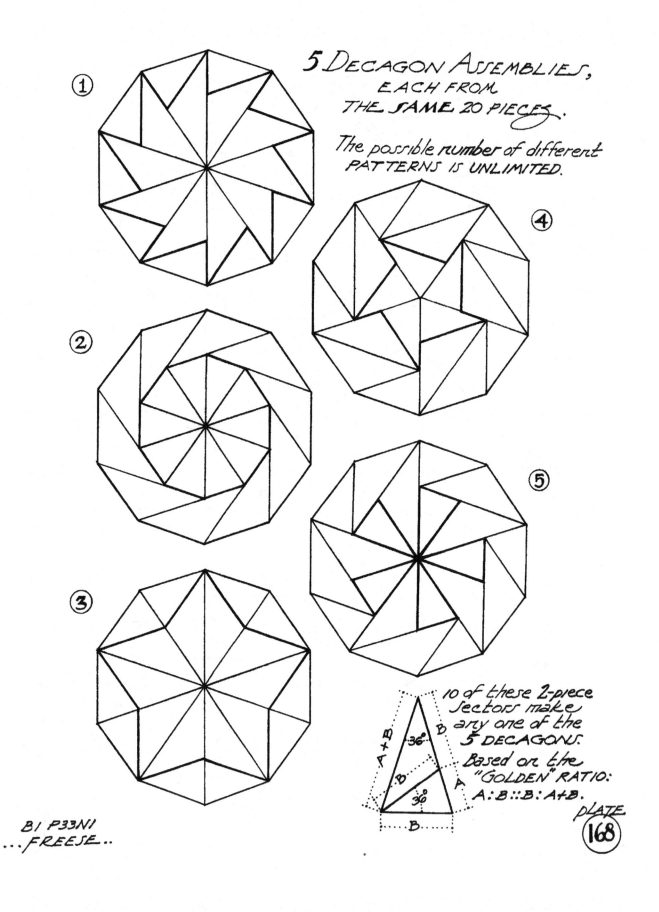

5 DECAGON ASSEMBLIES,
EACH FROM
THE SAME 20 PIECES.

The possible number of different
PATTERNS IS UNLIMITED.

① ② ③ ④ ⑤

10 of these 2-piece
sectors make
any one of the
5 DECAGONS.

Based on the
"GOLDEN" RATIO:
A:B::B:A+B.

A+B · 36° · B
B
A
36°
B

BI P33N1
...FREESE..

PLATE
168

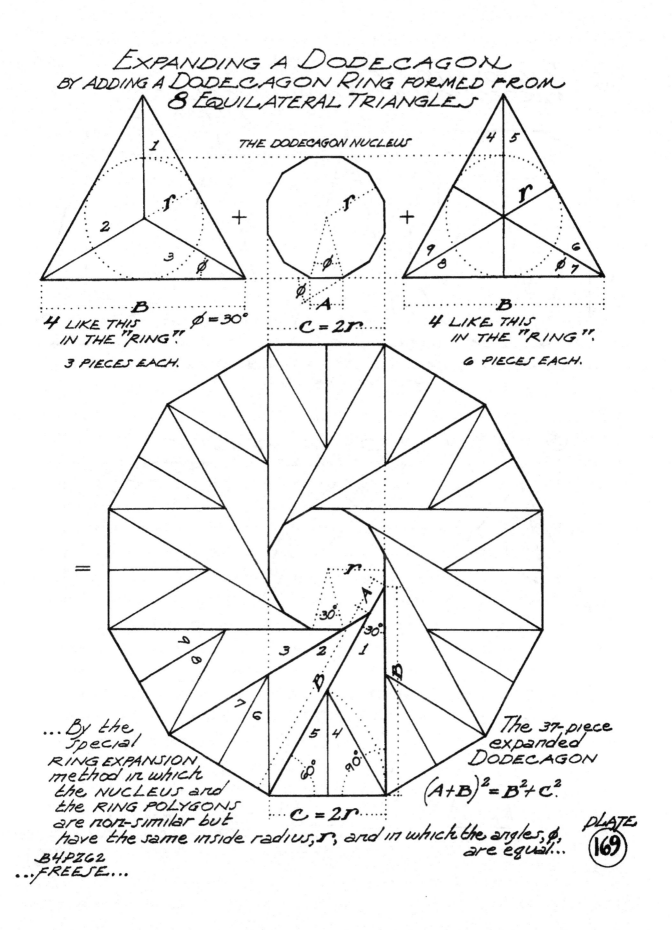

EXPANDING A DODECAGON
BY ADDING A DODECAGON RING FORMED FROM
8 EQUILATERAL TRIANGLES

THE DODECAGON NUCLEUS

$\phi = 30°$

$C = 2r$

B

4 LIKE THIS
IN THE "RING".

3 PIECES EACH.

B

4 LIKE THIS
IN THE "RING".

6 PIECES EACH.

... By the
special
RING EXPANSION
method in which
the NUCLEUS and
the RING POLYGONS
are non-similar but
have the same inside radius, r, and in which the angles, ϕ,
are equal...

B4PZ62
...FREESE...

The 37-piece
expanded
DODECAGON

$(A+B)^2 = B^2 + C^2$

$C = 2r$

PLATE 169

① TRANSFORMATION OF A DODECAGON INTO 3 EQUAL ISOSCELES TRIANGLES

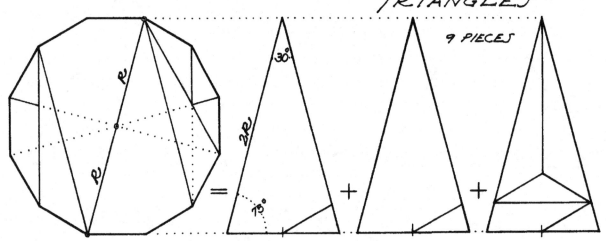

9 PIECES

FOR R=1, THE AREA OF DODECAGON = 3. ∴ THE AREA OF EACH ISOSCELES TRIANGLE = 1.

② DODECAGON EQUIVALENTS

3 PIECES

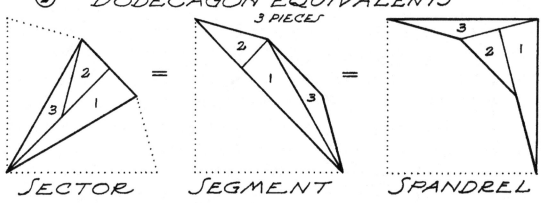

SECTOR   SEGMENT   SPANDREL

③ A PENTAGON EQUIVALENCE

6 PIECES

④ AN OCTAGON "TURNS OUT" TO BE A TETRASKELION. 5 PIECES

INSCRIBED TRIANGLE   THE FOUR SPANDRELS

B1P36C2
B1P28A2,3.
B4P266
B1P19A1
...FREESE...

PLATE 170

TRANSFORMATION OF REGULAR POLYGONS,
EACH INTO ITS
2 COMPONENT POLYGRAMS.

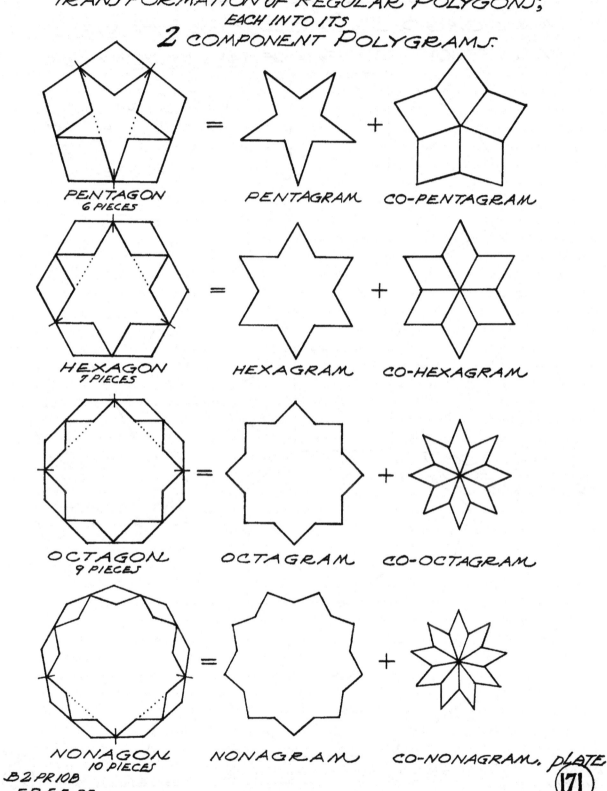

PENTAGON
6 PIECES
= PENTAGRAM + CO-PENTAGRAM

HEXAGON
7 PIECES
= HEXAGRAM + CO-HEXAGRAM

OCTAGON
9 PIECES
= OCTAGRAM + CO-OCTAGRAM

NONAGON
10 PIECES
= NONAGRAM + CO-NONAGRAM. PLATE

B2 PR 10B
...FREESE...

171

DISSECTION OF THE
DODECAGON, DECAGON & HEXAGON
INTO DIFFERENT POLYGONS OF _SAME LENGTH OF SIDES._

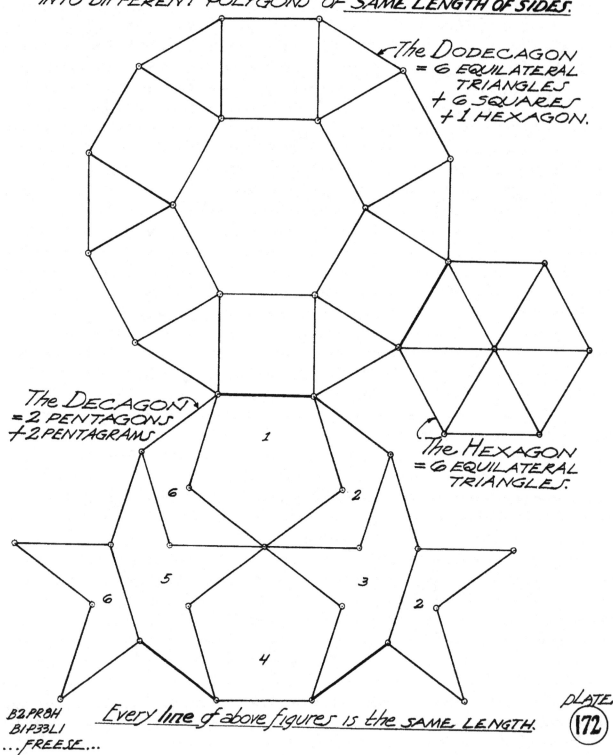

←The DODECAGON
= 6 EQUILATERAL
TRIANGLES
+ 6 SQUARES
+ 1 HEXAGON.

The DECAGON→
= 2 PENTAGONS
+ 2 PENTAGRAMS

The HEXAGON
= 6 EQUILATERAL
TRIANGLES.

1

6   2

6   5   3   2

4

Every line of above figures is the SAME LENGTH.

PLATE
(172)

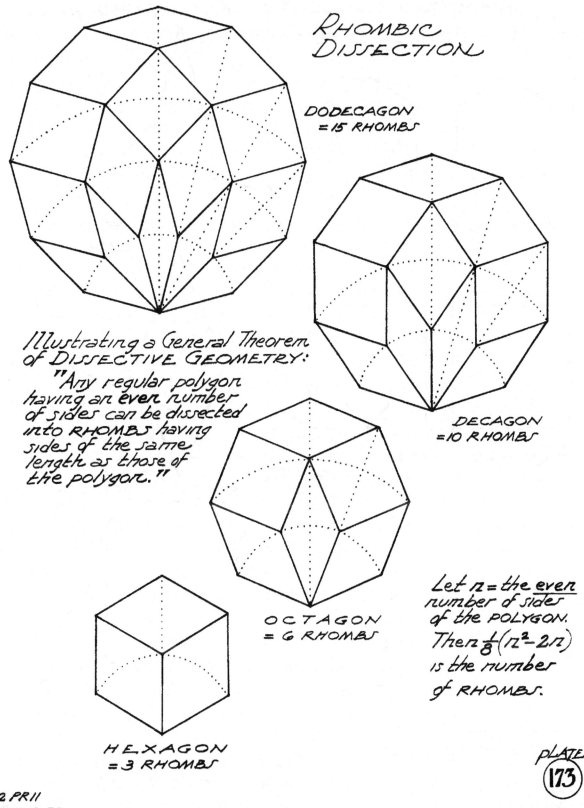

RHOMBIC DISSECTION

DODECAGON
= 15 RHOMBS

DECAGON
= 10 RHOMBS

Illustrating a General Theorem
of DISSECTIVE GEOMETRY:
"Any regular polygon
having an *even* number
of sides can be dissected
into RHOMBS having
sides of the same
length as those of
the polygon."

OCTAGON
= 6 RHOMBS

Let n = the *even*
number of sides
of the POLYGON.
Then $\frac{1}{8}(n^2-2n)$
is the number
of RHOMBS.

HEXAGON
= 3 RHOMBS

PLATE
173

B2 PR11
...FREESE...

TRANSFORMING CONVEX POLYGONS INTO CONCAVE POLYGONS

① A REGULAR HEXAGON, CUT LIKE THIS, MAKES A CONCAVE HEXAGON.
4 PIECES

② A SQUARE, CUT LIKE THIS, MAKES A CONCAVE OCTAGON.
5 PIECES

③ A SQUARE, CUT LIKE THIS, MAKES A CONCAVE DODECAGON.
5 PIECES

B1 P21A
B1 P21A2
...FREESE...

PLATE 174

A CONVEX DODECAGON
TRANSFORMED INTO
3 CONCAVE DODECAGONS

12 PIECES

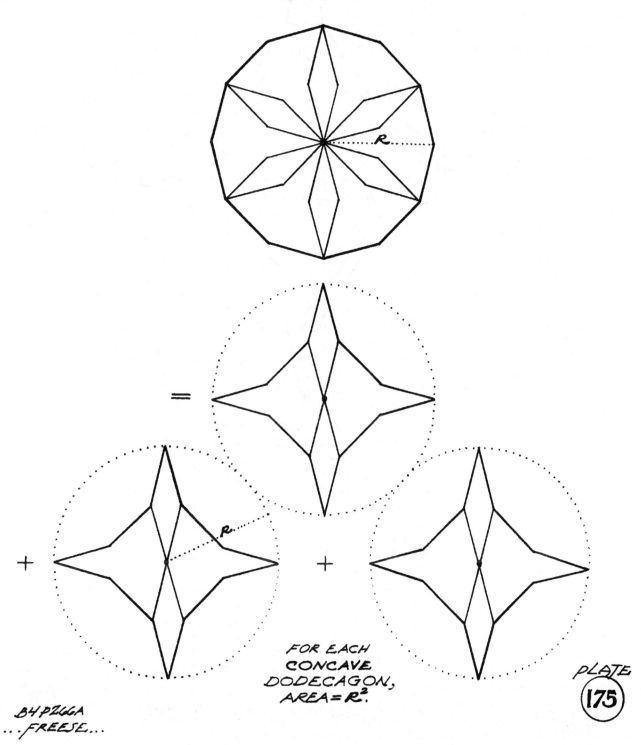

FOR EACH
CONCAVE
DODECAGON,
AREA = R^2

B4 P266A
...FREESE...

PLATE
175

Squaring the Concave Hexadecagon.

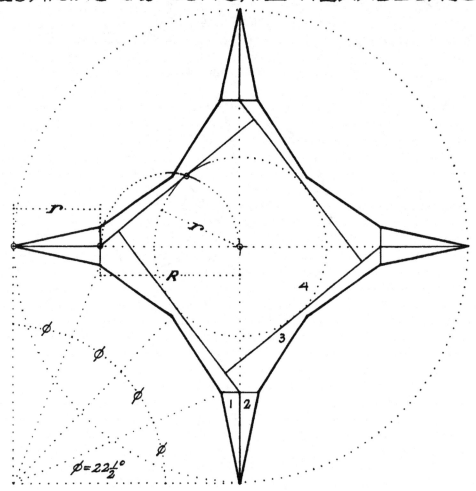

$\phi = 22\tfrac{1}{2}°$

13 PIECES

=

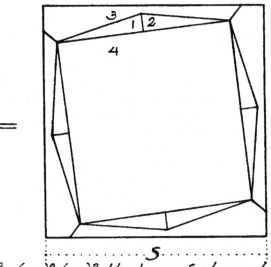

$S^2 = (2R)^2 - (2r)^2$, the basis of above layout.

PLATE
176

5 Decagons, 2 of which are concave ones, transformed into 2 Pentagons

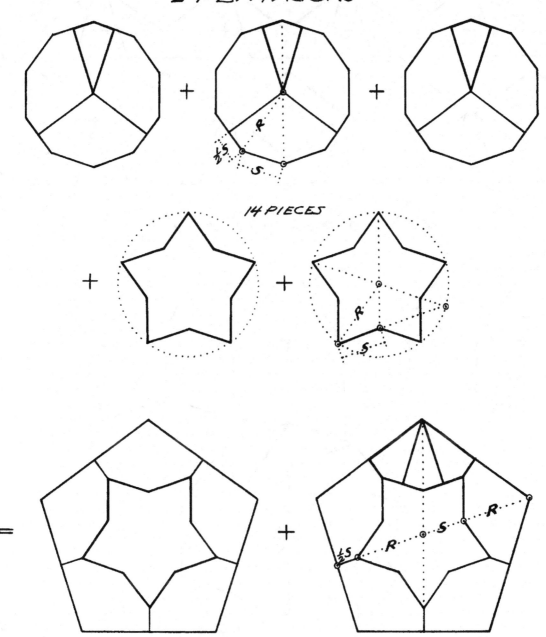

14 Pieces

Plate
177

B1 P33 J1
...Freese...

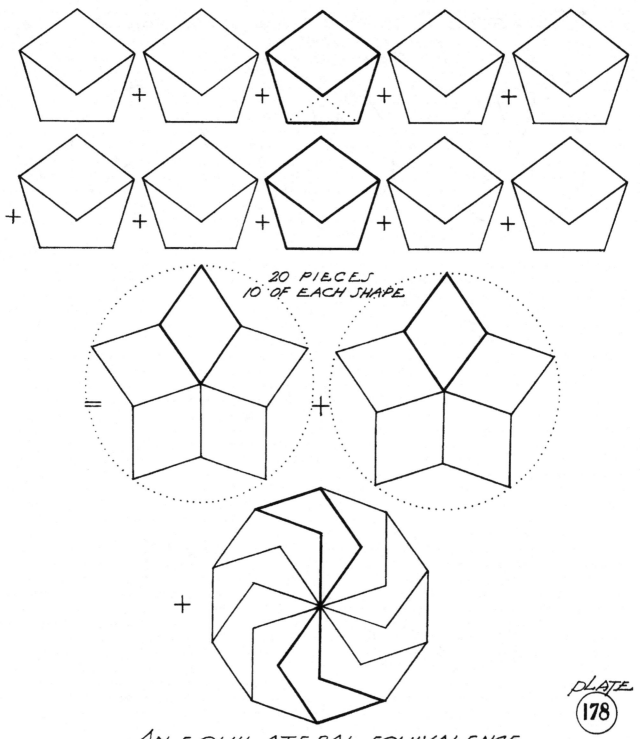

20 PIECES
10 OF EACH SHAPE

pLATE (178)

AN EQUILATERAL EQUIVALENCE.

10 PENTAGONS = 3 DECAGONS (2 OF WHICH ARE CONCAVE),
EVERY SIDE OF EACH PIECE AND FIGURE HAVING THE
SAME LENGTH.

B4PZ68A
...FREEZE...

Chapter 18

Mixed Polygons to One (Plates 179–188)

Freese was the first to consider dissecting two different regular polygons of equal area to a square. In Plates 179-183 he produced five different examples of such dissections. His general approach was to identify a convenient trapezoid whose area is the same as that of each polygon and whose shape is half of the square. He then dissected each polygon to the trapezoid and assembled the two trapezoids to complete the square.

In Plates 184-186 Freese extended this problem to three or four regular polygons to a square. In Plate 184 he partitioned the square into three sections of equal area, and then dissected each regular polygon into one of those sections. In Plate 185 he partitioned the square into two congruent rectangles, and then dissected a pair of figures into each rectangle. In Plate 186, he partitioned the square into two congruent trapezoids and two congruent squares, and then dissected each of four regular polygons to one of the sections from the square.

Finally in Plates 187 and 188, Freese dissected three regular polygons to three polygons that he assembled into a regular hexagon in one case, and similarly dissected three polygons into three polygons that he assembled into a regular dodecagon. For Plate 187, the three regular polygons were a triangle, a square, and a pentagon, while for Plate 188, the three regular polygons were a square, a hexagon, and an octagon.

The trick in these dissections is in choosing how to partition the large polygon so that the dissection uses fewer pieces. In Plate 180 Freese saved a piece by choosing a trapezoid that has a 3-piece dissection from an equilateral triangle. Gavin Theobald mimicked that approach in Figure I183a so that he could save a piece. Of course, Freese's inventiveness set the stage for such clever improvements. Can you figure out what's going on in Figure I187?

Plate 179: This plate, the first of five afore-mentioned consecutive examples, shows Freese's dissection of a regular hexagon and an equilateral triangle to a square. With one cut Freese converted the hexagon to a parallelogram, which he then converted to a trapezoid that is half of the square, using two more cuts. He next dissected the triangle to a congruent trapezoid in four pieces, and then assembled the square.

Plate 180: Similar to the dissection in the previous plate, Freese dissected an equilateral triangle and a regular dodecagon to a square. Again Freese dissected the polygons into congruent trapezoids that fit together to form the square. The dodecagon seems to require five pieces to convert into an appropriate trapezoid, but we can choose amongst an infinite number of such trapezoids. Freese identified a particular trapezoid that is half of the desired square and for which the dissection from an equilateral triangle uses only three pieces. Once he knew the particular trapezoid, he then adjusted the slope of the long, almost-vertical line segment in the dodecagon to get the dissection to the desired trapezoid. The next three of Freese's plates handle a dodecagon in similar fashion.

Plate 181: Freese dissected a regular pentagon and a regular dodecagon to a square, again by combining two dissections. He dissected the pentagon to a trapezoid in five pieces, and dissected the dodecagon to a congruent trapezoid in five pieces. For this dissection, however, Gavin Theobald found an imaginative dissection of the pentagon to an irregular hexagon that fills half the square and uses only four pieces. He completed the square by dissecting the dodecagon as earlier, but by substituting an irregular cut for the straight cut used in the previous two plates.

To dissect the pentagon to the irregular hexagon, Theobald sliced the cap off the top of the pentagon and slid it to the left of the remaining isosceles trapezoid to produce a trapezoid of the same height. He then made a cut from the midpoint of the right side to the long side (the top of the trapezoid), of length equal to one half of the square's side length. He then rotated the resulting small piece around what had been the midpoint. Finally, he cut a perpendicular to the long side thus created, of length of one half of the square's side length.

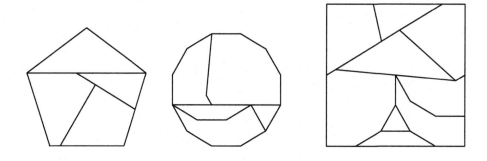

I181: Improved pentagon and dodecagon to a square [Gavin Theobald]

Plate 182: Freese dissected a regular hexagon and a regular dodecagon to a square by combining two dissections. As he did in Plate 179, he dissected the hexagon to a trapezoid that fills half of the square. And he dissected the dodecagon to a congruent trapezoid, using five pieces.

Plate 183: Freese dissected a regular octagon and a regular dodecagon to a square, again by combining two dissections. As in the three previous plates, he dissected the polygon other than the dodecagon, namely the octagon, to a trapezoid that fills up half of the square using six pieces. He then dissected the dodecagon to a congruent trapezoid, using five pieces.

Theobald found a way to improve the dissection by one piece, obtaining the 10-piece dissection in Figure I183a. For the octagon, he discovered a 3-piece T-strip element, which he then crossposed with a strip of trapezoids (half squares), as we see in Figure I183b, to get five pieces from the octagon.

Plate 184: Freese was probably the first to dissect the trio of an equilateral triangle, a regular hexagon, and a regular dodecagon, all of equal area, to a square. His approach was to split the square into two gnomons and a 1 × 3 rectangle. He used a T-strip crossposition to dissect the triangle to one gnomon. Next, he converted the hexagon to a rectangle and then applied a P-slide. Finally, he used the dissection of a dodecagon to three squares in Plate 114 to produce the second gnomon.

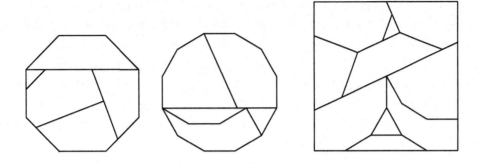

I183a: Improved octagon and dodecagon to a square [Gavin Theobald]

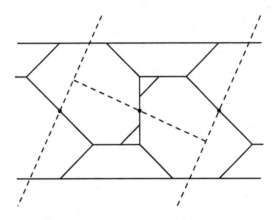

I183b: Crossposition for octagon to a half-square trapezoid

Freese's dissection has a blemish, namely a small, thin right triangle in the dissection of the equilateral triangle to one gnomon. Gavin Theobald removed this blemish and reduced the number of pieces by two, using two congruent irregular octagons rather than two gnomons. A 15-piece dissection (Figure I184) is the result.

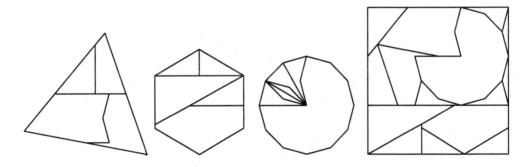

I184: Improved triangle, hexagon, and dodecagon to a square [Gavin Theobald]

Plate 185: It appears that Freese was the first to dissect a pair of equilateral triangles and a pair of regular pentagons, all of equal area, to a square. He dissected each pair into a rectangle twice as long as it is high, and then stacked one rectangle on top of the other to get the square.

Freese found symmetric dissections for each pair of polygons, saving four pieces over dissecting each of the polygons to a square. Yet Gavin Theobald found a nifty way to save two more pieces, by using more interesting strip dissections.

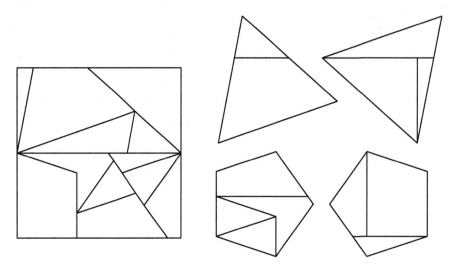

I185: Improved two triangle and two pentagons to a square [Gavin Theobald]

Plate 186: It appears that Freese was the first to dissect the quartet of an equilateral triangle, a regular pentagon, a regular hexagon, and a regular octagon, all of equal area, to a square. He used a dissection of each of the nonsquare polygons to a small square, packing them into the large square, and then swapped two pieces between the equilateral triangle dissection and the regular pentagon dissection. The swapped piece is the small triangular piece from the equilateral triangle dissection. Thus Freese saved one piece from a straightforward 20-piece dissection.

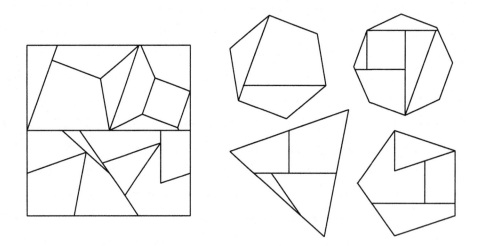

I186: Improved triangle, pentagon, hexagon, and octagon to a square [Gavin Theobald]

Gavin Theobald found a way to save two pieces from Freese's dissection, by first dissecting the triangle and pentagon into just eight pieces to form a (2×1)-rectangle, and then later dissecting the hexagon and octagon into just nine pieces to form another (2×1)-rectangle, as we see in Figure I186. What a double bonus! Note that for dissecting the octagon, Gavin used the (mirror image) of the dissected octagon in the upper left of Freese's Plate 89.

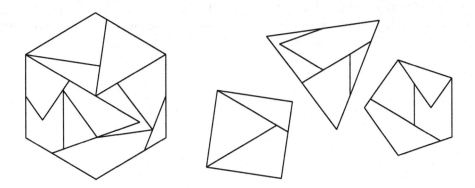

I187a: Improved triangle, square, and pentagon to a hexagon [Gavin Theobald]

Plate 187: Freese was the first to dissect a hexagon to the trio of an equilateral triangle, a square, and a regular pentagon, all of equal area. His approach was to dissect each figure to a $60°$-rhombus and then assemble the three rhombuses into the hexagon. He used four, three, and six pieces respectively, for a total of 13 pieces. With appropriate choices for these three dissections, as well as gluing one piece to another and cutting an indentation in one piece that accommodates the extension of the other piece, Gavin Theobald reduced this total to 11 pieces, as in Figure I187a.

The key is to first dissect the pentagon to a rhombus using a strip dissection. The result winds up in the lower left rhombus of the hexagon, where Gavin merged two piece from the pentagon with a trapezoidal piece from the equilateral triangle. To make space for the enlarged piece, its neighbor in the hexagon loses the trapezoid at the same time that its neighbor in the triangle gains the trapezoid. Figure I187b shows the crossposition that helps in dissecting the equilateral triangle. The dissection of the square in Plate 187 remains unchanged.

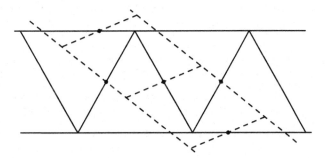

I187b: Crossposition of triangle for triangle, square, and pentagon to a hexagon

Plate 188: Freese was the first to dissect the trio of a square, a regular hexagon, and a regular octagon, all of equal area, to a regular dodecagon. To accomplish this Freese identified a third of the regular dodecagon and dissected the other three figures to such a third. For the square there is an easy 3-piece dissection to that third of a dodecagon. For the regular hexagon, Freese crossposed two P-strips as we see in Figure 188. He obtained the first P-strip from a regular hexagon by cutting along a diagonal to remove an isosceles triangle. He obtained the second P-strip from one third of the dodecagon by cutting along a diagonal to remove an isosceles triangle. For the regular octagon, Freese took the dissection of a regular octagon to a square in the lower right corner of Plate 89 and slid the square down so that the lowest corner of the square coincided with the lower right corner of the octagon. He then overlaid the resulting dissected square with the dissected square in the upper right corner of Plate 188.

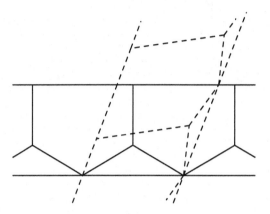

I188: Crossposition for regular hexagon to one third of dodecagon

This concludes the chapter on dissecting different regular polygons of equal area to a large regular polygon. The basic approach is to choose how to partition the large polygon into sections of equal area. For dissections of two different regular polygons to a square, Freese chose as one of the two polygons a regular dodecagon for four of the dissections. This allowed him to cut that dodecagon into a variety of different trapezoids, and he chose a trapezoid that he could dissect in few pieces to the other regular polygon. The regular hexagon fared well in Plate 182. Dissecting three different polygons is more problematic. After improving Freese's Plate 184 by one piece, I wondered if further improvement is possible. Then Gavin Theobald rose to the challenge and found the improvement! Perhaps it is possible to also improve Plate 188.

This chapter may have been our last big challenge, because the next chapter (the one before the Conclusion) is in large measure a group of special cases that has less to do with finding minimal dissections and more to do with identifying triangles that have relationships between angles or triangles that have integral-length sides. Even so, who can complain about triangles with integral side lengths, two hingeable dissections, an improved dissection, pairs of Pythagorean triples, and Chebyshev polynomials? So don't unfasten your seat belt—not yet!

A HEXAGON
PLUS AN EQUILATERAL TRIANGLE of SAME AREA DISSECTED TO FORM A SQUARE.

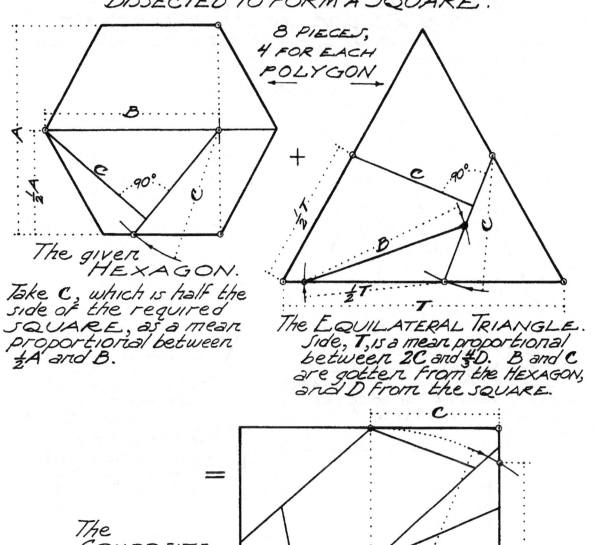

8 PIECES, 4 FOR EACH POLYGON

The given HEXAGON.

Take **C**, which is half the side of the required SQUARE, as a mean proportional between ½A and B.

+

The EQUILATERAL TRIANGLE. Side, **T**, is a mean proportional between 2C and ⁴⁄₃D. B and C are gotten from the HEXAGON, and D from the SQUARE.

=

The COMPOSITE SQUARE formed by the 8 pieces of the above 2 different equiareal POLYGONS.

2C = SIDE OF THE REQUIRED SQUARE. C is gotten from the HEXAGON.

PLATE
179

B4 P270A
...FREESE...

AN EQUILATERAL TRIANGLE
PLUS A DODECAGON of SAME AREA
DISSECTED TO FORM A SQUARE

8 PIECES

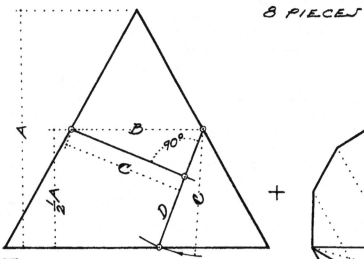

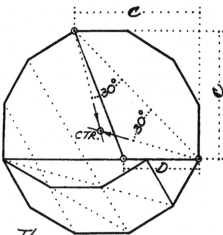

+

The given EQUILATERAL TRIANGLE.   C, half the side of the required SQUARE, is a mean proportional between ½A and B.

The DODECAGON that has the SAME AREA as the EQUILATERAL TRIANGLE from which C & D are taken.

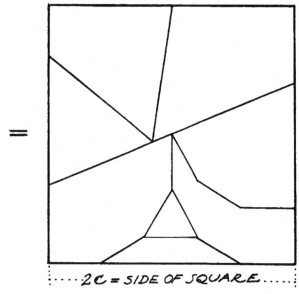

=

····· 2C = SIDE OF SQUARE ·····

The resultant COMPOSITE SQUARE formed from the 8 pieces of the two non-similar equiareal POLYGONS

pLATE
180

B4 PZ69
...FREESE...

A PENTAGON
PLUS A DODECAGON of <u>SAME AREA</u>
DISSECTED TO FORM A SQUARE.

10 PIECES,
5 FOR EACH POLYGON
THAT MAKES UP THE
SQUARE

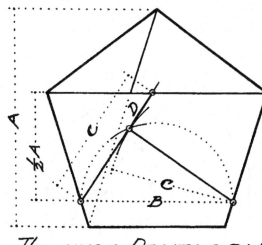

The given PENTAGON.
C, half the side of the
required SQUARE, is
a mean proportional
between ½A and B.

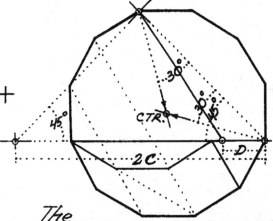

+

The
DODECAGON that has
the SAME AREA as
the PENTAGON
from which
C & D are taken.

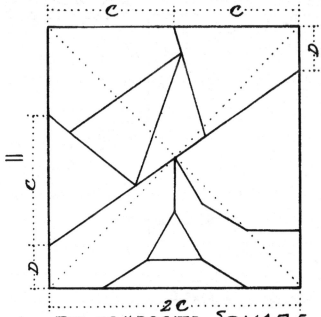

THE COMPOSITE SQUARE
RESULTING FROM ASSEMBLING
THE 10 ABOVE-DETERMINED PIECES

PLATE
181

A HEXAGON
PLUS A DODECAGON of SAME AREA
DISSECTED TO FORM A SQUARE.

9 PIECES

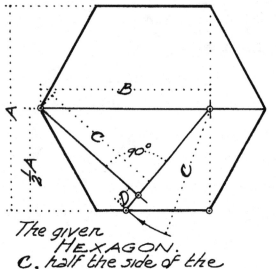

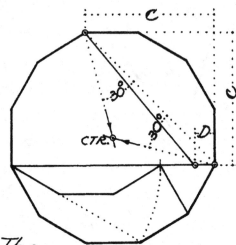

+

The given
HEXAGON.
C, half the side of the
required SQUARE, is a
mean proportional
between ½A and B.

The
DODECAGON that
has the SAME AREA as
the HEXAGON from
which C & D are taken.

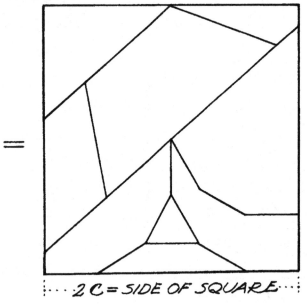

=

2 C = SIDE OF SQUARE

The
COMPOSITE
SQUARE formed by
the 9 pieces of the
above two polygons.

B4 P271
...FREESE...

PLATE
182

AN OCTAGON
PLUS A DODECAGON of SAME AREA
DISSECTED TO FORM A SQUARE

11 PIECES

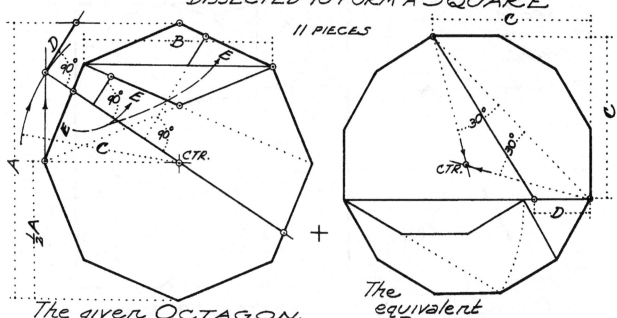

The given OCTAGON.
C, half the side of the required SQUARE, is a mean proportional between $\frac{1}{2}A$ and B.

+

The equivalent DODECAGON determined by the dimensions C & D taken from the OCTAGON.

=

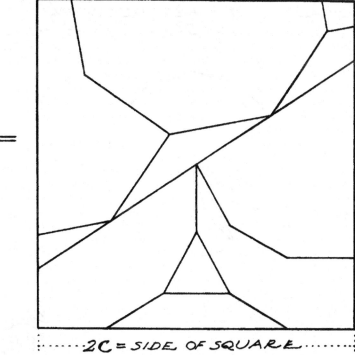

2C = SIDE OF SQUARE

The COMPOSITE SQUARE formed from the 11 pieces of the OCTAGON and DODECAGON.

PLATE
183

B4P272
...FREESE...

AN EQUILATERAL TRIANGLE
PLUS A HEXAGON
PLUS A DODECAGON
dissected
to form a
SQUARE.

17 PIECES.

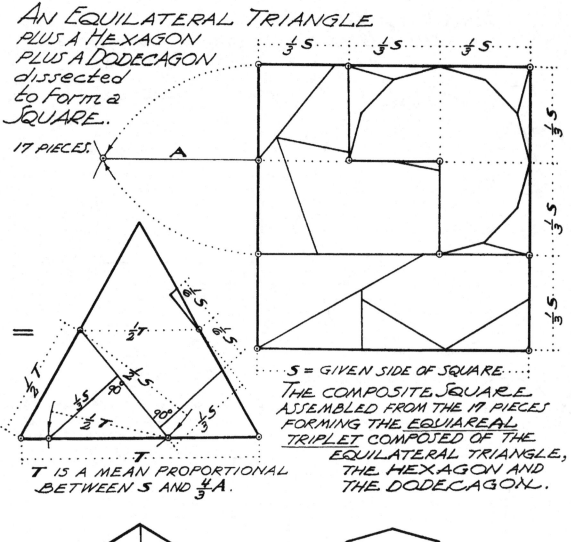

A

$\frac{1}{3}S$ $\frac{1}{3}S$ $\frac{1}{3}S$

$\frac{1}{3}S$

$\frac{1}{3}S$

$\frac{1}{3}S$

=

$\frac{1}{2}T$ $\frac{1}{3}S$ $\frac{1}{3}S$

$\frac{1}{3}T$ $\frac{1}{3}S$ 90° $\frac{1}{3}S$

90° $\frac{1}{2}T$ $\frac{1}{3}S$

T

T IS A MEAN PROPORTIONAL
BETWEEN S AND $\frac{4}{3}$A.

S = GIVEN SIDE OF SQUARE

THE COMPOSITE SQUARE
ASSEMBLED FROM THE 17 PIECES
FORMING THE EQUIAREAL
TRIPLET COMPOSED OF THE
EQUILATERAL TRIANGLE,
THE HEXAGON AND
THE DODECAGON.

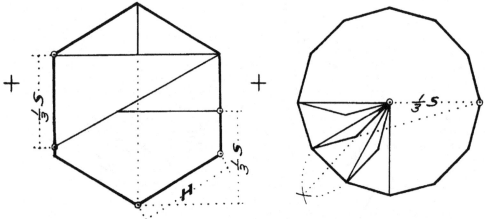

+ $\frac{1}{3}S$ $\frac{1}{3}S$ + $\frac{1}{3}S$

H

H IS A MEAN PROPORTIONAL
BETWEEN S AND $\frac{2}{9}$A.

PLATE
184

B4 PZ74
...FREESE...

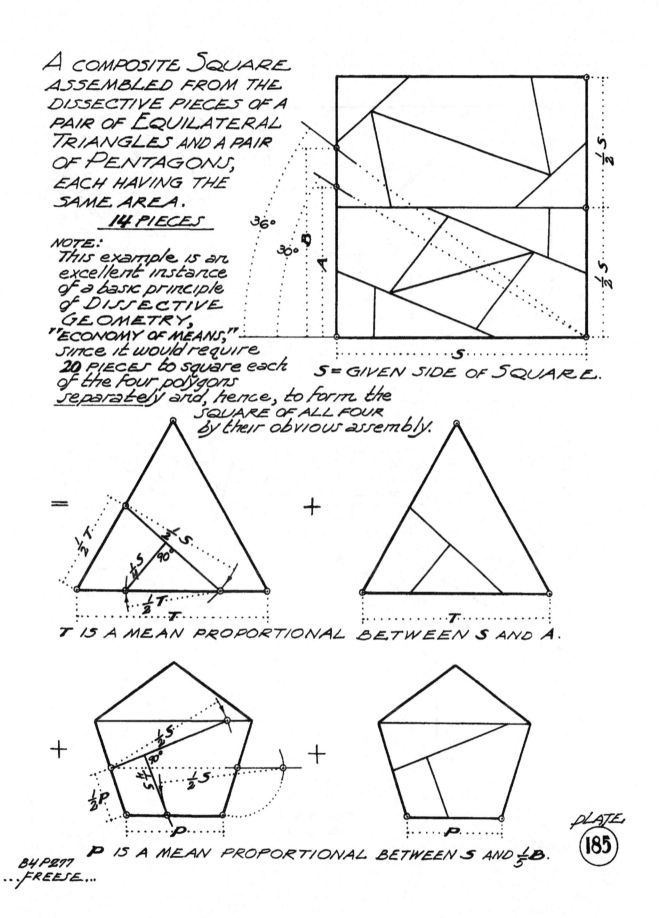

A COMPOSITE SQUARE ASSEMBLED FROM THE DISSECTIVE PIECES OF A PAIR OF EQUILATERAL TRIANGLES AND A PAIR OF PENTAGONS, EACH HAVING THE SAME AREA.

__14 PIECES__

NOTE:
This example is an excellent instance of a basic principle of DISSECTIVE GEOMETRY, "ECONOMY OF MEANS," since it would require __20__ PIECES to square each of the four polygons __separately__ and, hence, to form the SQUARE OF ALL FOUR by their obvious assembly.

S = GIVEN SIDE OF SQUARE.

T IS A MEAN PROPORTIONAL BETWEEN S AND A.

P IS A MEAN PROPORTIONAL BETWEEN S AND ⅕B.

PLATE
185

B4 P277
...FREESE...

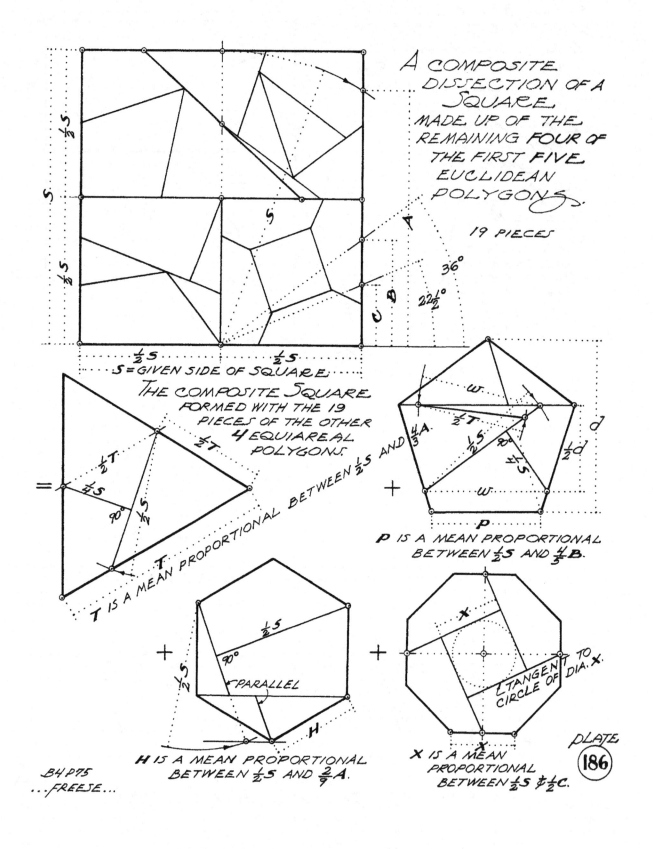

A COMPOSITE DISSECTION OF A SQUARE.
MADE UP OF THE REMAINING FOUR OF THE FIRST FIVE EUCLIDEAN POLYGONS.

19 PIECES

36°

22½°

S = GIVEN SIDE OF SQUARE.

½S ½S

THE COMPOSITE SQUARE FORMED WITH THE 19 PIECES OF THE OTHER 4 EQUIAREAL POLYGONS.

=

½T ½T

½S

90°

½S

T

T IS A MEAN PROPORTIONAL BETWEEN ½S AND ¾A.

+

w

½T

4/3 A

½S

90°

½S

w

P

d

½d

P IS A MEAN PROPORTIONAL BETWEEN ½S AND ⁴⁄₅B.

+

½S

90°

½S

PARALLEL

H

H IS A MEAN PROPORTIONAL BETWEEN ½S AND 2/9 A.

+

X

∠ TANGENT TO CIRCLE OF DIA. X.

X

X IS A MEAN PROPORTIONAL BETWEEN ½S & ½C.

PLATE 186

B4 P75
...FREESE...

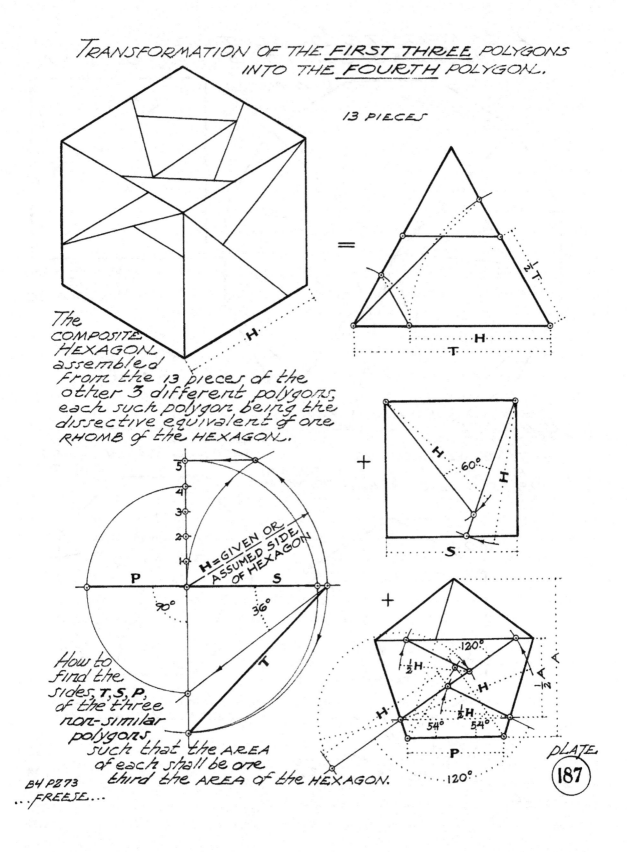

TRANSFORMATION OF THE <u>FIRST THREE</u> POLYGONS
INTO THE <u>FOURTH</u> POLYGON.

13 PIECES

The
COMPOSITE
HEXAGON
assembled
from the 13 pieces of the
other 3 different polygons,
each such polygon being the
dissective equivalent of one
RHOMB of the HEXAGON.

$\frac{1}{2}$ T

H

T

=

+

H   60°   H

S

How to
find the
sides, **T, S, P,**
of the three
non-similar
polygons
such that the AREA
of each shall be one
third the AREA of the HEXAGON.

H = GIVEN OR
ASSUMED SIDE
OF HEXAGON

P   S

90°   36°

T

5
4
3
2
1

+

120°

$\frac{1}{2}$H

$\frac{1}{2}$A

A

H

H

$\frac{1}{2}$H

54°   54°

P

120°

B4 P273
...FREESE...

PLATE
187

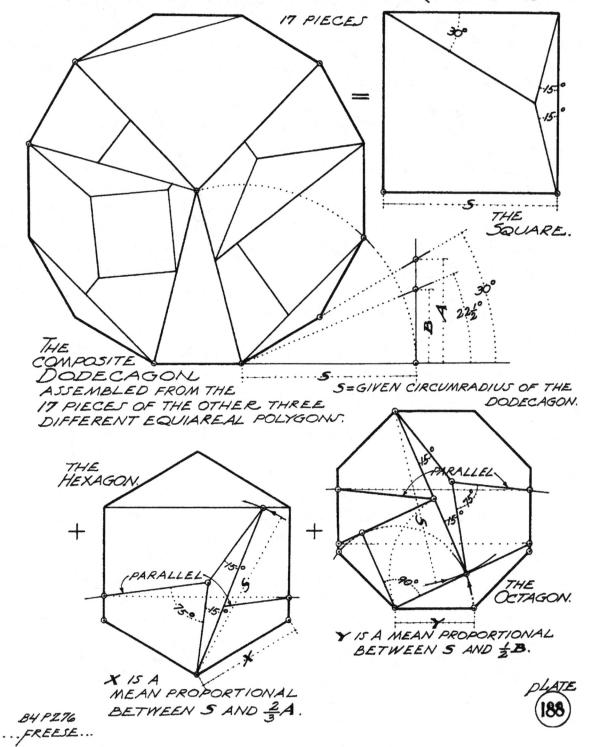

A COMPOSITE DISSECTION
OF THE DODECAGON INTO 3 OTHER POLYGONS

17 PIECES

30°

·15°

·15°

THE
SQUARE.

30°
22½°
B A

THE
COMPOSITE
DODECAGON
ASSEMBLED FROM THE
17 PIECES OF THE OTHER THREE
DIFFERENT EQUIAREAL POLYGONS.

S = GIVEN CIRCUMRADIUS OF THE
DODECAGON.

THE
HEXAGON.

15°

PARALLEL

75° 15°

15°

PARALLEL

15°

15°

90°

THE
OCTAGON.

Y IS A MEAN PROPORTIONAL
BETWEEN S AND ½ B.

X IS A
MEAN PROPORTIONAL
BETWEEN S AND ⅔ A.

B4 PZ.76
...FREESE...

PLATE
188

Chapter 19

Special Triangles (Plates 189–200)

In his final set of plates, Freese surveyed dissections of triangles that have integral side lengths or triangles whose angles are integral multiples of some basic angle. He explored the properties of various angles in the dissections of isosceles triangles in Plates 189-190, 193, and 200. He also explored the properties for classes of triangles with integral side lengths in Plates 191-192, and 194-199.

Although Freese focused more on special properties than on dissections in this chapter, there is nonetheless an opportunity to hinge his dissection in Plate 194 and to improve his dissection in Plate 196.

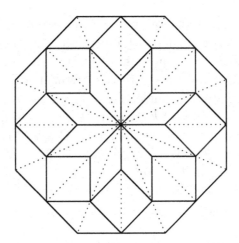

I189: A regular octagon structure from a dissection of a right triangle [GNF]

Plate 189: Freese's plate for the "multiple dissection of right triangles into isosceles triangles" demonstrates the rhombic structure of regular polygons with an even number of sides. We can use his dissection of a right triangle with an angle of 22.5° to fill out a regular octagon with 24 rhombuses of the same side length, as we see in Figure I189. Similarly, we can use his dissection of a right triangle with small angle of 18° to fill out a regular decagon with 40 rhombuses of the same side length. Freese's dissection of a right triangle with small angle of 15° fills out a regular dodecagon with 60 rhombuses of the same side length.

Inside Figure I189 we can easily identify eight small regular octagons whose structure matches the regular octagon in Freese's Plate 173. And the structure illustrated in Figure I189 is clearly present in Freese's dissection of four octagons to one octagon in Plate 95. We could use this same type of structural property to dissect regular decagons and regular dodecagons, though Freese did not use those structures in his decagon and dodecagon dissections.

Plate 190: In this plate, Freese investigated the cases in which he could dissect an isosceles triangle to some fixed number n of isosceles triangles. He further placed a restriction that each resulting isosceles triangle must have the pair of equal sides be the same length as the pair of resulting equal sides from any other resulting isosceles triangle. Without this last restriction, there would be more cases. For $n = 2$, there would also be an isosceles triangle with an apex angle of $108°$. For $n = 3$, there would be one symmetrically distinct way to dissect an isosceles right triangle. Also for $n = 3$, there would be four symmetrically distinct ways to dissect an isosceles triangle with an apex angle of $36°$. Similarly, there would be even more ways to dissect isosceles triangles to four or five isosceles triangles.

Plate 191: In this plate Freese elaborated on the two "3-in-1 dissections" that he had shown in the previous plate. Here he required the lengths of all sides of the isosceles triangle to be integral. Freese based triangles 1 and 2 on the Pythagorean identities $3^2 + 4^2 = 5^2$ and $7^2 + 24^2 + 25^2$. He based triangles 3 and 4 on the identities $5^2 + 12^2 = 13^2$ and $119^2 + 120^2 = 169^2$. He based triangles 6 and 7 on isosceles triangles that have one angle twice another. Similarly, he based triangles 8 and 9 on isosceles triangles that have one angle three times another. Note that triangle 6 has an interior edge labeled 6 that should be labeled 4.

Plate 192: In this plate Freese showed how to dissect an equilateral triangle of side length 167 to four sets of three congruent scalene triangles with integer side lengths plus an equilateral triangle of side length 19. The general puzzle that he addressed is to find, for any specified equilateral triangle of integral side length, a dissection to k sets of three congruent scalene integral triangles, for some integer k, plus one integral equilateral triangle. For that puzzle, it makes sense to ask for such a dissection for the smallest possible value of k, given that the resulting triangles are then primitive.

A way to approach these problems is to use the Law of Cosines, which becomes $c^2 = a^2 + b^2 - ab$ when the angle opposite side c is $60°$. We can generate all integral solutions to this equation by taking $a = m^2 - n^2$, $b = 2mn - n^2$, and $c = m^2 - mn + n^2$ for all pairs of positive integers m and n with $m > n$. We then generate all suitable triples of integers up to some particular value, and search backwards through our table of solutions to find an appropriate decomposition.

Plate 193: Of Freese's four "problems" in dissective geometry the first two have neat solutions, as we shall see. Problem 1 is a follow-up to his four examples in Plate 189. In that plate, Freese cut each of the lower three right triangles to N isosceles triangles. Freese then argued that the smallest of the resulting angles is $\phi = 90°/(N+1)$. Problem 1 in Plate 193 is just the example with $N = 6$. Thus $\phi = 90°/7 = 12\frac{6}{7}°$.

Problem 2 resolves as follows. Freese dissected the large isosceles triangle to five smaller isosceles triangles. Let each isosceles triangle sitting atop the lowest one have equal angles of size α, and an apex of size $180° - 2\alpha$. Then the two top isosceles triangles have equal angles of size 2α and a third angle of size $180° - 4\alpha$. And the lowest isosceles triangle will have an apex angle of 6α and equal angles of size $90° - 3\alpha$. But now there is a straight angle of size $\alpha + 6\alpha = 7\alpha$. Thus $\alpha = 180°/7$. Finally $\phi = \alpha + 90° - 3\alpha = 90° - 2\alpha = 38\frac{4}{7}°$.

Problems 3 and 4 appear more impressive than they really are. Let ϕ be the measure of each of the equal angles of an isosceles triangle. Then let c be a constant such that $\tan^3 \phi + \tan \phi = c$. If we apply Cardano's formula for the solution to a cubic equation, we get a value of $\tan \phi$. Applying the arctan function gives us the value ϕ. For problem 3, following this procedure for the cubic equation gives us $\phi \approx 54.048°$. The same approach will give a value for ϕ in Problem 4.

The dissection for Problem 3 is of some interest, because Freese claimed in Plate 8 that we can dissect an isosceles triangle to a square using the same technique as when the isosceles triangle is actually equilateral. We see the example for Problem 3 in Figure I193, which gives a hinged dissection.

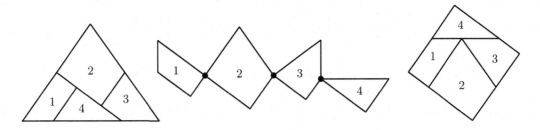

I193: Hingeable dissection of isosceles triangle in Problem 3

Plate 194: Freese seems to have been the first to find a 4-piece rational dissection of an oblique triangle to a regular hexagon. The particular triangle has an angle of 60°. Freese apparently did not realize that his 4-piece dissection is hingeable, which we verify in Figures I194a and I194b.

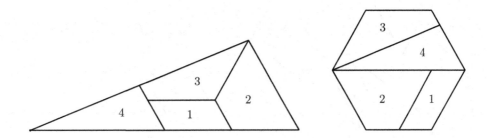

I194a: 4-piece rational dissection of oblique triangle to a regular hexagon

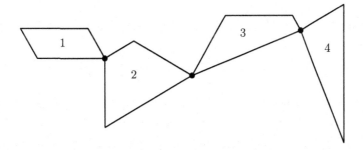

I194b: Hinged pieces for rational oblique triangle to a regular hexagon [GNF]

Plate 195: Freese was the first to dissect ten congruent regular hexagons to a triangle with side lengths of 6, 10, and 14. That triangle is related to the triangle in Plate 194 in the following way. If we slice an equilateral triangle of side length 3 from the triangle with side lengths 3, 7, and 8, we get a triangle with sides of length 3, 5, and 7, which is the triangle in this plate, but with each dimension halved.

Plate 196: Freese was the first to dissect ten congruent triangles of side lengths 15, 35, and 40 to a triangle of side lengths 60, 100, and 140. Note that the ten triangles are similar to the triangle from Plate 174, while the one large triangle here is similar to the triangle of Plate 195. Thus Plates 194-196 demonstrate related properties. All of the side lengths in Freese's 17-piece dissection are integral.

We can improve Freese's dissection by laying out the ten small triangles as suggested by the dotted (and solid) lines in the large triangle in Figure I196. Equilateral triangles of side length 15 and 20 stick out beyond the boundary of the large triangle, and there is a deficiency in the large triangle in the shape of an equilateral triangle of side length 25. Performing the implied Pythagorean-like dissection, and making just one more cut in addition to those indicated by the dotted edges, completes the resulting 15-piece dissection.

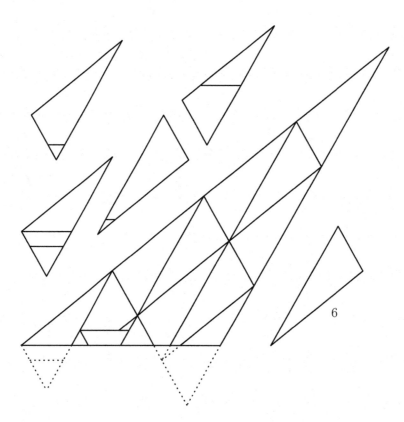

I196: Improved ten $(15, 35, 40)$-triangles to a $(60, 100, 140)$-triangle [GNF]

Plate 197: In this plate, Freese gave four related dissections of pairs of integral right triangles. Each pair of triangles has one triangle similar to a 20 : 21 : 29 right triangle and the other similar to a 12 : 35 : 37 right triangle. In each pair of right triangles, the triangles are of equal area. In the

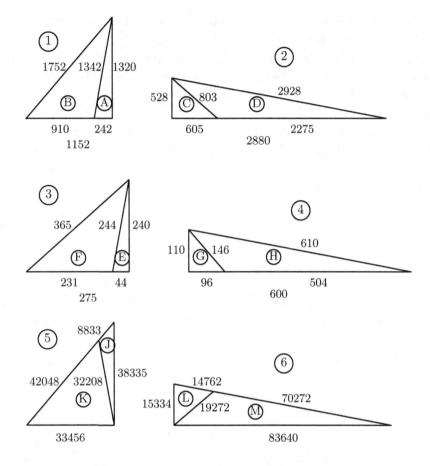

I197: Another example of special pairs of integral right triangles [GNF]

first two pairs of right triangles, Freese dissected each right triangle to an integral right triangle and an integral oblique triangle, where the two resulting right triangles are of equal area and a resulting right triangle from each of the pair is similar to the right triangle of the other of the pair. In the remaining two pairs of right triangles, Freese dissected each right triangle to two oblique triangles, where the resulting two triangles from one right triangle are similar to the resulting two triangles from the other right triangle, and each triangle from one right triangle has area equal to that of the dissimilar triangle from the other right triangle.

This set of examples is not of the form of geometric dissections per se, because Freese did not rearrange the pieces from one dissected figure to form the other figure. I do not know whether he discovered the phenomenon himself, or had found it in an article or a book. The key to making this example work seems to be identifying a pair of Pythagorean triples for which the corresponding right triangles have the same area. I wondered if this example was unique, but after a quick search identified another pair of Pythagorean triples whose corresponding right triangles have the same area, namely 48 : 55 : 73 and 22 : 120 : 122. Indeed, we can identify the integral triangles that form a solution similar to what we see in Plate 197. We see the first three pairs in Figure I197. The computation of these is not so difficult, but is perhaps a bit tedious. Readers may wish to determine the fourth pair of right triangles in this example themselves.

Plate 198: Freese analyzed two examples of integral right triangles which he dissected to two non-similar right triangles and an integral oblique triangle that contains one angle that is an integral multiple of another angle. He based his first example on right triangles similar to those with side lengths 3, 4, and 5 for the first and 7, 24, 25 for the second.

Applying the Law of Sines to the oblique triangle with edge lengths in the ratio of 11, 25, and 30, Freese obtained $(\sin a)/25 = (\sin(2a))/30$. Noting that $\sin a = 4/5$ and $\cos a = 3/5$ in his example, and applying $\sin(2a) = 2\sin(a)\cos(a)$, we confirm that $\sin(2a) = 24/25$ and indeed that $(\sin a)/25 = (\sin(2a)/30$.

In the second example, applying the Law of Sines to the oblique triangle with edge lengths in the ratio of 112, 125, and 195, Freese obtained $(\sin a)/125 = \sin(3a)/195$. Noting that $\sin a = 3/5$ in his example, and applying $\sin(3a) = -4\sin^3 a + 3\sin a$, we confirm that $\sin(3a) = 117/125$ and indeed $\sin(a)/125 = (\sin(3a))/195$.

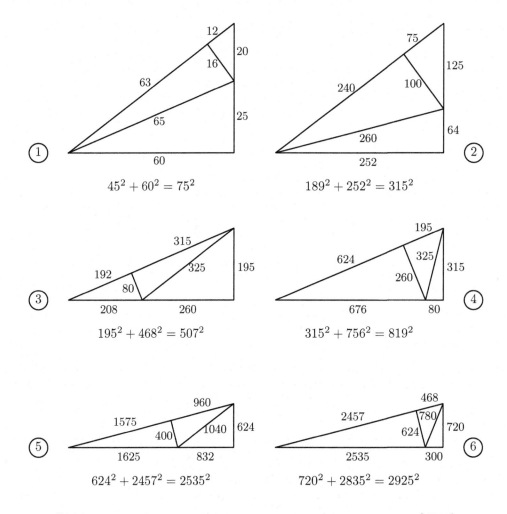

I199: Three other pairs of integral right triangles for Plate 199 [GNF]

Plate 199: This plate shows a nifty relationship that is shared among three Pythagorean triples. Freese observed that $3^2 + 4^2 = 5^2$, $5^2 + 12^2 = 13^2$, and $33^2 + 56^2 = 65^2$, and that there are right triangles with the corresponding side lengths. He then observed that there are six integral right triangles that dissect to triangles similar to the three triangles. Triangles 1 and 2 in Plate 199 have side lengths in the ratios of $3:4:5$, with their leftmost triangles having side lengths in the same ratio. Triangles 3 and 4 have side lengths in the ratios of $5:12:13$, with their leftmost triangles having side lengths in the same ratio. And triangles 5 and 6 have side lengths in the ratios of $33:56:65$, with their leftmost triangles having side lengths in the same ratio.

The example that Freese presented in this plate is one of an infinite number of such examples. We can generate such examples by taking any two different sets of Pythagorean triples, say $a^2 + b^2 = c^2$ and $A^2 + B^2 = C^2$ and generating two more as follows. (See Dickson 1929, which reproduces a discussion by Diophantus in his Chapter 4.) Multiply the lefthand sides together, and similarly for the righthand sides, yielding

$$a^2 A^2 + a^2 B^2 + b^2 A^2 + b^2 B^2 = c^2 C^2$$

We rearrange this identity to get two Pythagorean identities

$$(bB - aA)^2 + (aB + bA)^2 = (cC)^2$$

$$(aA + bB)^2 + (aB - bA)^2 = (cC)^2$$

If we choose $a = 3$, $b = 4$, $c = 5$ and $A = 5$, $B = 12$, $C = 13$, we produce for the first identity $33^2 + 56^2 = 65^2$, as in the plate. Choosing those same values, but now substituting into the second identity, we produce $63^2 + 16^2 = 65^2$. Figure I199 shows the six corresponding triangles for this second identity.

Another example with small integers would have $a = 3$, $b = 4$, $c = 5$, $A = 8$, $B = 15$, and $C = 17$. This produces the identities $36^2 + 77^2 = 85^2$ and $84^2 + 13^2 = 85^2$. Readers may enjoy finding the corresponding six triangles for each of those two resulting identities.

Plate 200: Freese gave a dissection of a certain triangle to 5 isosceles triangles that all have integral lengths. The isosceles triangle with the smallest base angles has side lengths of 11, 11, and 21, giving a base angle a where $\cos a = 21/22$. The base angles of the remaining 4 isosceles triangles are $2a$, $3a$, $4a$, and $5a$. The cosines of the base angles of the 4 other isosceles triangles are all rational, because $\cos(na) = T_n(\cos a)$, where T_n is the Chebyshev polynomial of the first kind:

$$T_n(x) = \sum_{m=0}^{\lfloor n/2 \rfloor} \binom{n}{2m} x^{n-2m} (x^2 - 1)^m$$

For example, the first four Chebyshev polynomials are $T_1(x) = x$, $T_2(x) = 2x^2 - 1$, $T_3(x) = 4x^3 - 3x$, and $T_4(x) = 8x^4 - 8x^2 + 1$.

This last chapter has focused on various properties of right triangles, isosceles triangles, and triangles with edges of integral length. So in this chapter Freese returned full circle to where he began, with triangles. The last four plates are a fascinating exploration of integral lengths in right and isosceles triangles.

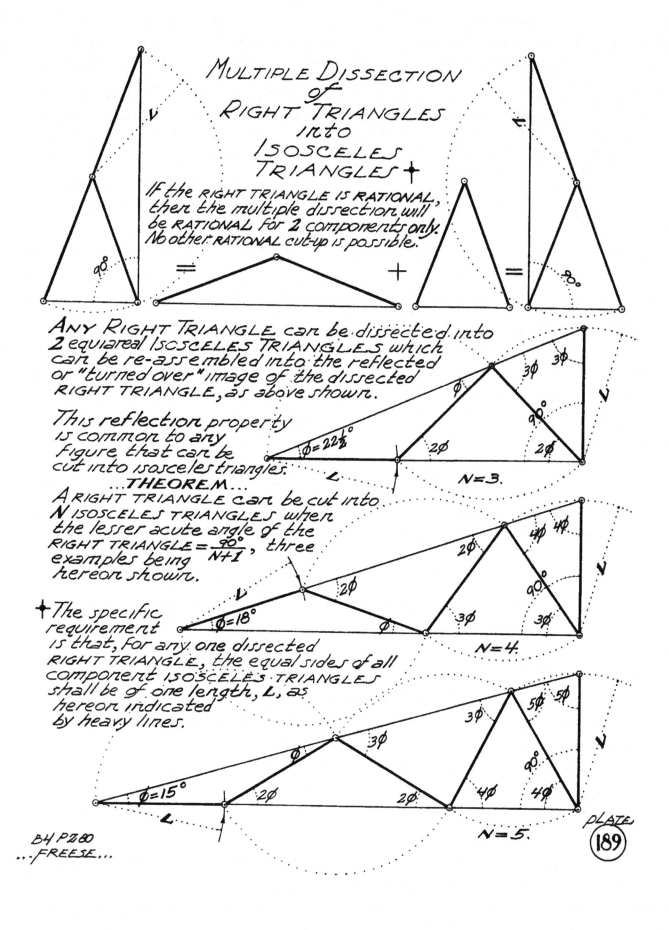

MULTIPLE DISSECTION
of
RIGHT TRIANGLES
into
ISOSCELES
TRIANGLES ✛

IF the RIGHT TRIANGLE IS RATIONAL,
then the multiple dissection will
be RATIONAL for 2 components only.
No other RATIONAL cut-up is possible.

90°

=   +   =

ANY RIGHT TRIANGLE can be dissected into
2 equiareal ISOSCELES TRIANGLES which
can be re-assembled into the reflected
or "turned over" image of the dissected
RIGHT TRIANGLE, as above shown.

This reflection property
is common to any
figure that can be
cut into isosceles triangles.

...THEOREM...
A RIGHT TRIANGLE can be cut into
N ISOSCELES TRIANGLES when
the lesser acute angle of the
RIGHT TRIANGLE = $\frac{90°}{N+1}$, three
examples being
hereon shown.

✛ The specific
requirement
is that, for any one dissected
RIGHT TRIANGLE, the equal sides of all
component ISOSCELES TRIANGLES
shall be of one length, L, as
hereon indicated
by heavy lines.

$\phi = 22\frac{1}{2}°$ ϕ 3ϕ 3ϕ $90°$ 2ϕ 2ϕ N=3.

$\phi = 18°$ 2ϕ ϕ 4ϕ 4ϕ $90°$ 3ϕ 3ϕ 2ϕ N=4.

$\phi = 15°$ ϕ 3ϕ 2ϕ 2ϕ 5ϕ 5ϕ $90°$ 4ϕ 4ϕ 3ϕ N=5.

B4 PZ80
...FREESE...

PLATE 189

MULTIPLE DISSECTION of ISOSCELES TRIANGLES into ISOSCELES TRIANGLES. ✝

2-IN-1 DISSECTIONS

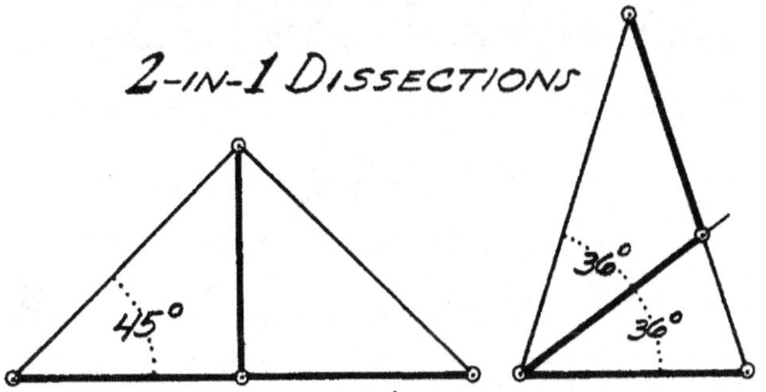

45°

36°
36°

These two are the ONLY ISOSCELES triangles that can be cut into 2 ISOSCELES triangles.

4-IN-1 DISSECTIONS

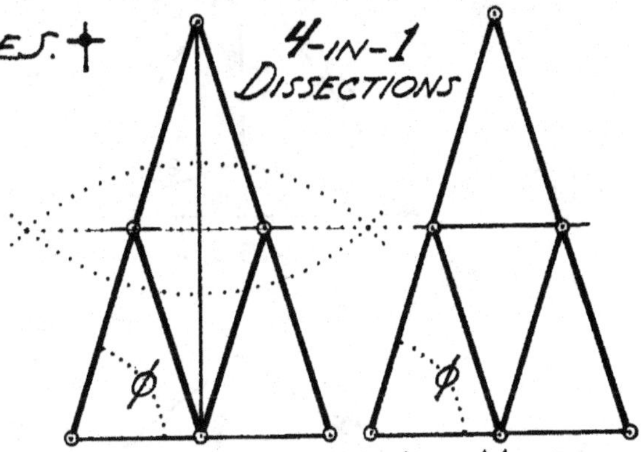

φ

φ

φ = any angle less than 90°.

3-IN-1 DISSECTIONS

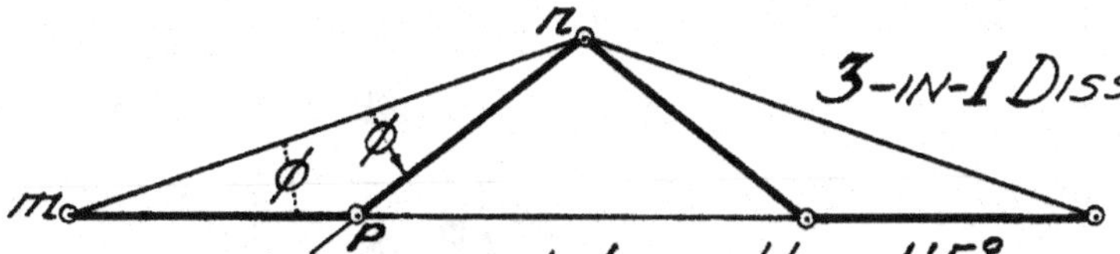

n

φ φ

m   P

φ = any angle less than 45°.

Note:
Point P can also be located by the perpendicular bisector of mn.

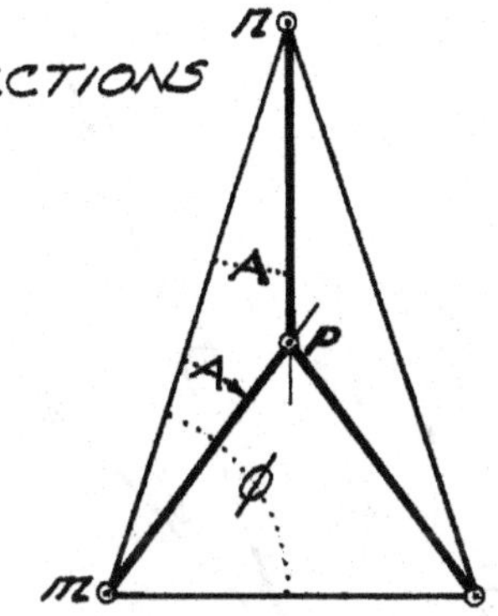

n

A

A   P

φ

m

φ = any angle more than 45°.

5-IN-1 DISSECTIONS

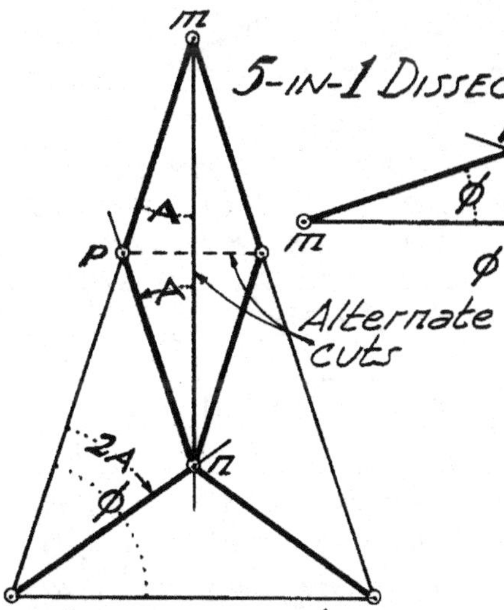

m

A

P   n

A

Alternate cuts

2A

φ   n

φ = any angle more than 60°.

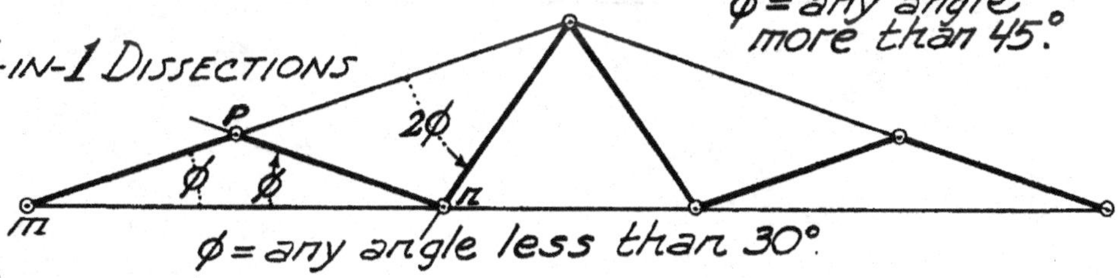

P

φ φ

m   2φ

n

φ = any angle less than 30°.

✝ The specific requirement is that, for any one dissected triangle, the equal sides of all component triangles shall be of one length, as hereon indicated by heavy lines.

PLATE
190

B4 PZ 79
...FREESE...

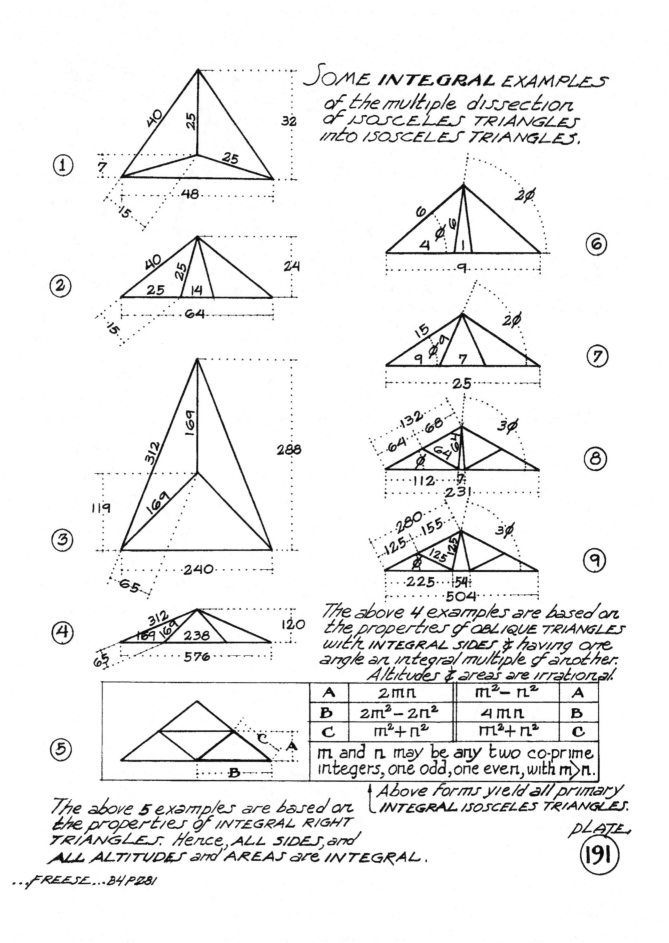

SOME *INTEGRAL* EXAMPLES
of the multiple dissection
of ISOSCELES TRIANGLES
into ISOSCELES TRIANGLES.

① 40 25 32 7 25 48 15

② 40 25 25 14 24 64 15

③ 169 312 169 119 288 240 65

④ 312 169 169 238 120 65 576

⑤ C A B

⑥ 6 6 4 1 9 2∅ ∅

⑦ 15 9 ∅ 9 7 2∅ 25

⑧ 132 68 64 ∅ 112 7 231 3∅

⑨ 280 155 125 125 125 ∅ 225 54 504 3∅

The above 4 examples are based on
the properties of OBLIQUE TRIANGLES
with INTEGRAL SIDES & having one
angle an integral multiple of another.
Altitudes & areas are irrational.

A	$2mn$	$m^2 - n^2$	A
B	$2m^2 - 2n^2$	$4mn$	B
C	$m^2 + n^2$	$m^2 + n^2$	C

m and n may be any two co-prime
integers, one odd, one even, with m>n.

Above forms yield all primary
INTEGRAL ISOSCELES TRIANGLES.

The above 5 examples are based on
the properties of INTEGRAL RIGHT
TRIANGLES. Hence, ALL SIDES, and
ALL ALTITUDES and AREAS are INTEGRAL.

PLATE.
⑲①

...FREESE...B4 P281

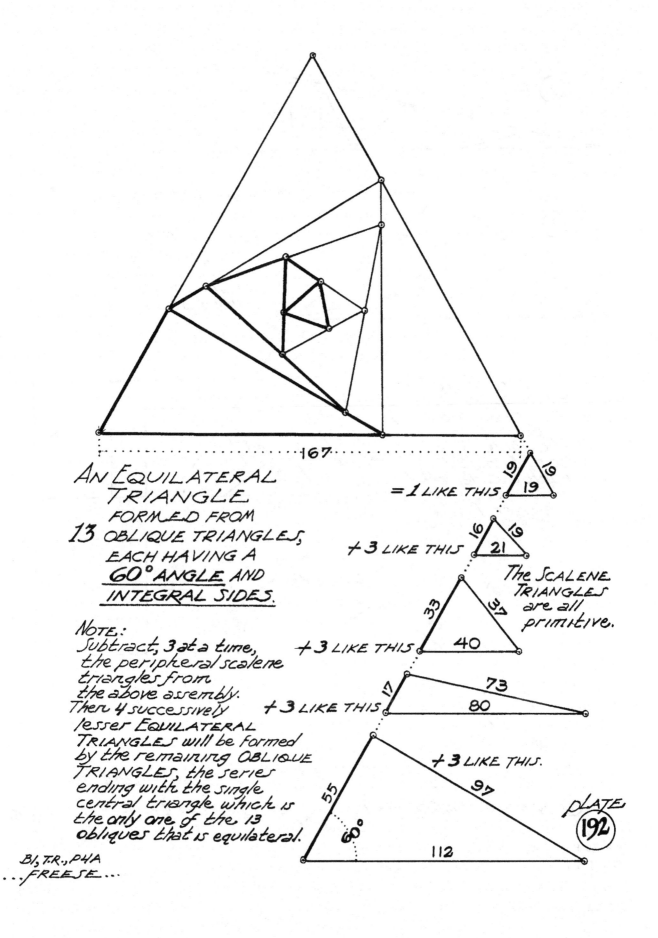

AN EQUILATERAL
TRIANGLE
FORMED FROM
13 OBLIQUE TRIANGLES,
EACH HAVING A
60° ANGLE AND
INTEGRAL SIDES.

= 1 LIKE THIS

+ 3 LIKE THIS

The SCALENE
TRIANGLES
are all
primitive.

+ 3 LIKE THIS

+ 3 LIKE THIS

+ 3 LIKE THIS.

NOTE:
Subtract, 3 at a time,
the peripheral scalene
triangles from
the above assembly.
Then 4 successively
lesser EQUILATERAL
TRIANGLES will be formed
by the remaining OBLIQUE
TRIANGLES, the series
ending with the single
central triangle which is
the only one of the 13
obliques that is equilateral.

B1, T.R., P4A
...FREESE...

PLATE
192

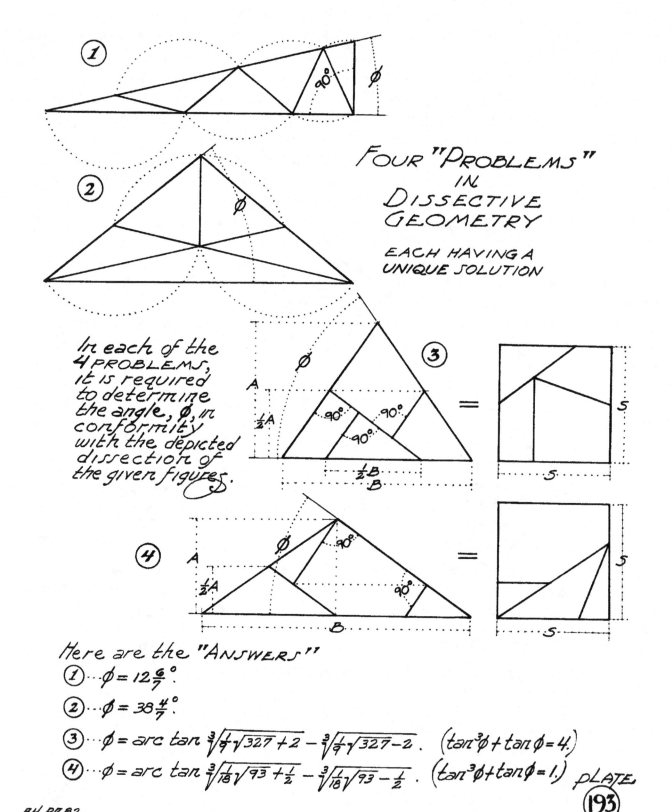

Four "Problems" in Dissective Geometry

Each having a unique solution

In each of the 4 PROBLEMS, it is required to determine the angle, φ, in conformity with the depicted dissection of the given figures.

Here are the "Answers"

① ... $\phi = 12\frac{6}{7}°$.

② ... $\phi = 38\frac{4}{7}°$.

③ ... $\phi = \arctan \sqrt[3]{\frac{1}{9}\sqrt{327}+2} - \sqrt[3]{\frac{1}{9}\sqrt{327}-2}$. $(\tan^3\phi + \tan\phi = 4.)$

④ ... $\phi = \arctan \sqrt[3]{\frac{1}{18}\sqrt{93}+\frac{1}{2}} - \sqrt[3]{\frac{1}{18}\sqrt{93}-\frac{1}{2}}$. $(\tan^3\phi + \tan\phi = 1.)$

PLATE 193

A RATIONAL 4-PIECE TRANSFORMATION OF AN OBLIQUE TRIANGLE WITH INTEGRAL SIDES INTO A REGULAR HEXAGON.

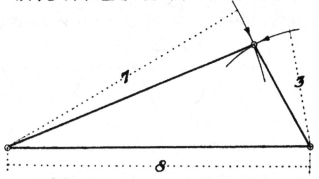

THE OBLIQUE TRIANGLE.

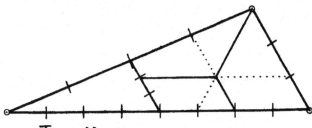

THE 4-PIECE DISSECTION,
in which all linear values are RATIONAL.

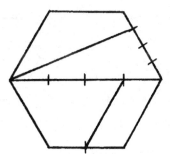

THE TRANSFORMATION
INTO A REGULAR HEXAGON.

Demonstration:

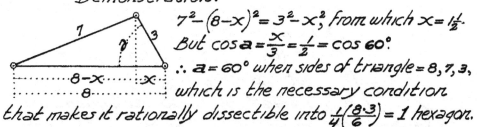

$7^2 - (8-x)^2 = 3^2 - x^2$, from which $x = 1\frac{1}{2}$.

But $\cos a = \dfrac{x}{3} = \dfrac{1}{2} = \cos 60°$.

$\therefore a = 60°$ when sides of triangle $= 8, 7, 3$,

which is the necessary condition

that makes it rationally dissectible into $\frac{1}{4}\left(\dfrac{8 \cdot 3}{6}\right) = 1$ hexagon.

PLATE
194

A RATIONAL TRANSFORMATION
OF 10 EQUAL REGULAR HEXAGONS
INTO
AN OBLIQUE SCALENE TRIANGLE WITH INTEGRAL SIDES.
18 PIECES

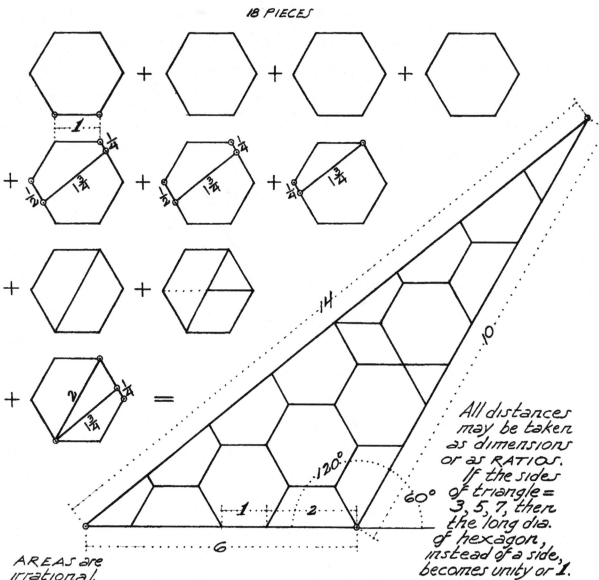

All distances
may be taken
as dimensions
or as RATIOS.
If the sides
of triangle =
3, 5, 7, then
the long dia.
of hexagon,
instead of a side,
becomes unity or **1**.

AREAS are
irrational.

Demonstration:
$14^2 - (6+x)^2 = 10^2 - x^2$, from which $x = 5$.
But $\cos a = \dfrac{x}{10} = \dfrac{5}{10} = \cos 60°$.
$\therefore a = 60°$, $b = 180° - 60° = 120°$, when sides
of triangle = 6, 10, 14, *or in that ratio.*
Hence, the triangle is rationally dissectible into $\dfrac{6 \cdot 10}{6} = 10$ hexagons.

R.T. PJ2
...FREESE...

PLATE
(195)

A UNIQUE ALL-INTEGRAL TRANSFORMATION OF
10 EQUAL OBLIQUE SCALENE TRIANGLES, HAVING
INTEGRAL SIDES AND A 60°ANGLE, INTO A SINGLE
SCALENE TRIANGLE HAVING INTEGRAL SIDES AND A 120°ANGLE.

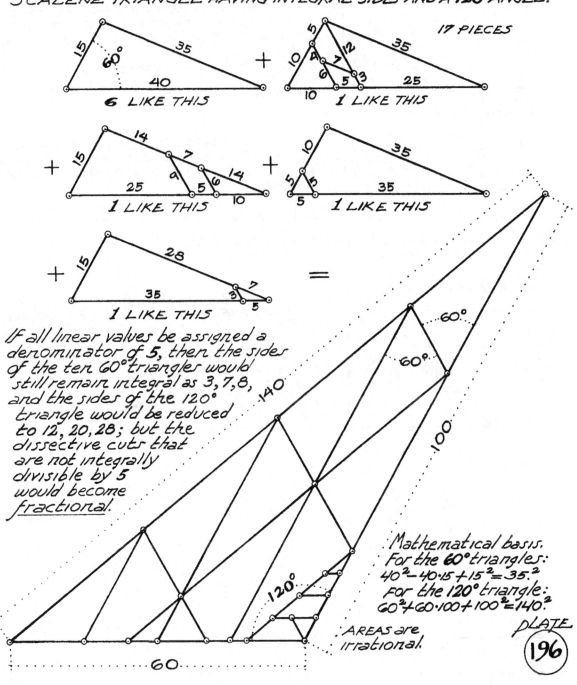

17 PIECES

6 LIKE THIS

1 LIKE THIS

1 LIKE THIS

1 LIKE THIS

1 LIKE THIS

If all linear values be assigned a
denominator of 5, then the sides
of the ten 60°triangles would
still remain integral as 3, 7, 8,
and the sides of the 120°
triangle would be reduced
to 12, 20, 28; but the
dissective cuts that
are not integrally
divisible by 5
would become
fractional.

Mathematical basis.
For the 60°triangles:
$40^2 - 40 \cdot 15 + 15^2 = 35^2$
For the 120° triangle:
$60^2 + 60 \cdot 100 + 100^2 = 140^2$

AREAS are
irrational.

PLATE
196

4 PAIRS OF NON-SIMILAR INTEGRAL EQUIAREAL RIGHT TRIANGLES, EACH PAIR DISSECTED TO FORM 2 DIFFERENT PAIRS OF NON-SIMILAR INTEGRAL RIGHT AND OBLIQUE TRIANGLES OF EQUAL AREA PER PAIR.

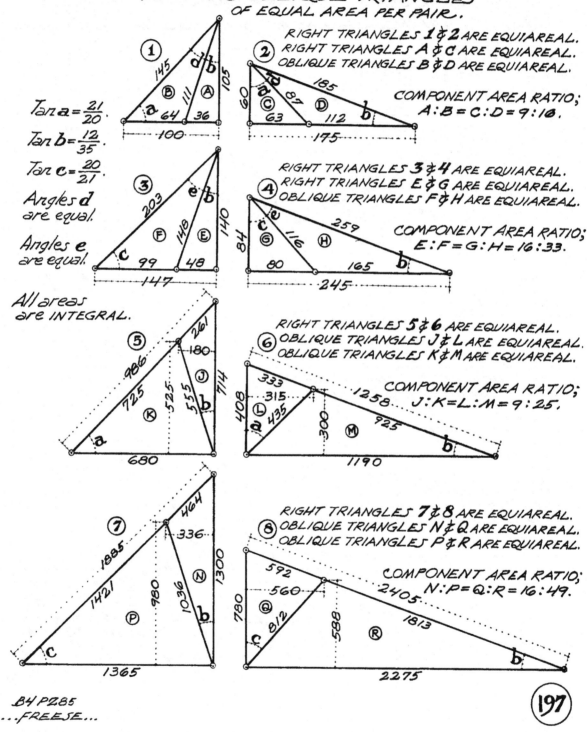

RIGHT TRIANGLES 1 & 2 ARE EQUIAREAL.
RIGHT TRIANGLES A & C ARE EQUIAREAL.
OBLIQUE TRIANGLES B & D ARE EQUIAREAL.

COMPONENT AREA RATIO;
A:B = C:D = 9:10.

$Tan\ a = \frac{21}{20}$.

$Tan\ b = \frac{12}{35}$.

$Tan\ c = \frac{20}{21}$.

Angles d are equal.

Angles e are equal.

All areas are INTEGRAL.

RIGHT TRIANGLES 3 & 4 ARE EQUIAREAL.
RIGHT TRIANGLES E & G ARE EQUIAREAL.
OBLIQUE TRIANGLES F & H ARE EQUIAREAL.

COMPONENT AREA RATIO;
E:F = G:H = 16:33.

RIGHT TRIANGLES 5 & 6 ARE EQUIAREAL.
OBLIQUE TRIANGLES J & L ARE EQUIAREAL.
OBLIQUE TRIANGLES K & M ARE EQUIAREAL.

COMPONENT AREA RATIO;
J:K = L:M = 9:25.

RIGHT TRIANGLES 7 & 8 ARE EQUIAREAL.
OBLIQUE TRIANGLES N & Q ARE EQUIAREAL.
OBLIQUE TRIANGLES P & R ARE EQUIAREAL.

COMPONENT AREA RATIO;
N:P = Q:R = 16:49.

197

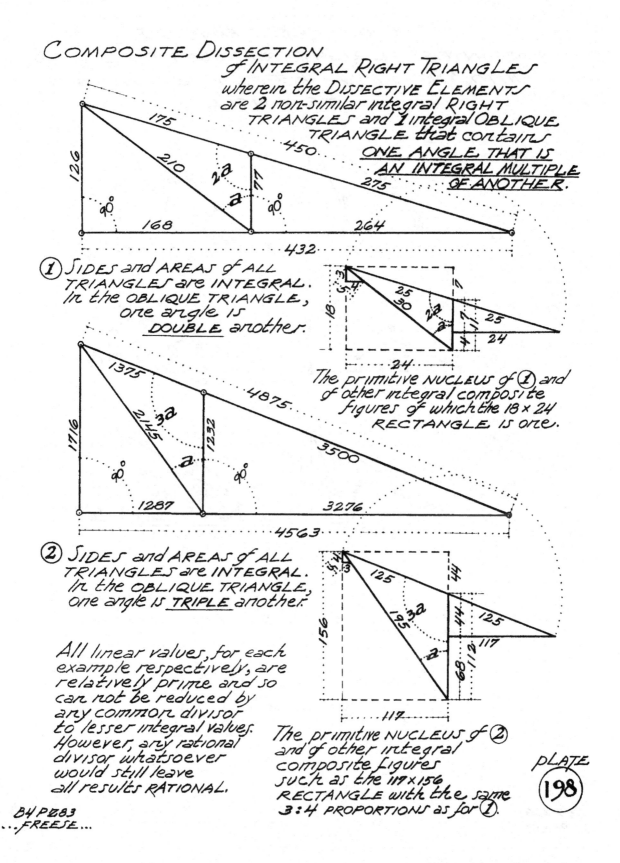

COMPOSITE DISSECTION of INTEGRAL RIGHT TRIANGLES wherein the DISSECTIVE ELEMENTS are 2 non-similar integral RIGHT TRIANGLES and 1 integral OBLIQUE TRIANGLE that contains ONE ANGLE THAT IS AN INTEGRAL MULTIPLE OF ANOTHER.

① SIDES and AREAS of ALL TRIANGLES are INTEGRAL. In the OBLIQUE TRIANGLE, one angle is DOUBLE another.

The primitive NUCLEUS of ① and of other integral composite figures of which the 18 × 24 RECTANGLE is one.

② SIDES and AREAS of ALL TRIANGLES are INTEGRAL. In the OBLIQUE TRIANGLE, one angle is TRIPLE another.

All linear values, for each example respectively, are relatively prime and so can not be reduced by any common divisor to lesser integral values. However, any rational divisor whatsoever would still leave all results RATIONAL.

The primitive NUCLEUS of ② and of other integral composite figures such as the 117×156 RECTANGLE with the same 3:4 PROPORTIONS as for ①.

B4 PZ83
...FREESE...

PLATE
198

6 DIFFERENT EXAMPLES
OF INTEGRAL RIGHT TRIANGLES DISSECTED
INTO 3 NON-SIMILAR UNEQUAL INTEGRAL RIGHT
TRIANGLES

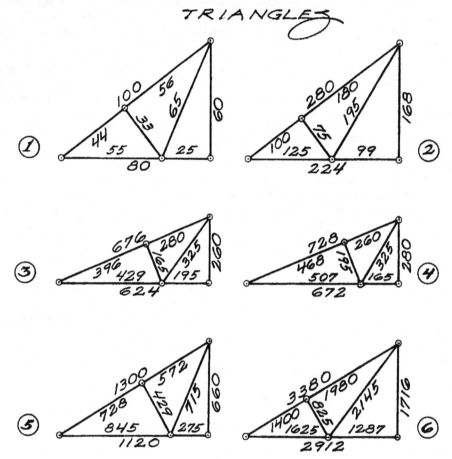

All triangles of the above 6 combinations are yielded by the following 3 PRIMITIVES and their inversions.

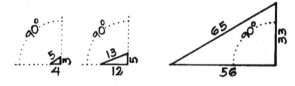

PLATE
199

B4 P286
...FREESE...

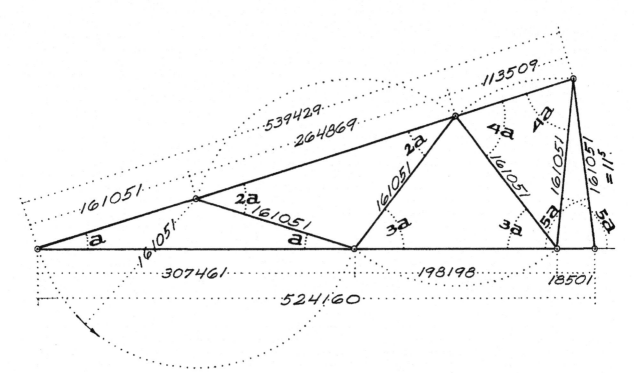

$$\cos a = \frac{21}{22}.$$

$$\cos 2a = \frac{199}{242}.$$

$$\cos 3a = \frac{819}{1331}.$$

$$\cos 4a = \frac{10319}{29282}.$$

$$\cos 5a = \frac{18501}{322102}.$$

AN OBLIQUE SCALENE TRIANGLE WITH INTEGRAL SIDES DISSECTED INTO 5 ISOSCELES TRIANGLES WITH INTEGRAL SIDES AND WITH BASE ANGLES INCREASING IN THE ARITHMETICAL PROGRESSION OF 1, 2, 3, 4, 5.

Any _INTEGRAL-SIDED_ OBLIQUE SCALENE TRIANGLE in which one angle is an _integral multiple_, N, of another, can be subdivided into N _INTEGRAL-SIDED ISOSCELES TRIANGLES_ provided that $\cos a > \cos \frac{90°}{N}$ ...... The above example, though of huge proportions, is the minimum one _for N=5,_ the _primitive nucleus_ being as follows:

PLATE
200
OF 200

Chapter 20

From the Past, Into the Future

We have come to the end of Freese's manuscript and its accompanying commentary. And we are now some sixty years beyond his "allotted participation in life," as Paul Peter Kiessig put it. Perhaps that is sufficient time to provide some historical heft to an assessment of his contributions. So what can we say?

From a life that he approached audaciously, that he explored unapologetically, and into which he sank his energy up until the very end, he distilled geometric marvels. He reached out to grasp an ancient art, envisioned impressive new problems and techniques, and attempted dissections that people had not previously contemplated. He had re-invigorated the technique of completing the tessellation, which had stagnated after that miraculous centuries-old dissection of a regular octagon to a square. He had breathed new life into Harry Hart's dissection of two regular polygons to one, by introducing a nifty variation of it.

Freese had dreamed up examples that pointed the way to an ever more ambitious program. With his early efforts of dissecting various many-sided regular polygons to a square, he had inspired Gavin Theobald decades later to confront those seemingly unapproachable problems with full force. With his fascination of hinged dissections, Freese re-energized my own focus on a kinetic art. He had conceived of strikingly rich sets of problems that are not yet fully explored, such as dissections of sets of disparate equal-area regular polygons to some larger regular polygon, and as sets of similar regular polygons with side lengths (or even areas) in arithmetic progression.

There is an elegance in Freese's framing of his inventive dissection problems. And in a day and age when we have come to rely on computer software to produce dissection diagrams, it is refreshing to be reminded by his detailed diagrams with dotted dimensions of how actually to measure and draw those geometric figures. (Indeed, less than fifteen years after Freese passed away, I started drafting my first geometric illustrations—with a ruler, a compass, and an India-ink marking pen!)

Gavin Theobald asked me about the mysterious codes that Freese had attached to each of his plates, usually in the lower left corner. Perhaps Freese had accumulated his dissections in four notebooks, designated as 'B1', 'B2', 'B3', and 'B4', though no such books were ever found. Referencing a particular book, maybe 'P' indicates a (page) number, 'PR' a (problem) number, and 'PZ' a (puzzle) number. There are other identifiers, such as 'PA', 'PS', 'PX', 'PY', and 'R.T.' (for Rational Triangles?). Could Freese have planned to erase all these codes once his plates were complete, but lost the energy to do so, as his life ebbed away?

Yes, it has been sixty years since Freese completed his manuscript. Even though many enthusiasts have advanced the field significantly since then, Freese's legacy continues to provide inspiration. So maybe, as his "BIOGRAPHY" suggests, his spirit "Haint dead Yet." Recognizing that potentiality, and recalling what I have done for my previous books, I will maintain (into the future) a webpage identifying improved dissections related to the manuscript of Ernest Irving Freese.

Bibliography

Abbott, T. G., Abel, Z., Charlton, D., Demaine, E. D., Demaine, M. L., and Kominers, S. D. (2012). Hinged dissections exist, *Discrete & Computational Geometry* **47**, 1, pp. 150–186.

Abu'l-Wafā' al-Būzjānī. Kitāb fīmā yahtāju al-sāni' min a' māl al-handasa (On the Geometric Constructions Necessary for the Artisan), Mashdad: Imam Riza 37, copied in the late 10th or the early 11th century. Persian manuscript.

Anonymous. Fī tadākhul al-ashkāl al-mutashābiha aw al-mutawāfiqa (Interlocks of Similar or Complementary Figures), Paris: Bibliothèque Nationale, ancien fonds. Persan 169, ff. 180r–199v.

Anonymous (1951). Solutions to problems in Eureka No. 13, *Eureka: The Journal of the Archimedeans*, 13, p. 23, R. F. Wheeler's solution is to the second of the 'Two Dissection Problems'.

Baetens, B. (1997). Opmerking van de heer bert baetens, *Wiskunde en Onderwijs* **23**, pp. 318–319.

Ball, W. W. R. (1939). *Mathematical Recreations and Essays*, eleventh edn. (MacMillan and Co., London), revised by H.S.M. Coxeter.

Bolyai, F. (1832). *Tentamen juventutem* (Typis Collegii Reformatorum per Josephum et Simeonem Kali, Maros Vasarhelyini).

Bosboom, J., Demaine, E. D., Demaine, M. L., Lynch, J., Manurangsi, P., Rudoy, M., and Yodpinyanee, A. (2016). Dissection with the fewest pieces is hard, even to approximate, in *Discrete and Computational Geometry and Graphs: 18th Japan Conference 2015, LNCS 9943*, pp. 37–48.

Bradley, H. C. (1921). Problem 2799, *American Mathematical Monthly* **28**, pp. 186–187.

Bradley, H. C. (1930). Problem 3048, *American Mathematical Monthly* **37**, pp. 158–159.

Brodie, R. (1891). Professor Kelland's problem on superposition, *Transactions of Royal Society of Edinburgh* **36**, **part II**, 12, pp. 307–311 + plates 1 & 2, see Fig. 12, pl. 2.

Busschop, P. (1876). Problèmes de géométrie, *Nouvelle Correspondance Mathématique* **2**, pp. 83–84.

Cardano, G. (1663). *Hieronymi Cardani Mediolanensis, philosophi ac medici celeberrimi*, Vol. III: De Rerum Varietate (Lugdani, Sumptibus Ioannis Antonii Hugvetan & Marci Antonii Ravaud), originally published in 1557. Other titles: Works. Opera omnia. See p. 248 (mislabeled p. 348).

Chaplin, J. E. (2012). *Round About the Earth: Circumnavigation from Magellan to Orbit* (Simon & Schuster, New York).

Cheney, W. F. (1933). Problem E4, *American Mathematical Monthly* **40**, pp. 113–114.

Cundy, H. M. and Langford, C. D. (1960). On the dissection of a regular polygon into n equal and similar parts, *Mathematical Gazette* **44**, p. 46.

Cundy, H. M. and Rollett, A. P. (1952). *Mathematical Models* (Oxford).

Dickson, L. E. (1929). *Introduction to the Theory of Numbers* (University of Chicago Press, Chicago), volume II, p. 225.

Dudeney, H. E. Perplexities, Monthly puzzle column in *The Strand Magazine* from 1910 to 1930.

Dudeney, H. E. Puzzles and prizes, Column in the *Weekly Dispatch*, April 19, 1896–March 27, 1904.

Dudeney, H. E. (1907). *The Canterbury Puzzles and other curious problems* (W. Heinemann, London), revised edition printed by Dover Publications in 1958.

Dudeney, H. E. (1917). *Amusements in Mathematics* (Thomas Nelson and Sons, London), revised edition printed by Dover Publications, 1958.

Dudeney, H. E. (1926). *Modern Puzzles and How to Solve Them* (C. Arthur Pearson, London).

Dudeney, H. E. (1931a). *A Puzzle-Mine* (Thomas Nelson and Sons, London).

Dudeney, H. E. (1931b). *Puzzles and Curious Problems* (Thomas Nelson and Sons, London).

Elliott, C. S. (1982–1983). Some new geometric dissections, *Journal of Recreational Mathematics* **15**, 1, pp. 19–27.

Elliott, C. S. (1985–1986). Some more geometric dissections, *Journal of Recreational Mathematics* **18**, 1, pp. 9–16.

Fourrey, E. (1907). *Curiosités Géométriques* (Vuibert et Nony, Paris).

Franck, H. A. (1910). *A Vagabond Journey Around the World* (Garden City Publishing, Garden City, New York).

Frederickson, G. N. (1972a). Polygon assemblies, *Journal of Recreational Mathematics* **5**, 4, pp. 255–260.

Frederickson, G. N. (1972b). Several star dissections, *Journal of Recreational Mathematics* **5**, 1, pp. 22–26.

Frederickson, G. N. (1974). More geometric dissections, *Journal of Recreational Mathematics* **7**, 3, pp. 206–212.

Frederickson, G. N. (1978–1979). Several prism dissections, *Journal of Recreational Mathematics* **11**, 3, pp. 161–175.

Frederickson, G. N. (1997). *Dissections Plane & Fancy* (Cambridge University Press, New York).

Frederickson, G. N. (2002). *Hinged Dissections: Swinging and Twisting* (Cambridge University Press, New York).

Frederickson, G. N. (2006a). *Piano-Hinged Dissections: Time to Fold!* (A K Peters, Wellesley, MA).

Frederickson, G. N. (2006b). Reflecting well: Dissections of two regular polygons to one, *Mathematics Magazine* **79**, pp. 87–95.

Frederickson, G. N. (2015). Folding pseudo-stars that are cyclicly hinged, in *Bridges Baltimore 2015: Mathematics, Music, Art, Architecture, Culture*, pp. 1–8.

Gardner, M. (1958). Problem E1309, *American Mathematical Monthly* **65**, 3, p. 205.

Gerwien, P. (1833). Zerschneidung jeder beliebigen Anzahl von gleichen geradlinigen Figuren in dieselben Stücke, *Journal für die reine und angewandte Mathematik (Crelle's Journal)* **10**, pp. 228–234 and Taf. III.

Goldberg, M. (1952). Problem E972: Six piece dissection of a pentagon into a triangle, *American Mathematical Monthly* **59**, pp. 106–107.

Graves, R. P. (1889). *Life of Sir William Rowan Hamilton*, Vol. III (Hodges, Figgis, & Co., Dublin), contains letter from Augustus deMorgan to W. R. Hamilton, September 5, 1855. See pp. 499–503.

Grinspan, J. (2014). Anxious youth, then and now, Op-ed column in the *New York Times*, January 1, p. 19.

Grinspan, J. (2015). D.I.Y. education before YouTube, Op-ed column in the *New York Times*, July 12, p. SR4.

Hadwiger, H. and Glur, P. (1951). Zerlegungsgleichheit ebener polygone, *Elemente der Mathematik* **6**, pp. 97–106.

Hart, H. (1877). Geometrical dissections and transpositions, *Messenger of Mathematics* **6**, pp. 150–151.

Hearn, R. A., Demaine, E. D., and Frederickson, G. N. (2003). Hinged dissection of polygons is hard, in *Canadian Conference on Computational Geometry*, pp. 98–102.

Jackson, J. (1821). *Rational Amusement for Winter Evenings* (London), subtitle: "A Collection of above 200 Curious and Interesting Puzzles and Paradoxes relating to Arithmetic, Geometry, Geography, &c.". See p. 84 and plate I, Fig. 6.

Kelland, P. (1855). On superposition, *Transactions of the Royal Society of Edinburgh* **21**, pp. 271–273 and plate V.

Kelland, P. (1864). On superposition. Part II, *Transactions of the Royal Society of Edinburgh* **33**, pp. 471–473 and plate XX.

Kordemskii, B. A. (1956). *Mathematical Know-how (Matematicheskaia smekalka)* (Gos. izd-vo tekhniko-teoret. lit-ry, Moscow).

Kordemsky, B. A. (1972). *The Moscow Puzzles* (Charles Scribner's Sons, New York), translation by Albert Parry.

Langford, C. D. (1956). To pentasect a pentagon, *Mathematical Gazette* **40**, p. 218.

Langford, C. D. (1960). Tiling patterns for regular polygons, *Mathematical Gazette* **44**, pp. 105–110.

Langford, C. D. (1967). Polygon dissections, *Mathematical Gazette* **51**, pp. 139–141.

Lemon, D. (1890). *The Illustrated Book of Puzzles* (Saxon, London).

Leonardo da Vinci (1973). *Il Codice Atlantico* (Giunti-Barbera, Firenze), facsimile reproduction of the Codex Atlanticus.

Lindgren, H. (1951). Geometric dissections, *Australian Mathematics Teacher* **7**, pp. 7–10.

Lindgren, H. (1953). Geometric dissections, *Australian Mathematics Teacher* **9**, pp. 17–21; 64.

Lindgren, H. (1956). Problem E1210: A dissection of a pair of equilateral triangles: Solution, *American Mathematical Monthly* **63**, pp. 667–668.

Lindgren, H. (1958). Problem E1309: Dissection of a regular pentagram into a square, *American Mathematical Monthly* **65**, pp. 710–711.

Lindgren, H. (1961). Going one better in geometric dissections, *Mathematical Gazette* **45**, pp. 94–97.

Lindgren, H. (1962). Dissecting the decagon, *Mathematical Gazette* **46**, pp. 305–306.

Lindgren, H. (1964). *Geometric Dissections* (D. Van Nostrand Company, Princeton, New Jersey).

Lowry, M. (1814). Solution to question 269, [proposed] by Mr. W. Wallace, in T. Leybourn (ed.), *Mathematical Repository*, Vol. III (W. Glendinning, London), pp. 44–46 of Part I.

Loyd, S. Weekly puzzle column in *Tit-Bits*, starting October 3, 1896, and continuing into 1897.

Loyd, S. Puzzle column in Sunday edition of *Philadelphia Press*, Feb. 23-June 29, 1902.

Loyd, S. (1914). *Cyclopedia of Puzzles* (Franklin Bigelow Corporation, New York).

Lucas, E. (1883). *Récréations Mathématiques*, Vol. 2 (Gauthier-Villars, Paris), second of four volumes. Second edition (1893) reprinted by Blanchard in 1960.

Macaulay, W. H. (1914). The dissection of rectilineal figures, *Mathematical Gazette* **7**, pp. 381–388.

Macaulay, W. H. (1919). The dissection of rectilineal figures, *Messenger of Mathematics* **48**, pp. 159–165.

Mott-Smith, G. (1946). *Mathematical Puzzles for Beginners and Enthusiasts* (Blakiston Co., Philadelphia), reprinted by Dover Publications, New York, 1954.

Ozanam, J. (1778). *Récréations Mathématiques et Physiques* (Claude Antoine Jombert, fils, Paris), see figures 123–126 and pages 297–302. This is material added by Jean Montucla, who is listed as a reviser under the pseudonym of M. de Chanla. See also pages 127–129 of *Recreations in Mathematics and Natural Philosophy*, by Jacques Ozanam, London, Thomas Tegg, 1840, translated from Montucla's edition by Charles Hutton, with additions.

Paterson, D. (1989). T-dissections of hexagons and triangles, *Journal of Recreational Mathematics* **21**, 4, pp. 278–291.

Perigal, H. (1873). On geometric dissections and transformations, *Messenger of Mathematics* **2**, pp. 103–105.

Richmond, H. W. (1944). Note 1704: Solution of a geometrical problem, *Mathematical Gazette* **28**, 278, pp. 31–32.

Rosenbaum, J. (1947). Problem E721: A dodecagon dissection puzzle, *American Mathematical Monthly* **54**, p. 44.

Scodel, J. (1991). *The English Poetic Epitaph: Commemoration and Conflict from Jonson to Wordsworth* (Cornell University Press, Ithaca, NY).

Siddons, A. W. (1932). 1020. Perigal's dissection for the theorem of Pythagoras, *Mathematical Gazette* **16**, 217, p. 44.

Sipser, M. (2013). *Introduction to the Theory of Computation, Third Edition* (CENGAGE Learning, Boston).

Steinhaus, H. (1960). *Mathematical Snapshots, 2nd edition* (Oxford University Press, New York), see p. 11.

Taylor, H. M. (1905). On some geometrical dissections, *Messenger of Mathematics* **35**, pp. 81–101.

Theobald, G. (2004). Geometric dissections, (http://www.gavin-theobald.uk).

Tilson, P. G. (1978–1979). New dissections of pentagon and pentagram, *Journal of Recreational Mathematics* **11**, 2, pp. 108–111.

Timmermans, E. (2005). Pythagoreïsche dissecties, *Pythagoras*, pp. 28–32.

Timmermans, E. (2007). Pythagorese dissections, *Cubism For Fun*, 73, pp. 12–13.

Valens, E. G. (1964). *The Number of Things* (Dutton, New York).

Varsady, A. (1989). Some new dissections, *Journal of Recreational Mathematics* **21**, 3, pp. 203–209.

Wallace, W. (ed.) (1831). *Elements of Geometry*, eighth edn. (Bell & Bradfute, Edinburgh), first six books of Euclid, with a supplement by John Playfair.

Wheeler, A. H. (1935). Problem E4, *American Mathematical Monthly* **42**, pp. 509–510.

Index

Interlocks of Similar or Complementary Figures,
  44, 53, 105

Abū'l-Wafā, 4, 57, 147, 148
Abbott, Timothy, *7*
Abel, Zachary, *7*
acrostic, *23*
Airy, George Biddle, *148*
Allen, Harris, *22*
arithmetic progression, *49, 65*

Baetens, Bert, *62*
Ball, W. W. Rouse, *2, 26*
Bennett, Geoffrey Thomas, *3*
Bolyai, Farkas, *1*
Bosboom, Jeffrey, *34*
Bradley, Harry, *46, 100*
Brodie, Robert, *67*
Burns, Silas, *14*
Busschop, Paul-Jean, *57, 85*

capped regular n-gon, *170*
Cardano, Girolamo, *56*
central rectangle of a regular polygon, *107, 122*
Charlton, David, *7*
Chebyshev polynomial, *187*
Chen, Tai, *125*
Cheney, William, *65*
co-polygram, *109*
Collison, David, *46*
completing the tessellation, *3, 32, 107, 125, 127,*
  156, 159, 172
concave hexagon, *165*
Conway, John, *39*
Coxeter, H. S. M., *2, 27*
Cundy, H. Martyn, *2, 28, 33, 128, 166*
customized strip, *120*
customized strip dissection, *117*
cyclic hinging, *27, 72, 116*

decagram, *34*
Demaine, Erik, *7, 34*
Demaine, Martin, *7, 34*
Diophantus, *187*
dissecting regular polygons to squares of certain
  area ratios, *92, 107, 121, 130*
dodecagram, *34*
double P-slide, *148*
Dudeney, Henry E., *2, 3, 5, 8, 27, 41, 47, 55–57,*
  148, 156

Elliott, C. Stuart, *50, 53, 55, 76*
Escott, E. B., *55*

Flagler railway, *13, 16, 17*
Fourrey, Emile, *2, 57*
Freese, Philippa (Cuny), *22*
Freese, William (Bill), *12*
Freese, William Henry, *16*
Freese, Winifred, *11, 13, 29*

Gardner, Martin, *9, 75*
general grid method, *31*
Gerwien, Karl, *1*
Ginsburg, Jekuthiel, *28*
Glur, Paul, *4, 6, 33*
Goldberg, Michael, *67*

Hadwiger, Hugo, *4, 6, 33*
Hanegraaf, Anton, *63, 115, 128*
Hart, Harry, *40, 68, 70, 110, 113, 115, 118, 120,*
  137, 139
Hatzipolakis, Antreas, *39*
Hearn, Robert, *34*
Hemmings, Edward C., *14*
Hill, A. E., *8*
hinged dissection, *5–8, 27, 32, 34, 38–42, 47–49,*
  51, 53, 55, 57, 62, 69, 71–73, 77, 79, 83, 86, 90,
  92, 94, 95, 97, 99–103, 105, 111, 115, 116, 118,
  120, 126, 131, 133, 148, 155–157, 159, 161, 165,

167, 172, 183
hollow square, *61*
Hunt and Burns, *14*
Hunt, Sumner, *14*

identity $(\sin \frac{\pi}{k})^2 + (\cos \frac{\pi}{k})^2 = 1$, *40, 72, 73*

Jackson, John, *1*

Kelland, Philip, *4, 33, 57, 148*
Kibbe, Vanessa, *12*
Kiessig, Paul Peter, *13*
Kominers, Scott, *7*
Kordemsky, Boris, *165*

Langford, C. Dudley, *9, 69, 107, 115, 116, 121,*
     127, 129, 166
Law of Cosines, *51, 147, 165, 182*
Law of Sines, *51, 186*
Leech, John, *65*
Lemon, Don, *55*
Leonardo da Vinci, *151*
Lindgren, Harry, *3, 5, 9, 39, 40, 43, 47, 49, 54,*
     67, 69, 73, 75, 85–88, 90, 94, 95, 99–101, 105,
     113, 117, 119, 125, 127, 128, 130, 132–134, 157,
     165, 169
Lowry, John, *1*
Loyd, Sam, *2, 55, 149, 153*
Lucas, Édouard, *2*
Luke, Dorman, *12, 27, 33*
Lynch, Jason, *34*

Macaulay, William, *2*
Manurangsi, Pasin, *34*
McElroy, Charles W., *5, 41*
Montucla, Jean, *1*
Mott-Smith, Geoffrey, *2, 8, 27, 57*

normal grid method, *31*
NP-hard, *34*

octagram, *34*
offset grid method, *31*
Ozanam, Jacques, *1*

P-slide method, *32*
P-strip technique, *31*
partially hinged dissection, *166*
Paterson, David, *47*
Perigal, Henry, *2, 3, 57, 59, 148, 150, 151*
photos
     Dudeney, Henry Ernest, *2*
     Freese's house, *13*
     Freese, Ernest Irving, *11, 24*

Freese, Ernest with bicycle, *19*
Freese, William Henry, *21*
Freese: Ernest, Bill, Winifred, *12*
Ginsburg, Jekuthiel, *28*
Hunt, Sumner P., *15*
Lindgren, Harry, *10*
Reid, Robert, *30*
Theobald, Gavin, *30*
Plato, *46*
pseudo-star, *34, 51, 165*
PSPACE-hard, *34*
Pythagorean triples, *61, 182, 185, 186*

Q-slide, *68*

raking grid method, *31*
Reid, Robert, *70, 101, 115, 128, 129, 134, 169*
rhombic method, *31*
rhombic structure of a pentagon, *164*
Richmond, Herbert, *65*
ring expansion, *70, 110, 115*
Rollett, A. P., *2*
Rosenbaum, Joseph, *127*
Rudoy, Mikhail, *34*

Siddons, Arthur W., *57*
Singmaster, David, *5*
sliced grid method, *32*
star, *34*
Steinhaus, Hugo, *47*
stellated decagon, *77*
stellated nonagon, *34, 51*
step technique, *56*

T-strip technique, *31*
Taylor, Henry M., *2, 38, 40*
Thābit ibn Qurra, *62*
Theobald, Gavin, *39, 45, 51, 62, 68, 74, 87, 94,*
     113, 116–118, 122, 137, 138, 142, 144, 150, 151,
     153, 161, 162, 165, 172, 176–179, 189
Tilson, Philip, *75*
Timmermans, Edo, *149*
translational dissection, *4, 8, 32, 47, 50, 51, 53–59,*
     61, 65, 76, 78, 85–90, 92–95, 97, 99–101,
     107–111, 115, 118, 120, 121, 124, 125, 127–129,
     131, 133, 134, 148, 150, 155, 165, 167, 169, 171,
     172
triangular number, *170*
turning over pieces, *33*

unfurling hinged pieces, *51, 72, 77, 158, 159, 161*
unsupported claims, *37, 39, 41, 57, 67, 85, 90, 105,*
     125, 148, 167

Valens, Evans, 127

Wallace, William, 1
Wheeler, Albert, 65

Wotherspoon, George, 148

Yodpinyanee, Anak, 34

Printed in the United States
By Bookmasters